T0345155

Texts in Applied Mathematics 1

L. Sirovich

Introduction to
Applied Mathematics

With 133 Illustrations

Springer-Verlag
New York Berlin Heidelberg
London Paris Tokyo

Lawrence Sirovich
Division of
  Applied Mathematics
Brown University
Providence, RI 02912, USA

Editors

F. John
Courant Institute of
  Mathematical Sciences
New York University
New York, NY 10012
USA

J.E. Marsden
Department of
  Mathematics
University of California
Berkeley, CA 94720
USA

L. Sirovich
Division of
  Applied Mathematics
Brown University
Providence, RI 02912
USA

M. Golubitsky
Department of
  Mathematics
University of Houston
Houston, TX 77004
USA

W. Jäger
Department of
  Applied Mathematics
Universität Heidelberg
Im Neuenheimer Feld 294
6900 Heidelberg, FRG

Mathematics Subject Classification (1980): 30xx 35xx 42xx 34xx

Library of Congress Cataloging-in-Publication Data
Sirovich, L., 1933–
  Introduction to applied mathematics.
  (Texts in applied mathematics ; 1)
  Bibliography: p.
  Includes index.
  1. Mathematics—1961–   . I. Title. II. Series.
QA39.2.S525 1988   515   88-27821

Printed on acid-free paper.

Camera-ready copy provided by the author.
Printed and bound by R.R. Donnelley and Sons, Harrisonburg, Virginia.
Printed in the United States of America.

9 8 7 6 5 4 3 2 1

ISBN 0-387-96884-9 Springer-Verlag New York Berlin Heidelberg
ISBN 3-540-96884-9 Springer-Verlag Berlin Heidelberg New York

Series Preface

Mathematics is playing an ever more important role in the physical and biological sciences, provoking a blurring of boundaries between scientific disciplines and a resurgence of interest in the modern as well as the classical techniques of applied mathematics. This renewal of interest, both in research and teaching, has led to the establishment of the series: *Texts in Applied Mathematics (TAM)*.

The development of new courses is a natural consequence of a high level of excitement on the research frontier as newer techniques, such as numerical and symbolic computer systems, dynamical systems, and chaos, mix with and reinforce the traditional methods of applied mathematics. Thus, the purpose of this textbook series is to meet the current and future needs of these advances and encourage the teaching of new courses.

TAM will publish textbooks suitable for use in advanced undergraduate and beginning graduate courses, and will complement the *Applied Mathematics Sciences (AMS)* series which will focus on advanced textbooks and research level monographs.

Preface

The material in this book is based on notes for a course which I gave several times at Brown University. The target of the course was juniors and seniors majoring in applied mathematics, engineering and other sciences. In actual fact, the students ranged from occasional highly prepared freshmen to graduate students. The last category usually made up one third to one half of the class. Overall, I would say that the students found the contents of the book challenging and exacting.

My basic goal in the course was to teach standard methods, or what I regard as a basic *bag of tricks*. In my opinion the material contained here, for the most part, does not depart widely from traditional subject matter. One such departure is the discussion of discrete linear systems (and this is really just a return to classical material). Besides being interesting in its own right, this topic is included because the treatment of such systems leads naturally to the use of discrete Fourier series, discrete Fourier transforms, and their extension, the Z-transform. On making the transition to continuous systems we derive their continuous analogues, viz., Fourier series, Fourier transforms, Fourier integrals and Laplace transforms. A main advantage to the approach taken is that a wide variety of techniques are seen to result from one or two very simple but central ideas. Students appeared both to grasp and to appreciate this consolidation of concepts.

Related to this and a recurrent theme in this text is the idea of transforming a problem to another simpler problem. This in turn leads to the use of eigenfunction methods. Virtually every method developed here is also derived by an eigenfunction approach. Moreover, some weight is laid on this being a natural way to view and analyze problems. This then leads to the geometrical point of view and to the introduction of abstract spaces. Since I felt that this was a very desirable approach I went to some lengths to motivate these ideas and make learning them as painless as possible.

As the remarks thus far imply I have placed emphasis on presenting a variety of approaches and perspectives—as many as I deemed possible. This is in keeping with a general principle which I subscribe to, namely that a deeper understanding of a subject is gained by viewing it from as many aspects as possible.

There are two basic prerequisites for this course: linear algebra and ordinary differential equations. The latter on the level of, for example, the books by Braun and by Boyce and DiPrima. (A list of references appears at the end of the book.) It is also appropriate to mention a word about the first three chapters which cover basic topics in complex variable theory. If one views this as a course in applied complex analysis then the first three chapters are the underpinnings. This portion of the course was taught in roughly five weeks and since a broad range of topics are included some sacrifices were required. Consequently there was no intention of having this course replace the traditional complex variable course. If anything I contend that the standard material in complex variable theory will be better appreciated by the student after a course of this type.

Above all, this course is intended as being one which gives the student a *can-do* frame of mind about mathematics. Too many math courses give the impression that mathematics is a minefield and that unless one is very very careful disasters will befall them. My view and the one that I have tried to present in this book is diametrically opposed to this. Students should be given confidence in using mathematics and not be made fearful of it. Partly with this in mind I have forgone the theorem-proof format for a more informal style. Although I have endeavored to make the mathematics respectable, rigor has not been given a high priority. Finally a concerted effort was made to present an assortment of examples from diverse applications with the hope of attracting the interest of the student, and an equally dedicated effort was made to be kind to the reader.

Only the help of many people made the completion of this book possible. Madeline Brewster and Andria Durk prepared an earlier version and played an essential role in assembling the present version; Kate MacDougall painstakingly and patiently prepared this final version. My colleague and friend Jack Pipkin performed the experiment of teaching this material from an earlier version of the manuscript. His criticism (sometimes severe) often took root. I take pleasure in expressing sincere gratitude to them all. Finally no words can express my deep appreciation to Candace Kent who took the course, corrected my errors, mathematical and otherwise. Her many improvements appear throughout the text. The blemishes, flaws and errors that remain are due to me and are there in spite of the best efforts of all these people. Finally thanks, with mixed feelings, also go to the late Walter Kaufmann-Bühler for sweet-talking me into writing this book.

I dedicate this book to the memory of my mother, Libby, who was my first and best teacher.

L. S.
Saltaire
July, 1988

Contents

1

Complex Numbers

1.1 Complex Numbers

The concept of *imaginary* numbers occurs early in the discussion of algebraic equations. For example, the quadratic equation

$$x^2 + 1 = 0$$

has the solutions $x = \pm i$, where $i = \sqrt{-1}$. In general, hybrid forms, called *complex numbers*, are found containing both real and imaginary parts. For example, if

$$x^2 - 2x + 2 = 0,$$

then

$$x = 1 \pm i$$

are the solutions. Complex numbers, which extend the *real* number system, are made necessary by the solution of algebraic equations with real coefficients. It is interesting to note that algebraic equations with complex coefficients have solutions which are complex—no further extension is necessary.

Complex numbers can be viewed as belonging to a two-space called the *complex plane* (see Figure 1.1). According to common convention, a typical complex number is denoted by the letter z, with

$$z = x + iy.$$

We also define the real and imaginary parts through

$$x = \operatorname{Re} z, \quad y = \operatorname{Im} z. \tag{1.1}$$

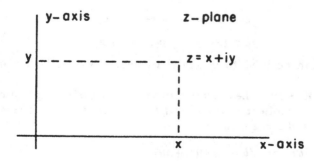

FIGURE 1.1.

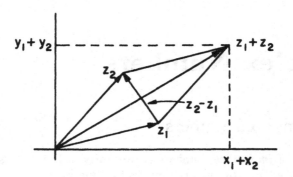

FIGURE 1.2.

The addition of the two complex numbers

$$z_1 = x_1 + iy_1, \quad z_2 = x_2 + iy_2$$

follows the rules of vector addition in two-space:

$$z = z_1 + z_2 = x_1 + x_2 + i(y_1 + y_2) = z_2 + z_1.$$

Both this operation and that of subtraction are indicated in Figure 1.2. The figure is familiar from analytical geometry and further explanation is not deemed necessary.

Complex numbers can be multiplied in the ordinary way and this differs from Cartesian two-vectors. In particular,

$$\begin{aligned} z_1 z_2 &= (x_1 + iy_1)(x_2 + iy_2) \\ &= (x_1 x_2 - y_1 y_2) + i(x_1 y_2 + x_2 y_1) = z_2 z_1, \end{aligned}$$

where $i^2 = -1$ has been used. Actually, the explicit appearance of i can be avoided by writing

$$z = (x, y)$$

and defining

$$z_1 + z_2 = (x_1 + x_2, y_1 + y_2),$$

$$z_1 z_2 = (x_1 x_2 - y_1 y_2, x_1 y_2 + x_2 y_1).$$

Such rules can be used to generate a complex arithmetic for use on a computer.

As usual, division is the operation inverse to that of multiplication. Thus z is called the quotient of a and b if $bz = a$. If we write $a = a_1 + ia_2$, $b = b_1 + ib_2$, and $z = x + iy$, then

$$\begin{aligned} bz &= (b_1 + ib_2)(x + iy) \\ &= (b_1 x - b_2 y) + i(b_2 x + b_1 y) = a_1 + ia_2. \end{aligned}$$

This complex equation can be put into the form of a matrix problem:

$$\begin{pmatrix} b_1 & -b_2 \\ b_2 & b_1 \end{pmatrix} \begin{pmatrix} x \\ y \end{pmatrix} = \begin{pmatrix} a_1 \\ a_2 \end{pmatrix}.$$

Thus solving $bz = a$ is equivalent to solving a 2×2 linear system, but not all 2×2 linear systems can be put into this complex form. It follows from the construction of complex numbers that in a complex equation the real and imaginary parts are separately equal. Thus we are led to two linear equations in x and y which, when solved, yields

$$x + iy = \frac{a_1 b_1 + a_2 b_2}{b_1^2 + b_2^2} + i\frac{-a_1 b_2 + a_2 b_1}{b_1^2 + b_2^2}$$

The denominator $b_1^2 + b_2^2$ is the squared distance of the complex number $b_1 + ib_2$ from the origin. More generally, if $z = x + iy$, its distance from the origin is denoted by

$$r = |z| = \mod z = (x^2 + y^2)^{1/2} \tag{1.2}$$

(mod, short for modulus). We can also write

$$|z|^2 = x^2 + y^2 = (x + iy)(x - iy) = z\bar{z}.$$

The last expression defines the complex conjugate; i.e., if $z = x + iy$, then $\bar{z} = x - iy$. (In certain instances the complex conjugate of z will also be denoted by z^*.) The conjugate is useful in representing the division of two complex numbers in terms of real and imaginary parts:

$$\frac{z_2}{z_1} = \frac{z_2 \cdot \bar{z}_1}{z_1 \cdot \bar{z}_1} = \frac{x_1 x_2 + y_1 y_2}{x_1^2 + y_1^2} + i\frac{-x_2 y_1 + x_1 y_2}{x_1^2 + y_1^2}.$$

We mention also that since $\bar{z}$ can be determined from z, $\bar{z}$ is really a function of z; i.e., formally $\bar{z} = \bar{z}(z)$. Figure 1.3 indicates the location of these quantities.

Polar Coordinates

Equation (1.2) defines the modulus of a complex number z. To complete the transformation from Cartesian to polar coordinates, we define the positive polar angle θ as being the angle measured counterclockwise from the positive real axis to the ray to z (see Figure 1.3.). For negative θ, measurement is made in the clockwise direction. The angle θ is also written as

$$\theta = \arg z,$$

where arg is short for argument.

It is clear that θ has the property

$$\tan \theta = \frac{y}{x},$$

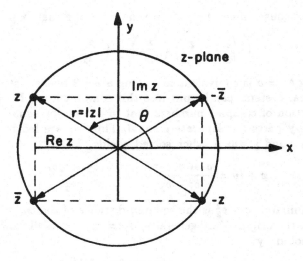

FIGURE 1.3.

which can be formally solved to give

$$\theta = \tan^{-1} \frac{y}{x}. \tag{1.3}$$

However, it should be recalled that the arctangent is customarily defined with its range as the open interval $(-\pi/2, \pi/2)$ (see Figure 1.4); therefore, some *fine print* is required along with (1.3). For example, if we consider the case where $\theta \in [-\pi, \pi]$, then θ is given by (1.3) for $x > 0$ and $\theta = \tan^{-1}(y/x) + \pi$ sign y for $x < 0$. (See Exercise 7.)

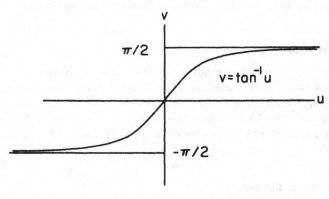

FIGURE 1.4.

1.2 Exponential Notation

The exponential function e^{at} for real a and t is defined by the convergent infinite series

$$e^{at} = \sum_{n=0}^{\infty} \frac{(at)^n}{n!} = \exp(at). \tag{1.4}$$

This can be regarded as the solution of the initial value problem

$$\frac{dw}{dt} = aw, \quad w(0) = 1.$$

If a is complex, the meaning of the series is not clear since we have not yet considered the idea of the convergence of a sum in the complex plane. For the moment we accept the above definition of the exponential as being valid for complex numbers (this will be justified in the next section).

If $a = i$ in (1.4), then formally

$$e^{it} = \sum_{n=0}^{\infty} \frac{i^n t^n}{n!} = 1 + it - \frac{t^2}{2} - \frac{it^3}{3!} + \frac{t^4}{4!} + \cdots$$

$$= \left(1 - \frac{t^2}{2!} + \frac{t^4}{4!} + \cdots\right) + i\left(t - \frac{t^3}{3!} + \cdots\right).$$

We recognize the sum in the first set of parentheses first as $\cos t$ and the second as $\sin t$. Thus we have shown

$$e^{it} = \cos t + i \sin t. \tag{1.5}$$

Further, from the differential equation, if

$$\frac{dw_1}{dt} = a_1 w_1, \quad \frac{dw_2}{dt} = a_2 w_2,$$

then cross-multiplication gives

$$\frac{dw_1 w_2}{dt} = (a_1 + a_2) w_1 w_2,$$

which demonstrates that

$$e^{a_1 t} = e^{a_2 t} = e^{(a_1 + a_2)t}$$

for complex a_1 and a_2. In particular,

$$e^z = e^{x+iy} = e^x e^{iy} = e^x(\cos y + i \sin y).$$

If we return to polar notation using (1.2) and (1.3), then

$$\begin{aligned} z &= x + iy + r \cos \theta + ir \sin \theta \\ &= r(\cos \theta + i \sin \theta), \end{aligned}$$

which from (1.5) states

$$z = x + iy = re^{i\theta}. \tag{1.6}$$

If this notation is applied to the product $z_1 z_2$, where $z_1 = r_1 e^{i\theta_1}$ and $z_2 = r_2 e^{i\theta_2}$, then

$$z_1 z_2 = (r_1 e^{i\theta_1})(r_2 e^{i\theta_2}) = r_1 r_2 e^{i(\theta_1 + \theta_2)},$$

from which we have

$$\arg z_1 z_2 = \arg z_1 + \arg z_2,$$

$$\operatorname{mod} z_1 z_2 = |z_1 z_2| = r_1 r_2 = (\operatorname{mod} z_1)(\operatorname{mod} z_2). \tag{1.7}$$

Therefore, *under multiplication, arguments add and moduli multiply.*

De Moivre's Formula

Consider z^n for integer n. If we use (1.6), then

$$z^n = (re^{i\theta})^n = r^n e^{in\theta} = r^n(\cos n\theta + i \sin n\theta);$$

or, if $|z| = r = 1$, then

$$e^{in\theta} = \cos n\theta + i \sin n\theta = (\cos \theta + i \sin \theta)^n, \tag{1.8}$$

which is known as *De Moivre's Formula* (or *Theorem*). If, for example, $n = 2$, this says

$$\begin{aligned}
\cos 2\theta + i \sin 2\theta &= e^{2i\theta} = (e^{i\theta})^2 = (\cos \theta + i \sin \theta)^2 \\
&= \cos^2 \theta - \sin^2 \theta + 2i \sin \theta \cos \theta,
\end{aligned}$$

the real and imaginary parts of which give the familiar trigonometric relations

$$\cos 2\theta = \cos^2 \theta - \sin^2 \theta$$

and

$$\sin 2\theta = 2 \sin \theta \cos \theta.$$

In general, De Moivre's Formula facilitates the demonstration of many trigonometric relations.

Roots of a Complex Number

If, for complex numbers w and z and integer n, we have that

$$w^n = z,$$

then w is said to be an nth root of z and is written as $z^{1/n}$. To find w, first write

$$w = Re^{i\Theta}, \quad z = re^{i\theta}$$

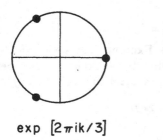

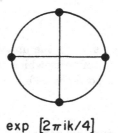

exp [2πik/3]   exp [2πik/4]

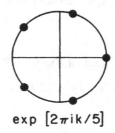

exp [2πik/5]

FIGURE 1.5.

so that

$$w^n = R^n e^{in\Theta} = re^{i\theta}.$$

On comparison of moduli and arguments, we obtain

$$R^n = r, \quad n\Theta = \theta + 2\pi N.$$

The term $2\pi N$ with integer N is included since arg is ambiguous up to an integer multiple of 2π; i.e., $\exp(i\theta) = \exp[i(\theta + 2N\pi)]$. If we solve for R and Θ, then

$$R = r^{1/n}$$

and

$$\Theta = \frac{\theta}{n} + 2\pi \frac{N}{n}, \quad N = 0, 1, \ldots, n-1.$$

These form the only choices for Θ since any other integer choice of N will yield a Θ which differs from one of the above by a multiple of 2π.

As a particular example, consider the nth roots of unity as determined by

$$w^n = 1.$$

From the above discussion, the n different roots are

$$w_k = e^{i2\pi k/n}, \quad k = 0, 1, \ldots, n-1.$$

Note that
$$|w_k| = 1,$$
so that all roots lie on the unit circle. Examples are shown in Figure 1.5.

Any nth root of unity which can generate the other roots by being raised to an integer power is referred to as a *primitive nth root of unity*. The quantity
$$\Omega_n = e^{2\pi i/n} \tag{1.9}$$
is always a primitive nth root of unity.

The notion of an arbitrary complex number to a complex power involves the definition of the logarithm, which we defer for the moment.

Exercises

1. Where is $\frac{1}{2}(z_1 + z_2)$ located in Figure 1.2?

2. What is the locus of $z_1 + t(z_2 - z_1)$ for t real?

3. Where is z if

   (a) $|z| = 5$?
   (b) $|z + 1| = 1$?
   (c) $\operatorname{Im} z = 2$?

4. Prove

   (a) $|z_1 + z_2| \le |z_1| + |z_2|$,
   (b) $|z_1 - z_2| \ge ||z_2| - |z_1||$,
   (c) $|1 - z| = |1 - \bar{z}|$,
   (d) $\overline{z_1 z_2} = \bar{z}_1 \bar{z}_2$,
   (e) $\overline{z_1/z_2} = \bar{z}_1/\bar{z}_2$.

   Give a geometrical construction in each case.

5. Write the following in terms of real and imaginary parts: (a) $1/(3+2i)$, (b) $(2+4i)/(1+i)$, (c) $(i/(1+i)) + ((1+i)/i)$, (d) $1 + (i/2) + (i/2)^2 + \cdots + (i/2)^6$, (e) $(1+i)(1+2i)(1+3i)$.

6. Solve $z = x + iy$, $\bar{z} = x - iy$ for x and y in terms of z and $\bar{z}$. (Thus every function of the plane $f(x,y)$ can be written as a function of z and $\bar{z}$. But since $\bar{z}$ is a function of z, every function of the plane is a function of z!)

7. Supply a rule to be used with (1.3) and Figure 1.3 to determine correctly $\theta = \arg z$ for $\theta \in [-\pi, \pi]$. Try the same for $\theta \in [0, 2\pi]$. Use the rule for $\theta \in [0, 2\pi]$ to put the following into polar form: (a) $1 + i$, (b) $-1 - 2i$, (c) $-1 + \sqrt{3}i$.

8. Sketch the loci of (a) arg $z = \pi/4$, (b) $|z| < 1$, (c) $|z - 1| = 2$.

9. Write each of the following in polar form:

    (a) $z = 1 - i\sqrt{3}$,

    (b) $z = -i$,

    (c) $z = 1 - i$,

    (d) $z = 1/(1 + i)$,

    (e) $z = i^{1/3}$,

    (f) $z = (1 + i)^4$.

10. (a) Prove

$$1 + z + z^2 + \cdots + z^n = \frac{1 - z^{n+1}}{1 - z}.$$

    (b) Use this to prove

$$1 + \cos\theta + \cdots + \cos n\theta = \frac{1}{2} + \frac{\sin[(n + \frac{1}{2})\theta]}{2\sin\theta/2}.$$

11. If $|z_1|, |z_2| \leq 1$, then show

$$|z_1 + z_2| \leq |1 + \bar{z}_1 z_2|.$$

When do we have equality?

12. If $z = re^{i\theta}$, $z_1 = r_1 e^{i\theta_1}$, $z_2 = r_2 e^{i\theta_2}$, find mod and arg of $1/z$, $\bar{z}_1/\bar{z}_2$.

13. Prove $\overline{z_1 z_2} = \bar{z}_1 \bar{z}_2$, $\overline{(z_1/z_2)} = \bar{z}_1/\bar{z}_2$.

14. Find the fourth roots of -9.

15. Find all primitive fifth roots of unity.

16. Solve $x^2 + 4 = 0$.

17. Find all values of $(1 - \sqrt{3}i)^{1/2}$.

18. Show that if $(\Omega_n)^k = W_k$, $k = 0, \ldots, n - 1$ (see (1.9)), then

$$W_1 + W_2 + \cdots + W_n = 0.$$

19. Find all roots of (a) $z^4 + 16 = 0$, (b) $z^8 - 2z^4 + 1 = 0$.

20. (a) If in

$$\frac{dw}{dt} = aw$$

the constant a and the variable w are complex, then this equation represents two real first order equations. What are these first order equations?

(b) Show that for the harmonic oscillator equation

$$\frac{d^2x}{dt^2} = \omega^2 x,$$

$w = x + i\omega(dx/dt)$ satisfies a single first order differential equation.

2

Convergence and Limit

2.1 Convergence and Limit

The question of convergence of a series has already arisen in the definition of the exponential, (1.4). To approach the question of *convergence* of a series

$$\sum_{n=0}^{\infty} b_n$$

we consider the partial sums

$$S_N = \sum_{n=0}^{N} b_n$$

and then define the sum of the series to be the *limit* of the sequence of partial sums, if its exists.

The concept of the limit of a sequence is easily thought of in geometrical terms. A sequence of reals $\{a_n\}$ is said to converge to a if there exists for any $\epsilon > 0$ an integer $N(\epsilon)$ such that

$$|a - a_n| < \epsilon$$

for all $n > N(\epsilon)$. If this is true, we write

$$\lim_{n \uparrow \infty} a_n = a \ \text{ or } \ a_n \to a.$$

Thus, if the elements of the sequence $\{a_n\}$ are viewed as points on the real axis, then, except for finitely many n, all the a_n lie in an arbitrarily small neighborhood of the limit a. On the other hand, a is said to be a *limit point* or a *point of accumulation* of the sequence $\{a_n\}$ if for infinitely many n the a_n lie in an arbitrarily small neighborhood of the limit a. The limit of a sequence is a limit point of the sequence, but if there is more than one limit point, then the sequence has no limit. The collection of all limit points is called the *limit set*. If, in addition, every deleted neighborhood of a point a (the *deleted neighborhood* of a point is the neighborhood of that point with that point deleted) contains a_n for some n, then a is said to be a *cluster point* of the sequence $\{a_n\}$. It should be noted that all cluster points are limit points but not vice versa. For example, if a point of a sequence is repeated infinitely often, it is a limit point but not a cluster point.

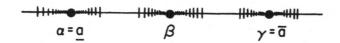

$$a = \underline{a} \qquad \beta \qquad \gamma = \overline{a}$$

FIGURE 2.1. A sequence $\{a_n\}$ on the real line with three limit points α, β, and γ. γ is the lim sup and α is the lim inf.

Intuitively it is clear that any bounded infinite collection of points contains at least one limit point (Bolzano–Weierstrass Theorem, see Hardy). In the case of real sequences, it is useful to distinguish two out of the many possible limit points. For a real sequence $\{a_n\}$, $\overline{a}$ is said to be the *upper limit* or *limit superior* (lim sup) if it is the largest limit point of the limit set. This is written as

$$\overline{\lim_{n\uparrow\infty}} a_n = \overline{a}. \tag{2.1}$$

In a similar manner, the smallest such limit point is denoted by $\underline{a}$, referred to as the *lower limit* or *limit inferior* (lim inf) and written as

$$\underline{\lim_{n\uparrow\infty}} a_n = \underline{a}. \tag{2.2}$$

See Figure 2.1.

Next we turn to the case of a sequence of complex numbers $\{a_n\}$ and apply the same definition; viz., a is the limit of the sequence if for any $\epsilon > 0$ there exists an integer $N(\epsilon)$ such that

$$|a - a_n| < \epsilon$$

for all $n > N(\epsilon)$. Now, however, the absolute value sign signifies the modulus, i.e., the distance function in the complex plane. The notions of cluster point and limit are clearly still applicable. But the ideas of lim sup and lim inf make no sense. It is further clear from simple geometrical considerations that Re $a_n \rightarrow$ Re a and Im $a_n \rightarrow$ Im a since these represent the projections on the real and imaginary axes (Figure 2.2).

Example. Consider the sequence $\{1 + (i/(1+n))\}$. This has just one limit point, 1, which is also a cluster point. The real part of the sequence also has 1 for a limit point but has no cluster points. The imaginary part of the sequence has zero for both a limit point and a cluster point. See Figure 2.3.

Cauchy Criterion

For a real sequence $\{a_n\}$, we know that a necessary and sufficient condition for convergence to a limit a is that

$$|a_n - a_m| < \epsilon$$

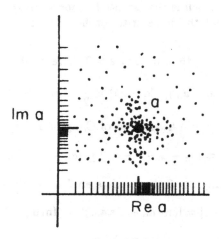

FIGURE 2.2.

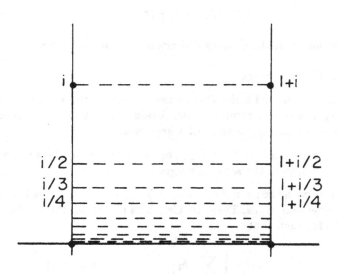

FIGURE 2.3.

for $m, n > N(\epsilon)$. This property, known as the *Cauchy Criterion* (see Hardy), we take as given. It states that terms of a sequence get closer as the indices get larger, which is the essential property of convergent sequences.

We now show that the same criterion holds true in the case of complex sequences.

Necessity: If $a_n \to a$, then for any $\epsilon > 0$ there exists an $N(\epsilon)$ such that

$$|a - a_m|, \quad |a - a_n| < \frac{\epsilon}{2}, \quad m, n > N.$$

Therefore (see Exercise 4 at the end of Chapter 1)

$$|a_n - a_m| = |a_n - a + a - a_m| \leq |a_n - a| + |a_m - a| < \epsilon.$$

Sufficiency: For any $\epsilon > 0$ there exists an $N(\epsilon)$ such that

$$\epsilon > |a_n - a_m| = [(\operatorname{Re} a_n - \operatorname{Re} a_m)^2 + (\operatorname{Im} a_n - \operatorname{Im} a_m)^2]^{1/2}$$

for all $m, n > N(\epsilon)$, from which we have

$$\epsilon > |\operatorname{Re} a_n - \operatorname{Re} a_m|, \quad \epsilon > |\operatorname{Im} a_n - \operatorname{Im} a_m|,$$

for all $m, n > N(\epsilon)$. From this final pair of inequalities we have the existence of a $\operatorname{Re} a$ and an $\operatorname{Im} a$ and hence the limit

$$a = \operatorname{Re} a + i \operatorname{Im} a$$

as a consequence of the Cauchy Criterion for real sequences.

Tests for Convergence

As in the case of real series and sequences, there follow a number of tests for convergence in the complex case. Again, we will use b_n to represent a term in a *series* and a_n, a term of a *sequence*.

Divergence Criterion: If, in $\sum_{n=0}^{\infty} b_n$, $b_n \not\to 0$ (i.e., b_n does not approach zero as $n \uparrow \infty$), then the series diverges.

Proof: Start with the partial sums $a_n = \sum_{k=0}^{n} b_k$, and assume the contrary, that the series converges. Then, by Cauchy's Criterion, there exists for any $\epsilon > 0$ an $N(\epsilon)$ such that

$$|a_m - a_n| = \left| \sum_{k=n+1}^{m} b_k \right| < \epsilon, \quad m > n > N(\epsilon).$$

Since the condition $m > n$ allows us to set $m = n+1$, we have an immediate contradiction.

If $b_n \to 0$ as $n \uparrow \infty$, nothing definite can be said. For example,

$$1 + \frac{1}{2} + \frac{1}{3} + \frac{1}{4} + \cdots + \frac{1}{N} = \sum_{n=1}^{N} \frac{1}{n}$$

diverges as $N \uparrow \infty$ while

$$1 + \frac{1}{2} + \frac{1}{2^2} + \cdots = 2.$$

(However, if $b_n \to 0$ as $n \uparrow \infty$ and the b_n alternate in sign, at least as $n \uparrow \infty$, then the series converges.)

Absolute Convergence Criterion: If $\sum_{n=0}^{\infty} |b_n| < \infty$ (i.e., the series converges *absolutely*), then $\sum_{n=0}^{\infty} b_n$ converges. (Recall that a series is *conditionally convergent* if $\sum_{n=0}^{\infty} b_n$ converges but $\sum_{n=0}^{\infty} |b_n|$ diverges.)

Comparison or M-Test: If $M_n \geq |b_n|$ for $n \geq N$ and

$$\sum_{n=0}^{\infty} M_n < \infty,$$

then $\sum_{n=0}^{\infty} b_n$ converges.

Proofs for the Absolute Convergence Criterion and M-Test are simple extensions of the real case and are left to the exercises.

Ratio Test: $\sum_{n=0}^{\infty} b_n$ converges if $|b_{n+1}/b_n| \leq r < 1$ for all n greater than some N and diverges if $|b_{n+1}/b_n| \geq 1$ for all n greater than some N. (We have $|b_{n+1}/b_n| \geq 1$ rather than $|b_{n+1}/b_n| > 1$ because limits are not used here.)

Proof: In the latter case $|b_n|$ is greater than some nonzero constant for infinitely many n, and we can apply the Divergence Criterion. In the former case, for $n > N$,

$$|b_n| = |b_N| \, |b_n/b_N| \leq |b_N| r^{n-N}$$

since

$$|b_{N+p}/b_N| = |b_{N+p}/b_{N+p-1}| \, |b_{N+p-1}/b_{N+p-2}| \cdots |b_{N+1}/b_N|.$$

Therefore $\sum_{n=0}^{\infty} |b_n|$ converges by the M-test, and, consequently, $\sum_{n=0}^{\infty} b_n$ converges by the Absolute Convergence Criterion.

Root Test: If $|b_n|^{1/n} < r < 1$ for all n greater than some N, then $\sum_{n=0}^{\infty} b_n$ converges. If $|b_n|^{1/n} \geq 1$ for infinitely many n greater than some N, then the series diverges.

The proof is left as an exercise.

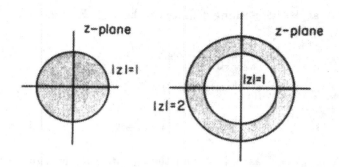

FIGURE 2.4.

2.2   Function of a Complex Variable. Continuity

Our concern is with functions defined on points of the complex plane. By these functions we mean rules for assigning a number, usually complex, to each point of some set in the complex plane. A convenient set of points to speak about is a *domain*, defined to be an *open connected set*. *Open* means that each point is an interior point (i.e., there exists a sufficiently small neighborhood of the point which belongs entirely to the set), and *connected* implies that any two points can be joined by a polygonal path lying entirely in the set. The set of all points z such that $|z| < 1$ is a domain, and so is the set of all points z such that $1 < |z| < 2$ (see Figure 2.4). The latter is said to be doubly connected while the first singly or simply connected. More generally, the connectivity is given by the count of necessary bounding curves of the domain.

A single-valued function of z in some domain D will be written

$$w = f(z).$$

This simply means that for every point $z = x + iy$ in D we obtain a complex number, w, which we write as $w = \phi + i\psi$. Since the value of $f(z)$ changes with position in the complex plane, as given by $z = x + iy$, we can write

$$w = f(z) = \phi(x, y) + i\psi(x, y).$$

This can be regarded as a mapping of the domain D in the z-plane into some set of points S in the w-plane (Figure 2.5).

Since we have defined a distance function in the complex plane (i.e., the modulus of the difference of two complex numbers), we can immediately extend the concept of *continuity* to complex functions. Suppose $f(z)$ is defined on a domain D. Then $f(z)$ is continuous at a point z_0 in D if for any $\epsilon > 0$ there exists a $\delta(\epsilon) > 0$ such that

$$|f(z) - f(z_0)| < \epsilon \quad \text{when} \quad |z - z_0| < \delta.$$

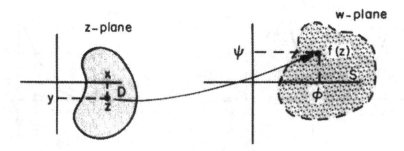

FIGURE 2.5.

This says that small neighborhoods map into small neighborhoods (Figure 2.6).

Suppose $w = f(z)$ is continuous in a domain D (i.e., $w = f(z)$ is continuous at all points in D) and maps D into S. If U is an open connected subset of S, then U is a domain by definition and we can speak of the inverse mapping

$$z = F(w) \quad \text{for} \quad w \in U;$$

and because of continuity, the set of points z forms an open subset of D. However, the inverse mapping F may be many-valued. For example,

$$w = z^2$$

is certainly a continuous function but both $z = 1$ and $z = -1$ map to $w = 1$.

To consider this problem further, write

$$z = re^{i\theta}$$

and then

$$w = z^2 = r^2 e^{2i\theta}.$$

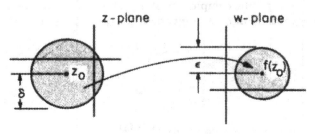

FIGURE 2.6.

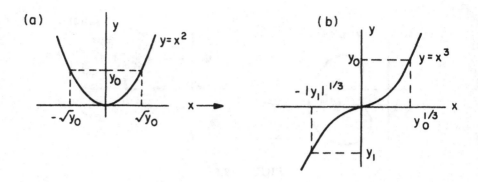

FIGURE 2.7.

It is clear that the upper half of the z-plane, namely

$$0 \le r < \infty, \quad 0 \le \theta < \pi,$$

maps into

$$0 \le |w| < \infty, \quad 0 \le \arg w < 2\pi,$$

which is the entire w-plane. Similarly, the entire lower half of the z-plane maps onto the entire w-plane. Therefore the inverse is double-valued. The case of a many-valued inverse will be treated later. If you consider the real case of, say, $y = x^2$, then, as a simple sketch can show, trouble appears in finding the inverse. There are two *branches* $x = \pm\sqrt{y}$, see Figure 2.7(a). Circumstances are really more subtle, however. To see this, consider $y = x^3$. With the help of a simple sketch (see Figure 2.7(b)), there is no trouble determining the inverse, viz., $x = y^{1/3}$; but if we allow x and y to be complex, multiple values again appear.

If, in the mapping $w = f(z)$, distinct points go into distinct points, the map is called *one-to-one* and the function is called *univalent*. In this case the inverse is single-valued. As an example of a univalent function, consider

$$w = az + b,$$

where a and b are specific complex constants and $a \ne 0$. The transformation w can be decomposed into its component parts. If we write

$$a = a_0 e^{i\theta_0}$$

and

$$z = re^{i\theta},$$

then

$$az = a_0 r e^{i(\theta + \theta_0)}.$$

Thus az can be regarded as a rotation of the z-plane by the angle θ_0 and a stretching by the factor a_0. The constant b simply translates each point of the rotated and stretched plane. Hence to each point of the z-plane we obtain a distinct point of the w-plane and vice versa, where the inverse is

$$z = \frac{1}{a}(w - b).$$

We complete this section with one final word about notation. Given the function

$$f(z) = \phi(x, y) + i\psi(x, y),$$

where ϕ and ψ are real-valued functions, we write

$$\overline{f(z)} = \phi(x, y) - i\psi(x, y).$$

To obtain this we merely change i to $-i$ wherever it appears. By contrast, we define

$$\overline{f}(z) = \overline{f(\bar{z})},$$

which simply says change i to $-i$ except when it occurs in z. For example, if

$$f(z) = i \sin z,$$

then

$$\overline{f}(z) = -i \sin z$$

and

$$\overline{f(z)} = -i \sin \bar{z}.$$

2.3   Sequences and Series of Functions

When speaking of sequences of complex functions $\{f_n(z)\}$, it will be tacitly assumed that the members $f_n(z)$ of the sequence are defined in some common domain.

By convergence of $\{f_n(z)\}$ to a function, say $f(z)$, in a domain D, we mean that for any z_0 belonging to D,

$$\lim_{n\uparrow\infty} f_n(z_0) = f(z_0).$$

The convergence is said to be *uniform* if for any $\epsilon > 0$ we can find an $N(\epsilon)$ independent of z so that

$$|f(z) - f_n(z)| < \epsilon, \quad n > N$$

for all z belonging to D. The passage to series is done in the usual way; viz., we consider partial sums

$$\sum_{n=0}^{N} g_n(z) = S_N(z)$$

and consider convergence of the sequence $\{S_N\}$.

An example of convergence that is not uniform is provided by the series

$$1 + z + z^2 + \cdots,$$

which converges to

$$\frac{1}{1-z}$$

in the disc $|z| < 1$. To obtain this limit, note that the partial sum is given by (Exercise 10(a), Chapter 1)

$$S_N = 1 + z + z^2 + \cdots + z^N = \frac{1 - z^{N+1}}{1 - z}.$$

Therefore

$$\left| \frac{1}{1-z} - S_N \right| = \frac{|z^{N+1}|}{|1-z|}.$$

This tends to zero for $N \uparrow \infty$ and any z such that $|z| < 1$. Why is the convergence nonuniform?

The various tests considered in Section 2.1 again apply. As an example, consider the

M-Test. Suppose $\{g_n(z)\}$ is defined in the domain D such that for $n > N$ and z in D

$$|g_n(z)| < M_n,$$

where the M_n are real positive constants such that $\sum_{n=0}^{\infty} M_n < \infty$. Then it is clear that $\sum_{n=0}^{\infty} g_n(z)$ is uniformly and absolutely convergent. For example, with the series $\sum_n (z/2)^n$, $|(z/2)^n| \leq 1/2^n$ for $n \geq 0$ and $|z| < 1$. Thus $M_n = 2^{-n}$ and, as we know, $\sum_n 2^{-n} = 2$.

Next suppose

$$\sum_{n=0}^{\infty} g_n(z) = g(z)$$

uniformly for z in D. If the terms of the series, $g_n(z)$, are continuous, then so is the limit of the series, $g(z)$. To see this, we examine the continuity of $g(z)$ at a point z_1 in D. Write

$$g(z_1) - g(z) = \left(g(z_1) - \sum_{n=0}^{N} g_n(z_1) \right) + \left(\sum_{n=0}^{N} g_n(z_1) - \sum_{n=0}^{N} g_n(z) \right)$$

$$+ \left(\sum_{n=0}^{N} g_n(z) - g(z) \right).$$

If we take the absolute value of both sides, we obtain

$$|g(z_1) - g(z)| \leq \left|g(z_1) - \sum_{n=0}^{N} g_n(z_1)\right| + \left|\sum_{n=0}^{N} g_n(z_1) - \sum_{n=0}^{N} g_n(z)\right|$$

$$+ \left|\sum_{n=0}^{N} g_n(z) - g(z)\right|.$$

It is clear that each term of the right-hand side can be made small. Formally, given an $\epsilon > 0$, we want to show there exists a $\delta(\epsilon) > 0$ so that the left-hand side is less than ϵ for $|z - z_1| < \delta(\epsilon)$. To accomplish this, we note that from the uniform convergence we do not have to worry about z and can find an $N_1(\epsilon)$ such that the first and third terms are each less than $\epsilon/3$ for any $N > N_1(\epsilon)$. Take $N = N_1 + 1$. Then, since $\sum_{n=0}^{N_1+1} g_n(z)$ is the finite sum of continuous functions, it too is continuous. Hence we can choose a $\delta(\epsilon) > 0$ so that for $|z - z_1| < \delta(\epsilon)$ the middle term is less than $\epsilon/3$, and this proves the result.

Power Series

In what follows we primarily will be concerned with power series. The sum

$$\sum_{n=0}^{\infty} a_n(z - z_0)^n \tag{2.3}$$

is said to be an (infinite) power series at z_0.

Basic Theorem for Power Series. *Every power series (2.3) has a radius of convergence R such that the series converges for $|z - z_0| < R$ and diverges for $|z - z_0| > R$. Furthermore,*

$$R = 1/\varlimsup_{n\uparrow\infty} |a_n|^{1/n}. \tag{2.4}$$

By $R = 0$ we mean that the series diverges for $z \neq z_0$.

Proof: Consider

$$(|a_n(z - z_0)^n|)^{1/n} = |a_n|^{1/n}|z - z_0|.$$

First we suppose

$$\varlimsup_{n\uparrow\infty} |a_n|^{1/n} = \infty$$

so that, by (2.4), $R = 0$. Then convergence of the series for $z = z_0$ is a trivial matter and divergence for $z \neq z_0$ is clear from the Divergence Criterion, where $|z - z_0| \neq 0$ and $|a_n(z - z_0)^n| \not\to 0$ as $n \uparrow \infty$.

Next suppose

$$\frac{1}{R} = \varlimsup_{n\uparrow\infty} |a_n|^{1/n} < \infty.$$

From this it follows that for any small $\epsilon > 0$ we can choose n sufficiently large such that

$$R - \frac{1}{|a_n|^{1/n}} < \epsilon$$

or

$$|a_n| < \frac{1}{(R-\epsilon)^n}.$$

Thus, if $|z - z_0| \leq R_1 < R$ and n is sufficiently large,

$$|a_n(z - z_0)^n| < \left(\frac{R_1}{R-\epsilon}\right)^n.$$

Hence, from the M-Test, the series converges for $|z - z_0| < R - \epsilon$. And since ϵ can be made arbitrarily small the series converges for $|z - z_0| < R$. On the other hand, the *Divergence Criterion* shows that the series diverges for $|z - z_0|/R > 1$.

The construction of the radius of convergence R in particular cases (see the Examples below and Exercise 7 at the end of this chapter) can be found from

$$R = \lim_{n \uparrow \infty} |a_n/a_{n+1}| \tag{2.5}$$

or

$$R = \lim_{n \uparrow \infty} 1/|a_n|^{1/n}, \tag{2.6}$$

depending on whether or not these limits exist and the ease of calculation of the ratio (2.5) of the nth root (2.6).

Examples

$$\frac{z}{1} - \frac{z^2}{2} + \frac{z^3}{3} - \frac{z^4}{4} + \cdots \quad (= \ln(1 + z)). \tag{2.7}$$

By any of the tests the radius of convergence is $R = 1$, so that the series (2.7) converges in the unit circle. As this example illustrates, no general statement can be made about convergence *on* the unit circle. For example, at $z = 1$ the series is

$$1 - \frac{1}{2} + \frac{1}{3} - \frac{1}{4} \pm \cdots,$$

which converges to $\ln 2$; while for $z = -1$ the series is

$$-\left(1 + \frac{1}{2} + \frac{1}{3} + \cdots\right),$$

which diverges. As is seen from the above, this is equivalent to trying to evaluate the logarithm of zero. The point $z = -1$ is in fact what halts the convergence of the series.

As another example, we go back to the series for the exponential

$$\exp(z) = \sum_{n=0}^{\infty} \frac{z^n}{n!}. \tag{2.8}$$

Any of the above criteria for the radius of convergence can be applied. The simplest is

$$R = \lim_{n\uparrow\infty} |a_n/a_{n+1}| = \lim_{n\uparrow\infty} \frac{1}{n!} \Big/ \frac{1}{(n+1)!} = \lim_{n\uparrow\infty}(n+1) = \infty.$$

The series converges in the entire z-plane.
    As a last example consider

$$1 + 2z + z^2 + (2z)^3 + z^4 + (2z)^5 + \cdots.$$

The coefficients a_n are

$$a_n = 1, 2, 1, 8, 1, 32, 1, 128, \ldots.$$

In this case the lim sup formula of (2.4) must be used:

$$R = 1/\overline{\lim_{n\uparrow\infty}} |a_n|^{1/n} = 1/\lim_{n\uparrow\infty}(2^n)^{1/n} = \frac{1}{2}.$$

This contrived series arises from the expansion, for small values of z, of

$$\frac{1}{1-z^2} + \frac{2z}{1-(2z)^2}.$$

The first term has singularities at $z = \pm 1$ and the second, at $z = \pm 1/2$. It is, of course, the second, smaller radius of convergence which dominates the calculation.
    Since the exponential (2.8) now has been made *legitimate* by the above discussion, we are at liberty to use it to define other functions. For example, we write

$$\cosh z = \frac{e^z + e^{-z}}{2}, \quad \sinh z = \frac{e^z - e^{-z}}{2} \tag{2.9}$$

and also

$$\cos z = \frac{e^{iz} + e^{-iz}}{2}, \quad \sin z = \frac{e^{iz} - e^{-iz}}{2i}. \tag{2.10}$$

In both cases we have extended to complex variables the definition given in the real case. Observe that the trigonometric and hyperbolic functions can now be linked as follows:

$$\cosh iz = \cos z, \quad \sinh iz = i \sin z. \tag{2.11}$$

Exercises

1. Prove the following tests of convergence: (a) Absolute Convergence Criterion, (b) M-test, (c) Root test.

2. For sequences $\{z_n\}$, determine whether the following are true or false and give reasons why:

   (a) z_n converges implies $|z_n|$ converges.

   (b) z_n converges implies $\arg z_n$ converges.

   (c) $\arg z_n$ and $|z_n|$ converge implies z_n converges.

   (d) $|z_n|$ converges implies z_n converges.

3. State if the following converge absolutely, converge conditionally, or diverge:

   (a) $\displaystyle\sum_{n=1}^{\infty} \frac{i^n}{n^2}$,

   (b) $\displaystyle\sum_{n=1}^{\infty} \frac{i^n}{n}$,

   (c) $\displaystyle\sum_{n=1}^{\infty} \frac{(1+i)^n}{n}$,

   (d) $\displaystyle\sum_{n=1}^{\infty} \frac{e^{2n}}{n^2}$.

4. Is the converse of the Ratio Test (i.e., if $\sum_{n=0}^{\infty} b_n$ is absolutely convergent, then $|b_{n+1}/b_n| \le r < 1$ for all n sufficiently large) true or false? Support your assertion.

5. In connection with the Basic Theorem for Power Series show that the radius of convergence R can be computed by (2.5) or (2.6), i.e., by

   (a) $\lim_{n\uparrow\infty} |a_n/a_{n+1}|$, if it exists, or

   (b) $\lim_{n\uparrow\infty} 1/|a_n|^{1/n}$, if it exists.

   [Hint: Make use of Ratio and Root Tests.]

6. Test the following series for convergence:

   (a) $\displaystyle\sum_{n=0}^{\infty} \frac{i^n}{2^n}$,

   (b) $\displaystyle\sum_{n=2}^{\infty} \frac{i^n}{\ln n}$,

   (c) $\displaystyle\sum_{n=1}^{\infty} \frac{(1+2i)^n}{n^2}$.

7. Find the radius of convergence R for each of the following series:

(a) $\displaystyle\sum_{n=1}^{\infty} \frac{z^n}{n^3}$,

(b) $\displaystyle\sum_{n=0}^{\infty} nz^n$,

(c) $\displaystyle\sum_{n=0}^{\infty} 2^n(z-1)^n$,

(d) $\displaystyle\sum_{n=1}^{\infty} \frac{z^n}{n^n}$,

(e) $\displaystyle\sum_{n=0}^{\infty} \frac{(z-i)^n}{4^n}$,

(f) $\displaystyle\sum_{n=1}^{\infty} (\ln n)^2 z^n$.

8. Prove the following properties of power series:

   (a) If $f(z)$ and $g(z)$ have power series in a common domain D (where for any $z_0 \in D$ the power series of $f(z)$ and $g(z)$ have the same radius of convergence R), then the function $(f+g)$ has a power series in D, which is given by the addition of the individual power series.

   (b) The power series of a function is unique.

   (c) Under the same hypothesis as (a), the product (fg) has a power series in D, which is given by the product of the individual power series.

9. For $0 \le r < 1$ sum the series

   (a) $\displaystyle\sum_{n=0}^{\infty} r^n e^{in\theta}$

   (b) $\displaystyle\sum_{n=-\infty}^{\infty} r^{|n|} e^{in\theta}$

   [Hint: These are geometrical series.]

10. Show

   (a) $\displaystyle \sinh z = \sum_{n=0}^{\infty} z^{2n+1}/(2n+1)! = -i \sin iz$,

   (b) $\displaystyle \cosh z = \sum_{n=0}^{\infty} z^{2n}/2n! = \cos iz$,

(c) $\sin z = \sin x \cosh y + i \cos x \sinh y$,

(d) $|\cos z|^2 = \cos^2 x + \sinh^2 y$,

(e) $\cosh z = \cosh x \cos y + i \sinh x \sin y$,

(f) $|\sinh z|^2 = \sinh^2 x + \sin^2 y$.

11. Show

(a) $\tanh \dfrac{z}{2} = \left(\dfrac{\sinh z/2}{\cosh z/2} \right) = \dfrac{\sinh x + i \sin y}{\cosh x + \cos y}$,

(b) $\left(\dfrac{ia - 1}{ia + 1} \right)^{ib} = \exp(-2b \cot^{-1} a)$.

12. Find the zeros of

(a) $\sin z$,

(b) $\cos z$,

(c) $\sinh z$,

(d) $\cosh z$.

3

Differentiation and Integration

3.1 Differentiation: Cauchy-Riemann Equations

To start the present discussion of differentiation, we recall the results of Exercise 6 at the end of Chapter 1, namely that the addition and subtraction of

$$z = x + iy, \quad \bar{z} = x - iy$$

yields

$$x = \frac{z + \bar{z}}{2}, \quad y = \frac{z - \bar{z}}{2i}. \tag{3.1}$$

Therefore any function $f = f(x, y)$ in the complex plane can be written as

$$f = f\left(\frac{z + \bar{z}}{2}, \frac{z - \bar{z}}{2}\right) = F(z, \bar{z}).$$

For example,

$$f = x^2 + y^2 = z\bar{z}.$$

An especially useful class of functions are those which do not explicitly depend on $\bar{z}$. More precisely, we want to examine functions of the plane possessing partial derivatives f_x and f_y for which

$$\frac{\partial f}{\partial \bar{z}} = 0. \tag{3.2}$$

The meaning of this derivative follows from (3.1) and the chain rule. Specifically,

$$\frac{\partial}{\partial \bar{z}} = \frac{\partial x}{\partial \bar{z}}\frac{\partial}{\partial x} + \frac{\partial y}{\partial \bar{z}}\frac{\partial}{\partial y} = \frac{1}{2}\left(\frac{\partial}{\partial x} + i\frac{\partial}{\partial y}\right),$$

since $\partial x/\partial \bar{z} = 1/2$ and $\partial y/\partial \bar{z} = i/2$. Thus, if we write $f = \phi + i\psi$ and substitute this expression for f in (3.2), we obtain

$$0 = \frac{\partial f}{\partial \bar{z}} = \frac{1}{2}(\phi_x + i\psi_x) + \frac{i}{2}(\phi_y + i\psi_y).$$

Separating this into real and imaginary parts of this equation yields

$$\frac{\partial \phi}{\partial x} = \frac{\partial \psi}{\partial y}, \quad \frac{\partial \phi}{\partial y} = -\frac{\partial \psi}{\partial x}. \tag{3.3}$$

These are referred to as the *Cauchy-Riemann equations. We define* $f = \phi + i\psi$ *to be analytic in a domain* D *if in it* f *is single-valued and satisfies the Cauchy-Riemann equations* (3.3).

From the above formalism, we can write $f(z)$ for an analytic function, and it would seem reasonable to write a derivative of it as df/dz. However, some ambiguity appears on closer inspection. For if we use the customary definition of derivative and write

$$\frac{df}{dz} = f'(z) = \lim_{\Delta z \to 0} \frac{f(z + \Delta z) - f(z)}{\Delta z} \tag{3.4}$$

the increment $\Delta z = \Delta x + i\Delta y = he^{i\theta}$ enters in the denominator. This segment may assume any orientation. Equation (3.4) thus appears to say that the evaluation of $f'(z)$ depends on θ. So, for (3.4) to make sense,

$$\lim_{\Delta z \to 0} \frac{f(z + \Delta z) - f(z)}{\Delta z} = \lim_{h \to 0} \frac{f(z + he^{i\theta}) - f(z)}{he^{i\theta}}$$

$$= \lim_{h \to 0} \frac{f(x + h\cos\theta + i(y + h\sin\theta)) - f(x + iy)}{h\cos\theta + ih\sin\theta}$$

should be independent of θ! This in fact is true in the case of analytic functions and is a consequence of the Cauchy–Riemann equations (3.3) (this is left to the exercises).

In the above discussion we have seen that if a function f is only dependent on z and f_x and f_y exist, then f satisfies the Cauchy–Riemann equations (3.3). Therefore, merely by inspection, we can say that functions such as $\sin z$, $\cosh z$, and z^n are analytic, and, if separated into real and imaginary parts, will satisfy the Cauchy–Riemann equations. Conversely, we may ask, after substituting (3.1) for x and y in

$$f = \phi(x, y) + i\psi(x, y),$$

whether or not f will be a function of z only. If f_x and f_y exist, the question is readily decided by the Cauchy–Riemann equations. For example, if

$$f = x^2 - y^2 + 2ixy,$$

then

$$\frac{\partial}{\partial x}(x^2 - y^2) = \frac{\partial}{\partial y}(2xy)$$

and

$$\frac{\partial}{\partial y}(x^2 - y^2) = -\frac{\partial}{\partial x}(2xy).$$

Hence (3.3) is satisfied, and, in fact, the substitution (3.1) gives $f = z^2$.

Exercises

1. Verify the Cauchy–Riemann equations for
   (a) $w = z^2$,
   (b) $w = e^z$,
   (c) $w = \sin z$.

2. Which of the following are analytic:
   (a) $f = xy + iy$,
   (b) $f = e^x \cos y + ie^x \sin y$,
   (c) $f = 1/(1 + z)$.

3. Prove that if $f(z)$ is analytic, then the limit in (3.4) is independent of the orientation of Δz.

4. For which values of the coefficients a and b are the following analytic:
   (a) $x + ayi + bxy$, (b) $x^2 + axyi + by^2$?

5. Show that the Cauchy–Riemann equations in polar coordinates, with $z = re^{i\theta}$, are
   $$\frac{\partial \phi}{\partial r} = \frac{1}{r} \frac{\partial \psi}{\partial \theta}, \qquad \frac{\partial \psi}{\partial r} = -\frac{1}{r} \frac{\partial \phi}{\partial \theta}.$$

6. Prove that if $f(z)$ and $g(z)$ are analytic on D, then $f + g$ and fg are analytic on D.

7. Show that the only real function which is analytic is a constant.

8. Show that $d\bar{z}/dz$ does not exist. [Hint: Form the differential quotient and show that the result depends on θ as $h \to 0$.]

3.2 Integration: Cauchy's Integral Theorem

The idea of a line integral in the z-plane is a simple variant of the line integral in the (x, y)-plane. Specifically, if Γ_{ab} is a simple path (not closed or self-crossing) between the points a and b in the z-plane, then by

$$\int_{\Gamma_{ab}} f(z)dz = \int_{\Gamma_{ab}} (\phi + i\psi)(dx + idy) = \int_{\Gamma_{ab}} [(\phi dx - \psi dy) + i(\phi dy + \psi dx)]$$

$$(3.5)$$

we mean the limit (if it exists) of the Riemann polygonal approximation

$$\sum_j f(z_j)(z_{j+1} - z_j) = \sum_j f_j \Delta z_j$$

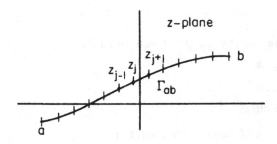

FIGURE 3.1.

as the maximum of the segmental lengths $|z_{j+1} - z_j|$ approaches zero (see Figure 3.1). It is clear that by separating the above expressions into their real and imaginary parts, we reduce them to real Riemann sums and integrals. In particular, we can write the polygonal approximation as

$$\sum_j f_j \Delta z_j = \sum_j [(\phi_j \Delta x_j - \psi_j \Delta y_j) + i(\phi_j \Delta y_j + \psi_j \Delta x_j)].$$

This reduces the complex Riemann sum to two real Riemann sums and the limit leading to the right-hand side of (3.5) is the same as the real case and requires no elaboration. (Note that (3.5) and the above polygonal approximation depend on the specific details of the path Γ_{ab}.)

Observe that if the curve Γ_{ab} is specified parametrically, say, by $z = z(t)$, $z(0) = a$, and $z(1) = b$, then (3.5) can be written as

$$\int_{\Gamma_{ab}} f(z)dz = \int_0^1 f(z)\frac{dz}{dt}dt = \int_0^1 [(\phi\dot{x} - \psi\dot{y}) + i(\phi\dot{y} + \psi\dot{x})]dt,$$

where the dot signifies differentiation with respect to t.

Stokes' Theorem in the plane states that for functions ϕ and ψ with continuous partial derivatives,

$$\oint_C (\phi dx + \psi dy) = \int_R \left(\frac{\partial\psi}{\partial x} - \frac{\partial\phi}{\partial y}\right) dx dy, \qquad (3.6)$$

where C is a simple (non-crossing) loop and R is the domain enclosed by it. As indicated by the arrow, C has the sense of direction which leaves R on its left as it is traversed. For example, if C is circle-like, its sense of direction is counterclockwise.

If $f = \phi + i\psi$ has continuous partial derivatives in a simply connected domain, then, from the expression for $df/d\bar{z}$ in Section 3.1,

$$2i \int_R \frac{\partial f}{\partial\bar{z}}dx dy = i \int_R \left(\frac{\partial\phi}{\partial x} - \frac{\partial\psi}{\partial y}\right) dx dy - \int_R \left(\frac{\partial\phi}{\partial y} + \frac{\partial\psi}{\partial x}\right) dx dy.$$

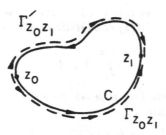

FIGURE 3.2.

Then if we apply Stoke's Theorem, (3.6), to the real and imaginary parts of the last relation,

$$i \int_R \left(\frac{\partial \phi}{\partial x} - \frac{\partial \psi}{\partial y} \right) dx dy - \int_R \left(\frac{\partial \phi}{\partial y} + \frac{\partial \psi}{\partial x} \right) dx dy$$

$$= i \oint_C (\psi dx + \phi dy) - \oint_C (-\phi dx + \psi dy)$$

$$= \oint_C (\phi + i\psi)(dx + idy) = \oint_C f(z) dz.$$

In the above, C has been assumed to be a simple loop lying in the domain D. This demonstration provides the extension of Stokes' Theorem to the complex plane. In summary, under the above-stated conditions we have shown

$$\oint_C f(z) dz = 2i \oint_R \frac{\partial f}{\partial \bar{z}} dx dy. \tag{3.7}$$

It should be recalled that Stokes' Theorem requires the domain to be simply connected.

From (3.7) we immediate have *Cauchy's Integral Theorem: If $f(z)$ is analytic on a simply connected domain D, then*

$$\oint_C f(z) dz = 0$$

for any simple loop C in D.

If z_0 and z_1 are any two points of C, we can split C into $\Gamma_{z_0 z_1}$ and $\Gamma'_{z_0 z_1}$, as shown in Figure 3.2, and write

$$\oint_C f(z) dz = \int_{\Gamma_{z_0 z_1}} f(z) dz - \int_{\Gamma'_{z_0 z_1}} f(z) dz = 0$$

or

$$\int_{\Gamma_{z_0 z_1}} f(z) dz = \int_{\Gamma'_{z_0 z_1}} f(z) dz. \tag{3.8}$$

This construction with the result, (3.8), then gives us *Cauchy's Integral Theorem (second form)*: If $f(z)$ is analytic on the simply connected domain D, then the value of

$$\int_{z_0}^{z_1} f(z)dz$$

is independent of the path for all paths lying entirely in D.

The converse is also true:

Morera's Theorem. *If in a simple domain D the value of*

$$\int_{z_0}^{z_1} f(z)dz$$

is independent of the path for all paths lying in D, then $f(z)$ is analytic. (See Ahlfors.)

Exercises

1. (a) Evaluate the integral

   $$\int_{1}^{e^{i\theta}} \frac{1}{z}dz$$

   by integrating along the circular arc $z = \cos\alpha + i\sin\alpha$, $0 \le \alpha \le \theta$.

   (b) Consider

   $$\int_{2}^{2e^{i\theta}} \frac{2z\,dz}{z^2 + 1},$$

   where the integration is along $|z| = 2$. Integrate by setting $z^2 + 1 = w$.

   [Hint: Carefully check the angular excursion in the w-plane.]

2. (a) Suppose $f = \phi + i\psi$ is analytic in some domain D. Prove that for any path $z_0 \to z$ in D

   $$f = \phi + i\int_{z_0}^{z}\left(-\frac{\partial\phi}{\partial y}dx + \frac{\partial\phi}{\partial x}dy\right) + i\psi_0,$$

   where $\psi_0 = \operatorname{Im} f(z_0)$.

   (b) Find an analytic function f for which $\operatorname{Re} f = x/(x^2 + y^2)$.

3. Prove that if C is a simple closed contour, then

   $$\frac{1}{2i}\oint_C \bar{z}\,dz$$

   is the area enclosed by C.

4. If f and g are analytic on D and C is a loop in D, then

$$\int_R \overline{f'}g\,dx\,dy = \frac{1}{2i} \oint_C \overline{f(z)}g(z)dz,$$

where R is the region enclosed by C. [Hint: $\overline{f'(z)} = \partial\overline{f}/\partial\overline{z}$.]

5. Show that

$$\oint z^n dz = \left\{ \begin{array}{ll} 2\pi i, & n = -1, \\ 0, & n \neq -1. \end{array} \right\}$$

The contour encloses the origin. [Hint: Take the contour to be a circle around the origin.]

6. Show that

$$\int_0^{1+i} z^* dz$$

depends on the path of integration. (Take two different convenient paths of integration.)

7. Evaluate each of the following by choosing a convenient path:

(a) $\int_0^{\pi+i} \cos(z/2)dz$,

(b) $\int_0^{i\pi} e^z\,dz$,

(c) $\int_{-1}^{i}(1+iz^2)dz$.

3.3 Differentiation and Integration of Power Series

We will have need of the following two theorems, both of which are virtually evident. (See Exercises 1 and 2 at the end of this section.)

If $f(z)$ is analytic on a domain D and a and b are two points that lie in D, then

$$\int_a^b f'(z)dz = f(b) - f(a) \tag{3.9}$$

for every path in D.

If $f(z)$ is analytic on a domain D and

$$F(z) = \int_a^z f(\zeta)d\zeta$$

with the path in D, then F is analytic and

$$\frac{dF}{dz} = f(z).$$

Consider the power series at z_0

$$f(z) = \sum_{n=0}^{\infty} a_n(z - z_0)^n, \tag{3.10}$$

which we assume has a radius of convergence $R > 0$. Although each term of the series (3.10) is analytic, we cannot yet say that $f(z)$ is analytic. This, in part, is what we want to show in this section.

The formal integration of the series (3.10) between any two points a and z lying in D (the path connecting a and z must lie in D as well) is

$$\sum_{n=0}^{\infty} a_n \int_a^z (\zeta - z_0)^n d\zeta = \sum_{n=0}^{\infty} \frac{a_n}{n+1}(\zeta - z_0)^{n+1}\big|_a^z$$

$$= \sum_{n=0}^{\infty} \frac{a_n}{n+1}(z - z_0)^{n+1} - \sum_{n=0}^{\infty} \frac{a_n}{n+1}(a - z_0)^{n+1}. \tag{3.11}$$

Since

$$\lim_{n \uparrow \infty} \left(\frac{|a_n|}{n+1} \right)^{1/(n+1)} = \lim_{n \uparrow \infty} |a_n|^{1/n}$$

(where $\lim n^{1/n} = 1$ since $n^{1/n} = e^{(1/n)\ln n}$), the radius of convergence of the formal series (3.11) is the same as that of (3.10). If we write

$$F(z) = \sum_{n=0}^{\infty} \frac{a_n(z - z_0)^{n+1}}{n+1}, \tag{3.12}$$

then (3.11) is $F(z) - F(a)$. In what follows, we avoid carrying the constant $F(a)$ by considering the integration from z_0. Next we write

$$f_N = \sum_{n=0}^{N} a_n(z - z_0)^n, \quad F_N = \sum_{n=0}^{N} \frac{a_n(z - z_0)^{n+1}}{n+1},$$

from which we see that

$$F_N(z) = \int_{z_0}^z f_N(z)dz.$$

We observe that each term in the series (3.10) is continuous and that for $|z - z_0| \le R_1 < R$ the series is uniformly convergent. Therefore, from an earlier assertion in Section 2.3, $f(z)$ is continuous, so that f and $f - f_N$ are integrable. For any path between z_0 and z in D,

$$\int_{z_0}^z (f - f_N)dz = \int_{z_0}^z f\,dz - F_N,$$

which implies

$$\left| \int_{z_0}^{z} f \, dz - F_N \right| = \left| \int_{z_0}^{z} (f - f_N) dz \right|.$$

If we take the modulus under the integration sign, then

$$\left| \int_{z_0}^{z} f \, dz - F_N \right| \le \int_{z_0}^{z} |f - f_N| |dz|.$$

Note that

$$|dz| = |dx + idy| = (dx^2 + dy^2)^{1/2}$$

is the differential of arc length. From the uniform convergence of the series, we can choose N sufficiently large so that for any $\epsilon > 0$ and for $|z - z_0| \le R_1 < R$, $|f - f_N| < \epsilon$. Hence

$$\left| \int_{z_0}^{z} f \, dz - F_N \right| \le \epsilon \int_{z_0}^{z} |dz|.$$

The last integral is simply the arc length, which we take to be a finite quantity. We have therefore proven that F_N converges to

$$\int_{z_0}^{z} f(z) dz$$

and hence that

$$F(z) = \int_{z_0}^{z} f(z) dz.$$

It is furthermore clear that $F(z)$ is independent of the path of integration because, as can be seen from (3.12), $F(z)$ depends only on the endpoints z_0 and z. It then follows from Morera's Theorem that $f(z)$ is analytic and from the second result mentioned at the outset of the section that $F(z)$ is also analytic with

$$F'(z) = f(z).$$

To consider the derivative of (3.10), we examine the result of formal differentiation

$$\sum_{n=0}^{\infty} n a_n (z - z_0)^{n-1}. \tag{3.13}$$

The radius of convergence is determined by

$$\lim_{n \uparrow \infty} \sqrt[n]{n |a_n|};$$

but as before we make use of the fact that $\lim n^{1/n} = 1$ and find that (3.13) also has R for a radius of convergence. By integrating (3.13), we reduce the present situation to the case discussed above. In particular, we have immediately that

$$f'(z) = \sum_{n=0}^{\infty} na_n(z - z_0)^{n-1}.$$

Both steps, integration and differentiation, can be repeated indefinitely since, after each step, we are left with essentially the same conditions. In summary, *a power series within the domain defined by its radius of convergence is an analytic function, which can be differentiated and integrated, term-by-term, any number of times without loss of analyticity and change of radius of convergence.*

We note in passing that comparison of (3.13) with (3.12) indicates that integration accelerates convergence while differentiation has the opposite effect—although for analytic functions this result is unimportant.

Exercises

1. Prove that if $f(z)$ is analytic on a domain D, then

$$\int_a^b f'(z)dz = f(b) - f(a)$$

   for every path in D. [Hint: Separate the integral into real and imaginary parts and use the Cauchy–Riemann equations.]

2. Prove that if $f(z)$ is analytic on a domain D and

$$F(z) = \int_a^z f(\zeta)d\zeta$$

   with the path in D, then

$$F'(z) = f(z).$$

   [Hint: Write $F = \Phi + i\Psi$ and use the fact that the integral of an analytic function is independent of the path of integration to show Φ and Ψ satisfy the Cauchy–Riemann equations.]

3.4   Cauchy's Integral Formula. Cauchy's Theorem in Multiply Connected Domains

Consider the doubly connected domain D pictured in Figure 3.3 (unhatched region). Cauchy's Integral Theorem can be extended to D by introducing a barrier B (which is just an artificial boundary), as indicated by the dashed

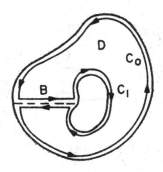

FIGURE 3.3.

line. In fact, if B is regarded as part of the boundary, the domain becomes simply connected and Cauchy's Integral Theorem applies directly. Consider, in particular, the closed loop formed by $-B + C_0 + C_i + B$. Then, by Cauchy's Integral Theorem, if $f(z)$ is analytic on D,

$$\int_{-B} f\, dz + \int_{C_0} f\, dz + \int_{C_i} f\, dz + \int_B f\, dz = 0.$$

Since f is single-valued (by definition), the first and last integrals cancel, and we are left with

$$\oint_{C_0} f\, dz = \oint_{-C_i} f\, dz$$

for *arbitrary* loops C_0 and C_i. We have therefore proven that in a doubly connected domain,

$$\oint_C f(z)\, dz = \text{constant}$$

for all simple paths C looping the inner boundary. In other words,

$$\oint_C f(z)\, dz$$

is either zero or a constant (which also may be zero), depending on whether C can be shrunken to zero or the boundary.

In general, if $f(z)$ is analytic on an $(N+1)$-tuply connected domain, then

$$\oint_{C_0} f\, dz + \oint_{C_1} f\, dz + \oint_{C_2} f\, dz + \cdots + \oint_{C_N} f\, dz = 0,$$

as is easily established by the introduction of barriers B_i, $i = 1, \ldots, N$ (see Figure 3.4). The paths C_i are such that D lies on the left as we move along

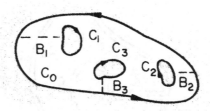

FIGURE 3.4.

them. An arbitrary loop integral

$$\oint_C f(z)dz$$

can then be represented in terms of the N constants

$$\oint_{C_i} f(z)dz, \quad i = 1, \ldots, N.$$

Exam,   …nsider the function

$$f(z) = \frac{1}{z - z_0}.$$

Clearly this is analytic in the doubly connected domain $z \neq z_0$. Then

$$\oint_C \frac{1}{z - z_0} dz = \oint_{\overline{C}} \frac{1}{z - z_0} dz,$$

where C and $\overline{C}$ are loops containing z_0 in their interiors and $\overline{C}$ is some convenient loop. A convenient choice of $\overline{C}$ is the unit circle centered at z_0; i.e., set $z = z_0 + e^{i\theta}$, $0 \leq \theta \leq 2\pi$. With this choice we have

$$\oint_{\overline{C}} \frac{1}{z - z_0} dz = \int_0^{2\pi} \frac{d(e^{i\theta})}{e^{i\theta}} = i \int_0^{2\pi} d\theta = 2\pi i.$$

Therefore

$$\oint_C \frac{1}{z - z_0} dz = 2\pi i \tag{3.14}$$

for any loop C containing z_0. By contrast the function

$$f(z) = \frac{1}{(z - z_0)^2},$$

which is also analytic in the doubly connected domain $z \neq z_0$, has a zero integral for any loop (not passing through z_0)—see Exercise 2 at the end of this section.

Next suppose $f(z)$ is analytic in a simply connected domain D containing the point z_0. It then follows that

$$\frac{f(z)}{z - z_0},$$

is analytic in the doubly connected domain $\hat{D}$ obtained from D by deleting the point z_0. Now suppose C loops the point z_0 and consider the integral

$$\frac{1}{2\pi i} \oint_C \frac{f(z)}{z - z_0} dz = \frac{1}{2\pi i} \oint_C \frac{f(z) - f(z_0)}{z - z_0} dz + \frac{f(z_0)}{2\pi i} \oint_C \frac{dz}{z - z_0},$$

where the last integral has been merely added and subtracted. But the coefficient of $f(z_0)$ in the last term was evaluated above, (3.14), and is equal to unity. To evaluate the first term on the right-hand side, observe that the value of the integral is independent of the path looping z_0. Therefore we can take the path to be a small circle around z_0, namely

$$z = z_0 + \delta e^{i\theta}, \quad 0 \leq \theta \leq 2\pi,$$

where $\delta > 0$ is small. Then changing the variable of integration from z to θ yields

$$\frac{1}{2\pi i} \oint_C \frac{f(z) - f(z_0)}{z - z_0} dz = \frac{1}{2\pi i} \int_0^{2\pi} \frac{f(z_0 + \delta e^{i\theta}) - f(z_0)}{\delta e^{i\theta}} i\delta e^{i\theta} d\theta$$

$$= \frac{1}{2\pi} \int_0^{2\pi} [f(z_0 + \delta e^{i\theta}) - f(z_0)] d\theta.$$

But the absolute value of the last expression can be made as small as we please by taking δ small, since f is continuous at the point z_0. Hence this integral vanishes in the limit $\delta \downarrow 0$, and we are left with

$$f(z_0) = \frac{1}{2\pi i} \oint_C \frac{f(z) dz}{z - z_0}$$

for any C in D looping z_0. Thus we have proven

Cauchy's Integral Formula. *If C is a loop in a simply connected domain on which $f(z)$ is analytic, then*

$$f(z) = \frac{1}{2\pi i} \oint_C \frac{f(\zeta) d\zeta}{\zeta - z} \tag{3.15}$$

for any point z lying inside C.

This formula underlies a remarkable property of an analytic function, namely that *it is completely determined in an interior region by its values on the boundary.* As a specific application of this idea we take C to be a circle centered at z and lying in the domain of analyticity, namely,

$$\zeta = z + re^{i\theta}, \quad 0 \leq \theta \leq 2\pi.$$

Under this change of variable (3.15) becomes

$$f(z) = \frac{1}{2\pi i} \int_0^{2\pi} \frac{f(z + re^{i\theta})ire^{i\theta}d\theta}{re^{i\theta}} = \frac{1}{2\pi} \int_0^{2\pi} f(z + re^{i\theta})d\theta. \qquad (3.16)$$

Equation (3.16) states

Mean Value Theorem. *An analytic function f at a point z is equal to the mean value of f around any circle which is centered at z and which only contains points of the domain D of analyticity of f.*

Conversely, *if a function possesses the mean value property (3.16) at every point in its domain, then the function is analytic.* (See, e.g., Nehari.)

Another immediate consequence of Cauchy's Integral Formula is that *analytic functions are infinitely differentiable.* This is seen by direct and repeated differentiation of (3.15), which yields

$$\frac{d^n f}{dz^n} = f^{(n)}(z) = \frac{n!}{2\pi i} \oint_C \frac{f(\zeta)}{(\zeta - z)^{n+1}}d\zeta. \qquad (3.17)$$

Liouville's Theorem

We have already encountered several functions which do not exhibit singularities in the finite portion of the plane. Examples are e^z, $\sinh z$, $\cos z$. Such functions are called *entire*. *Entire functions are everywhere analytic in the finite plane.* However, each of the examples just given diverges at infinity. This, in fact, is typically the case for entire functions.

Liouville's Theorem. *The only functions which are bounded everywhere are constants.*

Proof. Suppose $f(z)$ is entire and bounded, say $|f(z)| < M$. It can be represented by Cauchy's Integral Formula (3.15) and its derivatives, by (3.17). In particular

$$f'(z) = \frac{1}{2\pi i} \oint_C \frac{f(\zeta)d\zeta}{(\zeta - z)^2}.$$

Since $f(z)$ is analytic everywhere, we are at liberty to choose a loop C anywhere in the z-plane. We take the circle

$$\zeta = z + Re^{i\theta}, \quad 0 \leq \theta \leq 2\pi,$$

so that

$$f'(z) = \frac{1}{2\pi} \int_0^{2\pi} \frac{f(z + Re^{i\theta})}{Re^{i\theta}} d\theta.$$

If we take the absolute value of both sides of this equation and bring the absolute value inside the integral we obtain

$$|f'(z)| \le \frac{1}{2\pi R} \int_0^{2\pi} |f(z + Re^{i\theta})| d\theta \le \frac{M}{R}$$

since, by hypothesis, $|f(z)| < M$. Next let $R \uparrow \infty$, which demonstrates that

$$f'(z) = 0.$$

Hence Liouville's Theorem is proven.

A number of additional results flow from Cauchy's Integral Formula (3.15), and many of these form the subjects of the following section. We close this section with one consequence which plays a role in later work. We recall that if

$$f = \phi + i\psi$$

is analytic, then the Cauchy–Riemann equations are satisfied:

$$\frac{\partial \phi}{\partial x} = \frac{\partial \psi}{\partial y}, \quad \frac{\partial \phi}{\partial y} = -\frac{\partial \psi}{\partial x}.$$

Since we now know that f is infinitely differentiable, the same is also true of ϕ and ψ. Therefore, we can cross differentiate these two equations to eliminate either ϕ or ψ and obtain

$$\nabla^2 \phi = \left(\frac{\partial^2}{\partial x^2} + \frac{\partial^2}{\partial y^2} \right) \phi = 0 \quad (= \nabla^2 \psi). \tag{3.18}$$

This is a basic equation of partial differential equations and is called *Laplace's equation*. Thus the real and imaginary parts of an analytic function are harmonic—and because of the special relationship between ϕ and ψ, they are called *harmonic conjugates* of each other. The operator ∇^2 in (3.18) is referred to as the *Laplacian*. It can also be defined in n dimensions by

$$\nabla^2 = \frac{\partial^2}{\partial x_1^2} + \frac{\partial^2}{\partial x_2^2} + \cdots + \frac{\partial^2}{\partial x_n^2},$$

and functions, ϕ, satisfying

$$\nabla^2 \phi = 0$$

are called *harmonic*.

Example. The function

$$\phi = y^3 - 3x^2 y$$

is easily seen to be harmonic, i.e., satisfies (3.18). To find its harmonic conjugate, ψ, we use the Cauchy–Riemann relation,

$$\frac{\partial \phi}{\partial x} = -6xy = \frac{\partial \psi}{\partial y}.$$

On integrating this we obtain

$$\psi = -3xy^2 + f(x).$$

But

$$\frac{\partial \psi}{\partial x} = -\frac{\partial \phi}{\partial y}$$

then implies

$$-3y^2 + f'(x) = -3y^2 + 3x^2$$

and therefore

$$f = x^3 + k$$

where k is a constant. If we write $F(z) = \phi + i\psi$, then

$$F(z) = \phi + i\psi = i(z^3 + k).$$

Exercise 2 at the end of Section 3.2 gives an alternate construction and at the same time demonstrates the existence of a conjugate for each harmonic function.

Exercises

1. Prove that

$$\oint_C \frac{1}{(z-a)(z-b)} dz = 0$$

when C loops both a and b.

2. Show

$$\oint_C \frac{1}{(z-z_0)^n} dz = 0$$

for any integer $n \neq 1$.

3. Evaluate the following integrals:

(a) $\dfrac{1}{2\pi i} \oint_{|z-2|=1} \dfrac{e^z}{z-2} dz,$

(b) $\dfrac{1}{2\pi i} \oint_{|z|=4} \dfrac{\sin z}{z} dz,$

(c) $\dfrac{1}{2\pi i} \oint_{|z|=3/2} \dfrac{1}{(z^2+1)(z^2+4)} dz,$

(d) $\dfrac{1}{2\pi i}\displaystyle\oint_{|z|=1}\dfrac{e^z}{z^5}dz.$

4. Demonstrate that each of the following are harmonic and find the corresponding harmonic conjugate:

(a) $\phi = \ln(x^2 + y^2)^{1/2}$,

(b) $\phi = (x\sin y - y\cos y)\exp(-x)$,

(c) $\phi = 2x(1 - y)$,

(d) $\psi = \cos x \cosh y$,

(e) $\psi = y/(x^2 + y^2)$.

5. The Fundamental Theorem of Algebra states that if

$$P(z) = z^m + a_k z^{m-1} + \cdots + a_m = 0$$

is a polynomial of degree $m \geq 1$, then it has at least one root. Prove this. [Hint: Consider $1/P(z)$ and use Liouville's Theorem in a proof by contradiction.]

3.5 The Taylor and Laurent Expansions

Thus far we have seen that all convergent power series are analytic (see Section 3.3). However, we have also found that all analytic functions can be represented in integral form by means of the Cauchy Integral Formula (3.15). We are now in a position to show that *an arbitrary analytic function possesses a power series representation*. We start with Cauchy's Integral Formula (3.15),

$$f(z) = \frac{1}{2\pi i}\oint_C \frac{f(\zeta)d\zeta}{\zeta - z},$$

with the path C a circle of radius R, say, centered at z_0, and such that $f(z)$ is analytic in and on it (Figure 3.5). We then notice

$$\frac{1}{\zeta - z} = \frac{1}{\zeta - z_0 + z_0 - z} = \frac{1}{(\zeta - z_0)(1 - (z - z_0)/(\zeta - z_0))}.$$

For $|(z-z_0)/(\zeta-z_0)| < 1$ we can expand the last expression in a geometrical expansion:

$$\frac{1}{\zeta - z} = \frac{1}{\zeta - z_0}\left[1 + \frac{z - z_0}{\zeta - z_0} + \left(\frac{z - z_0}{\zeta - z_0}\right)^2 + \cdots\right] = \frac{1}{\zeta - z_0}\sum_{n=0}^{\infty}\left(\frac{z - z_0}{\zeta - z_0}\right)^n.$$

$$(3.19)$$

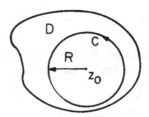

FIGURE 3.5.

The final expression is therefore convergent for $|(z - z_0)/(\zeta - z_0)| < 1$. If we multiply (3.19) by $f(\zeta)/2\pi i$, we obtain

$$\frac{1}{2\pi i} \frac{f(\zeta)}{\zeta - z} = \frac{1}{2\pi i} \frac{f(\zeta)}{\zeta - z_0} \sum_{n=0}^{\infty} \left(\frac{z - z_0}{\zeta - z_0} \right)^n,$$

which for $|z - z_0| \leq R_1 < R$ is a uniformly convergent series of continuous functions in ζ (z and z_0 are regarded as fixed at this stage of the discussion) and therefore can be integrated term-by-term (it is not deemed necessary to provide a proof of this as such a proof would only mimic that given in Section 3.3). We have therefore demonstrated that

$$f(z) = \frac{1}{2\pi i} \oint_C \frac{f(\zeta)}{(\zeta - z)} d\zeta = \sum_{n=0}^{\infty} \frac{(z - z_0)^n}{2\pi i} \oint_C \frac{f(\zeta) d\zeta}{(\zeta - z_0)^{n+1}}$$

$$= \sum_{n=0}^{\infty} \frac{f^{(n)}(z_0)(z - z_0)^n}{n!}, \qquad (3.20)$$

where in the last step we have made use of the representation of derivatives, (3.17). This series identifies the coefficients in a power series with the derivatives of the function at the center of the circle of convergence, and what we have is the infinite *Taylor expansion* of the function, $f(z)$. (The finite form with remainder is left as an exercise.)

From the derivation of (3.20), we also learn over what radius the expansion is valid and what in fact brings this convergence to a halt. Suppose $\hat{R}$ is the radius of convergence of the Taylor expansion (3.20). Then the claim is that the circle $|z - z_0| = \hat{R}$ must contain at least one singularity of the defining function f. For if this were not true (i.e., if f were analytic on $|z - z_0| = \hat{R}$), then f would also be analytic on a neighborhood of this circle, say, on the disk $|z - z_0| \leq \hat{R} + \epsilon$, $\epsilon > 0$. But this would lead to a contradiction since the above derivation of (3.20) would then give $\hat{R} + \epsilon$ as the radius of convergence. Therefore *the radius of convergence of an analytic function f about a point z_0 is the distance from z_0 to the nearest*

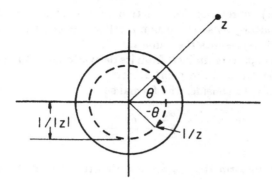

FIGURE 3.6.

singularity of f. This certainly explains why entire functions have power series representations everywhere convergent in the plane.

We return to the simple example

$$\frac{1}{1-z} = \sum_{n=0}^{\infty} z^n, \qquad (3.21)$$

which converges for $|z| < 1$ and which certainly illustrates the above rule about the radius of convergence. Consider the transformation

$$w = 1/z,$$

which, if we write $z = re^{i\theta}$, states

$$|w| = \frac{1}{r}, \quad \arg w = -\arg z = -\theta.$$

It clearly maps the exterior of the unit circle $|z| > 1$ into the interior of the unit circle $|w| < 1$ and vice versa. The mapping is illustrated in Figure 3.6. If we substitute this transformation into $1/(1-z)$, we obtain

$$-\frac{w}{1-w},$$

which, if expanded for $|w| < 1$, gives the convergent power series

$$-\sum_{n=1}^{\infty} w^n, \quad |w| < 1$$

and which, if transformed back to the z-plane, states

$$\frac{1}{1-z} = -\sum_{n=1}^{\infty} \frac{1}{z^n}, \quad |z| > 1. \qquad (3.22)$$

Equation (3.21) represents the function $1/(1 - z)$ inside the unit circle $|z| < 1$ by a convergent series in nonnegative powers. Equation (3.22), on the other hand, represents the same function for $|z| > 1$ by a convergent series in negative powers. In fact, from its derivation, (3.22) can be regarded as a power series around the *point at* ∞.

More generally, to consider the formal series

$$\sum_{n=1}^{\infty} \frac{a_n}{(z - z_0)^n}. \tag{3.23}$$

We can first transform it to a more conventional form by means of the mapping

$$w = \frac{1}{z - z_0}.$$

Then the series becomes

$$\sum_{n=1}^{\infty} a_n w^n,$$

with which we can associate a radius of convergence R by any of the rules given in Section 2.3 (see (2.4)–(2.6)). Then, if we return to the original form   ), we can say that it converges for

$$\frac{1}{|z - z_0|} < R = \frac{1}{\rho}$$

or for

$$|z - z_0| > \rho = \frac{1}{R},$$

i.e., outside a circle of radius ρ centered at z_0.

As another example consider

$$\frac{1}{1 - 2z},$$

which has a singularity at $z = 1/2$. Then, to find an expansion for $|z| > 1/2$, we write

$$\frac{1}{1 - 2z} = -\frac{1}{2z} \cdot \frac{1}{1 - 1/2z} = -\frac{1}{2z}\left(1 + \frac{1}{2z} + \frac{1}{4z^2} + \cdots\right) = -\sum_{n=1}^{\infty} \frac{1}{(2z)^n}. \tag{3.24}$$

If we recall that (3.21) is valid for $|z| < 1$ and that (3.24) is valid for $|z| > 1/2$, then their sum

$$\sum_{n=0}^{\infty} z^n - \sum_{n=1}^{\infty} \frac{1}{(2z)^n}$$

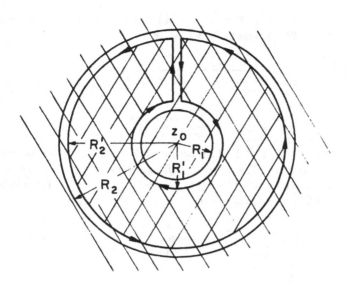

FIGURE 3.7.

is valid for (i.e., converges in the annulus) $1/2 < |z| < 1$. In fact, it represents the function $1/(1-z) + 1/(1-2z)$ in that region and we can write

$$\frac{1}{1-z} + \frac{1}{1-2z} = \sum_{n=0}^{\infty} z^n - \sum_{n=1}^{\infty} \frac{1}{(2z)^n}, \quad \frac{1}{2} < |z| < 1. \qquad (3.25)$$

The sum (3.25) is an example of a *Laurent expansion* and is typical of the representation that can be obtained for an analytic function in an annular domain. Specifically, let us suppose that $f(z)$ is analytic on the domain D given by

$$R_1 < |z - z_0| < R_2,$$

which is depicted in Figure 3.7 as the cross-hatched region. *We can represent $f(z)$ at any point z in D by the Cauchy Integral Formula in the form*

$$f(z) = \frac{1}{2\pi i} \oint_{|\zeta - z_0| = R_2' < R_2} \frac{f(\zeta)d\zeta}{\zeta - z} - \frac{1}{2\pi i} \oint_{|\zeta - z_0| = R_1' > R_1} \frac{f(\zeta)d\zeta}{\zeta - z}. \qquad (3.26)$$

The basis for this representation is indicated in Figure 3.7. We choose as a loop in Cauchy's Integral Formula (3.15), the heavily drawn curve. The contributions along the two vertical portions cancel one another since the function $f(z)$ is single-valued and we are left with (3.26). The strategy to adopt is now clear. In the first of the two integrals, we proceed with a conventional power series expansion, namely (3.19), and then obtain a form corresponding to (3.20). For the second integral we need a development analogous to (3.22). Specifically, we write

$$\frac{1}{\zeta - z} = \frac{1}{\zeta - z_0 + z_0 - z} = -\frac{1}{(z - z_0)(1 - (\zeta - z_0)/(z - z_0))}$$

$$= -\sum_{n=0}^{\infty} \frac{(\zeta - z_0)^n}{(z - z_0)^{n+1}} \tag{3.27}$$

since $|(\zeta - z_0)/(z - z_0)| < 1$. Both series, (3.19) and (3.27), converge; for in the first, ζ lies on the circle $|\zeta - z_0| = R'_2$ and $|z - z_0| < R'_2$, whereas in the second, ζ lies on the circle $|\zeta - z_0| = R'_1$ and $|z - z_0| > R'_1$. If we substitute (3.19) and (3.27) appropriately in (3.26), we obtain

$$f(z) = \sum_{n=0}^{\infty} (z - z_0)^n \frac{1}{2\pi i} \oint_{|\zeta - z_0| = R'_2} \frac{f(\zeta)}{(\zeta - z_0)^{n+1}} d\zeta$$

$$+ \sum_{m=0}^{\infty} \frac{1}{(z - z_0)^{m+1}} \frac{1}{2\pi i} \oint_{|\zeta - z_0| = R'_1} \frac{f(\zeta)}{(\zeta - z_0)^{-m}} d\zeta.$$

The first summation converges for

$$\left| \frac{z - z_0}{\zeta - z_0} \right| = \left| \frac{z - z_0}{R'_2} \right| < 1,$$

or better yet $|z - z_0| < R_2$. It therefore represents an analytic function in the domain that is marked by parallel lines of positive slope in Figure 3.7. Similarly, the second summation converges for $|z - z_0| > R_1$ and represents an analytic function in the domain that is marked by parallel lines of negative slope in Figure 3.7. The two sums are convergent in the intersection of the two domains, i.e., the cross-hatched portion of Figure 3.7.

At this point in the discussion, we remark that the value of

$$\oint_C \frac{f(\zeta)}{(\zeta - z_0)^n} d\zeta$$

for any integer n is independent of the path taken provided the path lies in D. We can therefore take the same path of integration for the integrals in both of the above summations. Also, if in the second summation we change the variable of summation to $-m = n + 1$, we can consolidate terms and write

$$f(z) = \sum_{n=-\infty}^{\infty} a_n (z - z_0)^n, \tag{3.28}$$

$$a_n = \frac{1}{2\pi i} \oint_C \frac{f(\zeta)}{(\zeta - z_0)^{n+1}} d\zeta, \quad n = \ldots, -1, 0, 1, 2, \ldots. \tag{3.29}$$

The contents of the above discussion are known as Laurent's Theorem, and an expansion of the form (3.28) is known as a Laurent expansion.

As a special case, consider functions analytic in some domain that includes the unit circle. We can then employ the Laurent expansion (3.28), with $z_0 = 0$ and with coefficients (3.29) evaluated on the unit circle

$$\zeta = e^{i\theta}, \quad 0 \le \theta < 2\pi.$$

Given this choice, we in fact have

$$a_n = \frac{1}{2\pi i} \int_0^{2\pi} \frac{f(e^{i\theta}) i e^{i\theta} d\theta}{e^{in\theta} e^{i\theta}} = \frac{1}{2\pi} \int_0^{2\pi} f(e^{i\theta}) e^{-in\theta} d\theta. \tag{3.30}$$

The function itself on the unit circle has the representation

$$f(e^{i\theta}) = \sum_{n=-\infty}^{\infty} a_n e^{in\theta}. \tag{3.31}$$

This function, $f(z)$, repeats itself as we move around the unit circle (or any circle on which it is defined). It is therefore a periodic function with period 2π. Equation (3.31) states that such a function can be represented by an infinite sum of harmonics of $\cos\theta$ and $\sin\theta$. Moreover, (3.30) states that coefficients of the series (3.31) are computed by simple *harmonic averages*. Equation (3.31) is an example of a Fourier series, and (3.30) is the formula for the Fourier coefficients. These ideas, as will be seen later in Chapter 5, can be extended to functions which are not analytic.

Although (3.28) and (3.29) give us the tools by which to develop the Laurent expansion of a function, we should try to avoid the labor involved in evaluating the coefficients, (3.29). Inspection is probably the best advice that can be given in regard to obtaining the Laurent expansion of a function.

Example. Consider the function

$$f(z) = \frac{1}{z(1+z^2)}.$$

This has singularities at $z = 0, i, -i$. In the annulus $0 < |z| < 1$, we easily have

$$f(z) = \frac{1}{z}(1 - z^2 + z^4 \mp \cdots)$$

$$= \frac{1}{z} + \sum_{k=1}^{\infty} (-)^k z^{2k-1}, \quad 0 < |z| < 1.$$

For $|z| > 1$ we again expand as a geometrical series to obtain

$$f(z) = \frac{1}{z^3(1 + 1/z^2)} = \frac{1}{z^3} \sum_{k=0}^{\infty} \left(\frac{-1}{z^2}\right)^k$$

$$= \sum_{k=0}^{\infty} \frac{(-)^k}{z^{2k+3}}.$$

Exercises

1. Find the finite Taylor expansion form of (3.20) with remainder for an analytic function. [Hint: Use the identity $(1 - z)(1 + z + z^2 + \cdots + z^{n-1}) = 1 - z^n$.]

2. Prove the following: If $f(z)$ is analytic in a domain which includes $|z - z_0| \le R$ and $|f(z)| \le M$ on $|z - z_0| = R$, then

$$|f^n(z_0)| \le \frac{Mn!}{R^n}, \quad n = 0, 1, 2, \ldots \quad \text{(Cauchy's Inequality).}$$

3. Find the Taylor expansion of

   (a) $1/z$ about $z = 1$,

   (b) e^z about $z = 1$,

   (c) $1/(z(z - 2))$ about $z = 1$.

   What is the radius of convergence in each case?

4. Find the (Laurent or Taylor) expansion of

   (a) $\dfrac{1}{(z - 1)(z - 2)}$ for $1 < |z| < 2$,

   (b) $\dfrac{1}{(z - 1)^2}$ for $|z| > 1$,

   (c) $\dfrac{1}{\sin z}$ for $0 < |z| < \pi$.

   [Hint: For Part (c), first take into consideration the fact that $\sin z$ has a simple zero at $z = 0$.]

5. Derive the binomial expansion

   $$(1 + z)^a = 1 + az + \frac{a(a - 1)}{1 \cdot 2} z^2 + \cdots = \sum_{k=0}^{\infty} \binom{a}{k} z^k$$

   for a real. What is the circle of convergence?

6. Find four terms in the Laurent expansions of

   (a) $\dfrac{\sinh z}{z^2}$, $|z| > 0$,

   (b) $\csc z$, $0 < |z| < \pi$,

   (c) $\dfrac{e^z}{z(1+z^2)}$, $0 < |z| < 1$.

7. Find the Laurent expansion of $1/z^2(1-z)$ for

   (a) $0 < |z| < 1$,

   (b) $|z| > 1$.

8. Find the Taylor expansion of

   (a) $\sin z$ about $z = \pi/4$,

   (b) e^z about $z = 1$,

   (c) $1/z^2$ about $z = 1$.

   What is the circle of convergence in each case?

9. Consider the Laurent expansion

$$\sum_{n=-\infty}^{\infty} a_n z^n.$$

   Indicate the annulus of convergence for each of the following choices of coefficients:

   (a) $a_n = 1$, $n \geq 0$, $a_n = 2^n$, $n < 0$;

   (b) $a_n = 2^n$, $n \geq 0$, $a_n = 1$, $n < 0$;

   (c) $a_n = 1/|n|!$ for all n.

3.6   Singularities of Analytic Functions

Thus far we have used the word *singularity* in a somewhat loose and informal manner. In this section, we give more precision to this term.

Isolated Singularities. Suppose $f(z)$ is analytic, except at some point z_0, in a domain D. The point z_0 is then called an isolated singularity of $f(z)$. Examples of such functions are

$$\frac{\sin z}{z}, \quad e^{1/z}, \quad \frac{\cos z}{z^2},$$

where in each case the origin is an isolated singularity.

Since, by definition $f(z)$ is analytic in a deleted neighborhood of z_0, it possesses a Laurent expansion (3.28) about the point z_0. There are three cases to distinguish:

(1) If the Laurent expansion possesses no negative powers, the singularity is said to be *removable*. For example,

$$\frac{\sin z}{z} = 1 - \frac{z^2}{3!} + \frac{z^4}{5!}$$

has a removable singularity at $z = 0$.

(2) If $f(z)$ has only a finite number of negative powers such that

$$f(z) = \sum_{n=-N}^{\infty} a_n (z - z_0)^n$$

with $0 < N < \infty$ and $a_{-N} \neq 0$, then f is said to have a *pole of order N* at z_0. For example,

$$\frac{\cos z}{z^2} = \frac{1}{z^2} - \frac{1}{2!} + \frac{z^2}{4!} - \cdots$$

has a pole of order two at the origin. Note that a pole of order one is also referred to as a *simple pole*. In general, with a pole of order N at z_0, we can write

$$f(z) = \frac{1}{(z - z_0)^N} g(z),$$

where $g(z)$ is analytic at z_0 (why?). Alternatively, we can write

$$f(z) = p(z) + q(z),$$

where

$$p(z) = \sum_{n=-N}^{-1} a_n (z - z_0)^n.$$

In this last case we refer to $p(z)$ as the *principal part*. The remainder $q(z)$ is analytic.

(3) If infinitely many negative powers are required in the summation, the point z_0 is said to be an *essential singularity*. For example,

$$e^{1/z} = 1 + \frac{1}{z} + \frac{1}{z^2 2!} + \frac{1}{z^3 3!} + \cdots$$

has an essential singularity at the origin.

It follows from the above discussion that if $f(z)$ has a pole of order N at the point z_0, then $1/f(z)$ has a zero of order N at that point. On the other hand, if $f(z)$ has a zero of order N at z_0, then $1/f(z)$ has a pole of order N at that point.

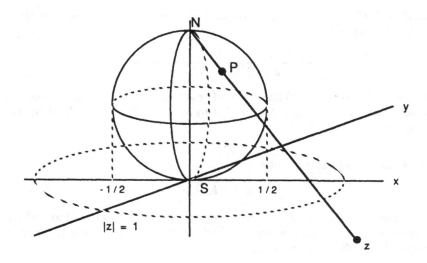

FIGURE 3.8.

Point at ∞. In our discussion thus far we have alluded to infinity as a single point. This is a useful convention and can be made precise in the following manner. Imagine the z-plane with a sphere of unit diameter resting on the origin (see Figure 3.8). It is convenient to refer to the point of tangency as the south pole, S, and the diametrically opposite point as the north pole, N. The z-plane is mapped onto the sphere by associating with each point z of the plane the point P of intersection of the straight line between z and N with the sphere. Under this mapping, called *stereographic projection*, the unit circle of the z-plane maps onto the equator, the interior of the unit circle maps onto the lower or southern hemisphere, and the exterior of the unit circle maps onto the upper or northern hemisphere. In particular, the neighborhood of the north pole is the image of points in the z-plane that are infinitely distant from the origin. The z-plane with ∞ included will be referred to as the *extended plane* and with ∞ deleted the *finite plane*.

This mapping allows us to treat infinity in the same way as points of the finite plane. Specifically, if $f(z)$ is analytic on $|z| > R$ but not necessarily analytic at ∞, it possesses a Laurent expansion (3.28), with $z_0 = 0$, of the form

$$f(z) = \sum_{n=-\infty}^{\infty} a_n z^n, \tag{3.32}$$

whose coefficients are given by

$$a_n = \frac{1}{2\pi i} \oint_{|z|=R} \frac{f(\zeta)}{\zeta^{n+1}} d\zeta \tag{3.33}$$

for $\overline{R} > R$. If (3.32) has no positive powers, then $f(z)$ is said to be analytic at ∞. For example

$$\frac{1}{z(z-1)(z-2)} = \frac{1}{z^3} \frac{1}{(1-1/z)(1-2/z)}$$

is analytic at infinity (and has a zero of order three). If (3.32) has N positive powers, with $a_N \neq 0$, then $f(z)$ is said to have a pole of order N at ∞. For example,

$$az + b$$

has a simple pole at ∞, while

$$az^2 + bz + c$$

has a pole of order two at ∞. In analogy with the finite case, if an infinite number of positive powers are required in (3.32), $f(z)$ is said to have an essential singularity at ∞; e.g.,

$$e^z = 1 + \frac{z}{1!} + \frac{z^2}{2!} + \frac{z^3}{3!} + \frac{z^4}{4!} + \cdots$$

has an essential singularity at ∞.

Branch Point Singularities. Branch points form another broad class of singularities and are associated with many-valued functions (see Section 2.2). For example, the origin is a branch point of the double-valued function $f(z) = \sqrt{z}$. We now treat two major groups of functions which possess this type of singularity and, through this treatment, describe what a branch point is and in what way it is a singularity.

Logarithm. We start by considering the exponential

$$e^w = z, \tag{3.34}$$

and, as the notation indicates, we will be interested in the inverse mapping $w = w(z)$. As in the real case, we denote the inverse by the logarithm:

$$w = \ln z. \tag{3.35}$$

It follows from the properties of the exponential, viz.,

$$e^{w_1} e^{w_2} = e^{w_1 + w_2} = z_1 z_2,$$

that

$$\ln z_1 z_2 = \ln z_1 + \ln z_2.$$

Also, it is immediate that $2n\pi i$, for any positive or negative integer n, can be added to the solution w, (3.35), of (3.34).

If we write $z = |z|e^{i\theta}$, then it follows from the properties of the logarithm that

$$w = \ln |z| + i\theta.$$

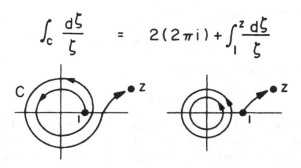

$$\int_C \frac{d\zeta}{\zeta} \;\; = \;\; 2(2\pi i) + \int_1^z \frac{d\zeta}{\zeta}.$$

FIGURE 3.9.

The imaginary part of w, $\theta = \arg z$, further underlines the ambiguity in the solution to (3.34); i.e., θ is known up to an integer multiple of 2π.

$\arg z$ is a monotonically increasing function of θ *and is not periodic* as we move counterclockwise around the unit circle. Thus, although it is clear that the origin is a singularity of $\ln z$, the difficulty would seem to be more severe. For example, if we reach $z = -1$ by moving counterclockwise, $\ln z = i\pi$, whereas reaching the same point in the clockwise direction gives $\ln z = -i\pi$.

Another way to illustrate this, and also a way to keep track of $\arg z$, is by making use of the relation

$$\frac{d \ln z}{dz} = \frac{1}{z},$$

which follows from differentiating (3.34) with respect to z. From this relation and from (3.9) we then have

$$\ln z = \int_1^z \frac{d\zeta}{\zeta}, \tag{3.36}$$

where we have taken unity as the lower limit of integration (the path of integration must not pass through the origin). Each time we circle the origin in the counterclockwise direction, the value of the integral is increased by $2\pi i$; and each time we circle the origin in the clockwise direction, the integral is decreased by $2\pi i$ (see Figure 3.9).

In order to keep track of where we are, that is, how many times we have encircled the origin, we create what is known as a *Riemann surface* for $\ln z$. We do this by imagining an infinite number of z-planes stacked one upon the other. Each of these z-planes is further imagined to be cut on its negative real axis. On one such plane or *sheet* we take

$$-\pi < \arg z < \pi.$$

On the sheet above this one we take

$$\pi < \arg z < 3\pi,$$

while on the sheet below it we take

$$-3\pi < \arg z < -\pi.$$

In general, above every plane $\arg z$ is incremented by 2π, and below every plane $\arg z$ is decremented by 2π. (The first of the above-mentioned sheets, where $-\pi < \arg z < \pi$, will be *distinguished* from the rest.) Next we attach all these planes along the cut, in spiral fashion, so that $\arg z$ is a continuous function as we circle the origin.

By the simple device of creating a Riemann surface for $\ln z$, we have made this function continuous and single-valued. In fact it is clear that $\ln z$ is analytic on its Riemann surface.

On each sheet of its Riemann surface $\ln z$ attains different values. On the *distinguished sheet* (i.e., $-\pi < \arg z < \pi$),

$$\ln z = \ln |z| + i \arg z.$$

This is referred to as the *principal branch* of the logarithm, and it is sometimes written as

$$w = \operatorname{Ln} z$$

to indicate that it is the principal branch. The function defined by the set of values of the logarithm on each of the other sheets is simply said to be a *branch* of the logarithm. Thus, on the branch above the principal branch (i.e., $\pi < \arg z < 3\pi$),

$$\ln z = \operatorname{Ln} z + 2\pi i.$$

The origin, which belongs to all sheets, is referred to as the *branch point* of the logarithm. A function is said to be singular at its branch point.

Finally, it should be noted that placing the cut, or *branch cut* as it is called, along the negative real axis was arbitrary. In doing this we are following an accepted convention. The branch cut can be taken along any simple curve between the origin and infinity which is also a branch point.

Power Functions. We next consider the function

$$f = z^{\alpha}, \tag{3.37}$$

which can be regarded as the inverse of $w^{1/\alpha} = z$. When it is written this way or as (3.37), we are faced with the idea, in general, of a complex number raised to a complex power—a notion not previously discussed. Consider the particular case

$$f^2 = z,$$

which leads to consideration of

$$f = \sqrt{z}.$$

Referring to the discussion given in Section 2.2, if we write $f = |f|e^{i\phi}$, we see that one circuit around the origin in the f-plane sends us twice around the z-plane. Therefore the pre-image (i.e., the inverse) of a point of the z-plane is ambiguous since it can be one of two values in the f-plane.

The present situation can be reduced to the already studied logarithm. To do this, observe that (3.34) raised to the power α yields

$$z^\alpha = (e^w)^\alpha = e^{\alpha w}.$$

From this result and (3.35), we are led to the definition of (3.37) as

$$f = z^\alpha = e^{\alpha \ln z}. \tag{3.38}$$

Note that the exponential itself is an entire function and hence single-valued. Thus the many-valuedness of z^α is reduced to the branch structure of $\ln z$. In particular, the principal branch of z^α is defined to be

$$e^{\alpha \ln z}.$$

Example. Let us consider $z^{1/2}$. Its principal branch according to the above discussion is

$$z^{1/2} = e^{\frac{1}{2}\ln|z| + \frac{1}{2}(\arg z)} = |z|^{1/2} e^{\frac{1}{2}\theta}, \quad -\pi < \theta = \arg z < \pi.$$

On all other branches

$$z^{1/2} = e^{\frac{1}{2}\ln|z| + \frac{i}{2}(\theta + 2\pi n)} = |z|^{1/2} e^{i\theta/2} e^{i\pi n}, \quad n = \pm 1, \pm 2, \ldots,$$

where θ is the location of arg z on the principal sheet. But $e^{i\pi n}$ is either plus or minus one depending on whether n is even or odd. This states that $z^{1/2}$ simply changes sign as we pass from one sheet to the next in either direction. Therefore $z^{1/2}$ has just two branches, and only a two-sheeted Riemann surface is necessary to make $z^{1/2}$ single-valued. The Riemann surface is constructed by considering two sheets, again cut along the negative real axis, and again attached as before. We are then left with two free unattached edges along which arg $z = 3\pi$ and arg $z = -\pi$. However,

$$|z|^{1/2} e^{i3\pi/2} = |z|^{1/2} e^{-i\pi/2};$$

that is, $z^{1/2}$ takes on the same values at the two free edges, and we therefore can attach these as indicated in Figure 3.10. In general, $z^{1/N}$ for an integer N has N branches and is single-valued on an N-sheeted Riemann surface.

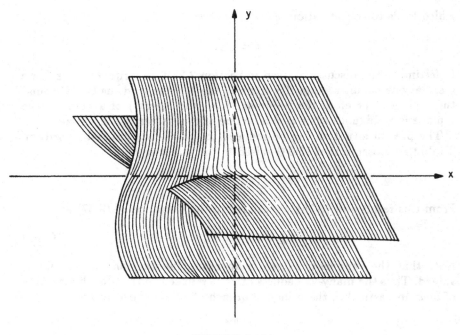

FIGURE 3.10.

As a result of this discussion, the idea of raising a complex number to a complex number is now computationally realizable. Consider α^β, where both α and β are complex. To evaluate this expression, first write

$$\alpha^\beta = e^{\beta \ln \alpha}.$$

Then substitute in the right-hand side $|\alpha| \exp(i \arg z)$ for α and $\beta_1 + i\beta_2$ for β, so that we are led to computable quantities.

Example. Consider the complex number $a = i^i$. From our discussion

$$i^i = e^{i \ln i}.$$

If we take $i = e^{i\pi/2}$, then

$$i^i = e^{i \ln e^{i\pi/2}} = e^{i(i\pi/2)} = e^{-\pi/2} \approx .2079.$$

Exercises

1. Locate in the extended z-plane *all* zeros and poles of

   (a) $\dfrac{z^2 - 1}{z^2 + 1}$,   (b) $\dfrac{z - 1}{z^3 + 1}$,   (c) $\dfrac{(z - 1)^2 (z + 2)^3}{z}$.

2. Find the Taylor series of $\ln z$ about $z = 1$. What is its radius of convergence?

3. Evaluate

$$(a) \; (1+i)^{2/3}, \qquad (b) \; i^{\sqrt{2}}, \qquad (c) \; \ln(1-i).$$

4. Consider $\sqrt{z}$ and place the branch cut on the positive real axis. Compare the resulting values with those that would be obtained if the branch cut were placed on the negative real axis, as was done in the text.

5. Show

(a) $\cos^{-1} z = i^{-1} \ln[z \pm i(1 - z^2)^{1/2}]$,

(b) $\sin^{-1} z = -i \ln(iz \pm \sqrt{1 - z^2})$,

(c) $\tanh^{-1} z = \frac{1}{2} \ln(1 + z/z - 1)$.

6. Find the Taylor expansion of $\sqrt{z}$ around $z = 2$. What is the radius of convergence?

7. Find the real and imaginary parts of z^z if $z = x + iy$.

8. Determine all values of (a) $1^{\sqrt{2}}$, (b) $(1 + i)^i$, (c) $\cos^{-1} i$, (d) $\ln i^{1/2}$.

9. Solve $\ln z = i\pi/2$ for z.

3.7 Residue Theory

Suppose $f(z)$ is analytic in a domain D except for z_0 in D. It then follows from Cauchy's Theorem that the integral of $f(z)$ around every loop in D enclosing z_0 has the same value. This value is known as the *residue* of $f(z)$ at z_0 and is denoted by

$$\text{Res}[f(z); z_0] = \frac{1}{2\pi i} \oint_{|z - z_0| = \epsilon} f(z) dz, \qquad (3.39)$$

where $|z - z_0| = \epsilon$ represents a small circle about z_0 including no other singularities. Since z_0 is an isolated singularity of $f(z)$, we have from Laurent's Theorem that

$$f(z) = \sum_{n=-\infty}^{\infty} a_n(z - z_0)^n,$$

and therefore substitution into (3.39) yields (see Exercise 2 at the end of Section 3.4)

$$\text{Res}[f(z); z_0] = a_{-1}. \qquad (3.40)$$

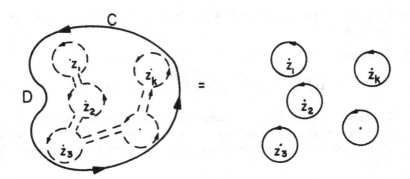

FIGURE 3.11.

More generally, we have

Residue Theorem. *Suppose $f(z)$ is analytic in some domain D and has only isolated singularities. If C is a loop which lies in D and which encloses the singularities $z_1, z_2, z_3, \ldots, z_k$, then*

$$\frac{1}{2\pi i} \oint_C f(z)dz = \sum_{n=1}^{k} \text{Res}[f(z); z_n]. \tag{3.41}$$

This follows from the graphical construction seen in Figure 3.11, in which the loop C is deformed until the curve represented by the broken line is obtained. Cancellations on the straight legs then reduce the integral in (3.41) to integrals on the small loops encircling each of the singularities, as shown on the right side of Figure 3.11.

Example. Consider

$$I = \oint_{|z|=2} \frac{ze^z}{(z^2 - 1)}dz.$$

The integrand has simple poles at $z = \pm 1$, both of which are enclosed by the loop $|z| = 2$. Therefore

$$I = 2\pi i \left\{ \text{Res}\left[\frac{ze^z}{z^2 - 1}; 1 \right] + \text{Res}\left[\frac{ze^z}{z^2 - 1}; -1 \right] \right\}.$$

Each of these residues are easy to calculate; e.g.,

$$\text{Res}\left[\frac{ze^z}{z^2 - 1}; 1 \right] = \frac{1}{2\pi i} \oint_{|z-1|=\epsilon} \frac{ze^z/(z+1)}{z-1}dz = \frac{e}{2}$$

by Cauchy's Integral Formula (3.15). (By $|z - 1| = \epsilon$ is meant a small circle surrounding $z = 1$.) Therefore the value of I is

$$I = 2\pi i \left[\frac{e}{2} + \frac{e^{-1}}{2} \right].$$

With each residue calculation diverse strategies are possible, and often one is more efficient than the others. We leave some of the general rules for the exercises and illustrate residue calculations by a number of examples.

Example. Consider

$$\text{Res}\left[\frac{e^z}{(z-1)^2};1\right] = \frac{1}{2\pi i}\oint_{|z-1|=\epsilon}\frac{e^z}{(z-1)^2}dz.$$

We can calculate this directly by expanding e^z in the neighborhood of $z = 1$; i.e.,

$$e^z = ee^{z-1} = e\left\{1 + \frac{z-1}{1!} + \frac{(z-1)^2}{2!} + \cdots\right\}.$$

Then we substitute this expansion for e^z in the above integral and locate a_{-1} $(= e^1)$. Even simpler is the fact that the integral represents the first derivative of e^z evaluated at $z = 1$ (see (3.17)), from which the result is obtained immediately.

Example. Consider

$$J = \oint_{|z|=3}\frac{e^z}{(z-1)^2(z-2)}dz$$

$$= 2\pi i\left\{\text{Res}\left[\frac{e^z}{(z-1)^2(z-2)};1\right] + \text{Res}\left[\frac{e^z}{(z-1)^2(z-2)};2\right]\right\}.$$

In this case a *partial fraction decomposition* is useful. (This is discussed directly following this example.) Recall that this is obtained by writing

$$\frac{1}{(z-1)^2(z-2)} = \frac{A}{z-1} + \frac{B}{z-2} + \frac{C}{(z-1)^2}.$$

To evaluate the constants A, B, and C, a variety of procedures suggest themselves. For example, if we multiply both sides by $z-2$ and set z equal to 2, then we find

$$B = 1.$$

Next we multiply both sides by $(z-1)^2$ and set z equal to 1, which gives us

$$C = -1.$$

Finally we can use residue theory itself to find A. The residue at $z = 1$ of the right-hand side is A and of the left-hand side using (3.17) is

$$\frac{d}{dz}\left(\frac{1}{z-2}\right)\Big|_{z=1} = -1 = A.$$

Even more simply, we can substitute the value 1 for B and the value -1 for C so that

$$\frac{1}{(z-1)^2(z-2)} = \frac{A}{z-1} + \frac{1}{z-2} - \frac{1}{(z-1)^2}.$$

Since this is true for all z, we can substitute any convenient value, say $z = 0$, to find $A = -1$. Therefore returning to the integral which we have denoted by J, the integrand can be written in the form

$$\frac{e^z}{(z-1)^2(z-2)} = \frac{e^z}{z-2} - \frac{e^z}{z-1} - \frac{e^z}{(z-1)^2}.$$

From this we obtain

$$J = 2\pi i \left\{ \text{Res}\left[\frac{e^z}{z-2}; 2\right] - \text{Res}\left[\frac{e^z}{z-1}; 1\right] - \text{Res}\left[\frac{e^z}{(z-1)^2}; 1\right] \right\}.$$

Each term of the right hand may be evaluated by inspection, thereby giving us

$$J = 2\pi i\{e^2 - e - e\} = 2\pi i\{e^2 - 2e\}.$$

Partial Fractions. We pause now to give a simple proof of the general *partial fraction decomposition.* Suppose we have the *rational function*

$$f(z) = \frac{Q(z)}{P(z)}, \tag{3.42}$$

where the degree of the polynomial Q is less than the degree of the polynomial P. Denote the distinct zeros of P by

$$z_1, \ldots, z_k$$

and denote the principal part (see the discussion on poles in Section 3.6) of $f(z)$ about each of these points by

$$p_1(z), p_2(z), \ldots, p_k(z),$$

respectively. Then

$$q(z) = f(z) - \sum_{n=1}^{k} p_n(z)$$

has removable singularities at $z_1, \ldots, z_k$. Therefore $q(z)$ is analytic in the finite plane and vanishes at infinity (since f and each p_n vanishes as $|z| \uparrow \infty$). Therefore, by Liouville's Theorem, $q(z)$ is identically equal to zero and

$$f(z) = \sum_{n=1}^{k} p_n(z), \tag{3.43}$$

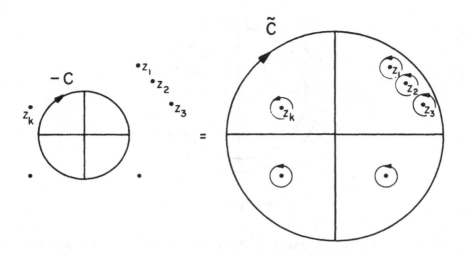

FIGURE 3.12.

which is just the partial fraction decomposition.

Residue of Infinity. A concept which is useful in computations is that of the residue at ∞. This by definition is

$$\text{Res}[f(z); \infty] = -\frac{1}{2\pi i} \oint_C f(z)dz, \qquad (3.44)$$

where $f(z)$ is assumed to be analytic in the deleted neighborhood of ∞ and C (traced in the counterclockwise direction) loops all finite singularities of f. (A little thought and appeal to Figure 3.8 indicates the appropriateness of the minus sign in this definition.)

Since ∞ is assumed to be an isolated singularity, we can expand $f(z)$ as a Laurent expansion centered at $z = 0$:

$$f(z) = \sum_{n=-\infty}^{\infty} a_n z^n, \quad R < |z| < \infty, \qquad (3.45)$$

where the circle $|z| = R$ is large enough so that it encloses all singularities of f. It then follows directly from (3.33) and (3.44), with C taken to be $|z| = \overline{R} > R$, that

$$\text{Res}[f(z); \infty] = -a_{-1}. \qquad (3.46)$$

As an application, consider

$$-\frac{1}{2\pi i} \oint_C f(z)dz,$$

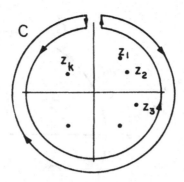

FIGURE 3.13.

where C is the loop indicated on the left in Figure 3.12 and where $f(z)$ possesses only isolated singularities at z_1, z_2, $z_3, \ldots, z_k$ outside C (this set may include $z = \infty$). Nothing need be said about the function inside the loop. Next, $-C$ can be distorted as indicated on the right-hand side of Figure 3.12. From this we have

$$-\frac{1}{2\pi i} \oint_C f(z)dz = \sum_{n=1}^{k} \text{Res}[f(z); z_n] + \text{Res}[f(z); \infty].$$

As another application, suppose $f(z)$ has only isolated singularities at z_1, $z_2, \ldots, z_k$ and at $z = \infty$. Then, for the loop C indicated in Figure 3.13,

$$\frac{1}{2\pi i} \oint_C f(z)dz = 0.$$

We observe, as we did in many earlier arguments, that since $f(z)$ is single-valued the two contributions on the vertical portions cancel, and we are left with two loop integrals. If we contract the inner loop to the origin, we obtain

$$\sum_{n=1}^{k} \text{Res}[f(z); z_n] + \text{Res}[f(z); \infty] = 0. \tag{3.47}$$

Example. Evaluate

$$\frac{1}{2\pi i} \oint_{|z|=2} \frac{z^{n-1}}{z^n - 1} dz,$$

where n is a positive integer. The straightforward approach is to contract the contour and pick up the residues (i.e., the simple pole contributions) at each of the n roots of unity—all of which lie on $|z| = 1$ (see Section 1.2). However, the result is immediate if we use the second of the above applications, (3.47). We simply find the residue at infinity using (3.46), where the circle $|z| = 2$ encloses the unit circle and thus all finite singularities of the function $z^{n-1}/(z^n - 1)$. In fact $a_{-1} = 1$ in (3.45) and in (3.46), and the value of the integral is 1.

Example. Consider

$$\frac{1}{2\pi i} \oint_{|z|=3} \frac{dz}{(z^2 + 1)(z^2 + 2)(z^2 + 3)(z - 4)}.$$

The integrand has poles at $z = \pm i, \pm i\sqrt{2}, \pm i\sqrt{3}$ and at $z = 4$; $z = \infty$ is a zero of order seven of the integrand. The first of the above applications turns this example into a trivial problem, since only the residue at $z = 4$, which is outside $|z| = 3$, need be calculated. The answer is

$$-\frac{1}{17 \times 18 \times 19}.$$

Evaluation of Real Integrals. A variety of real integrals can be evaluated by means of residue theory. We consider several representative cases.

   Case I:

$$I = \frac{1}{2\pi} \int_0^{2\pi} R(\cos \theta, \sin \theta) d\theta, \tag{3.48}$$

where R is a rational function of $\sin \theta$ and $\cos \theta$. The underlying strategy in the evaluation of this integral is first to transform it to an integral along the unit circle in the z-plane. On the unit circle we can write

$$z = e^{i\theta},$$

which gives

$$z + \frac{1}{z} = e^{i\theta} + e^{-i\theta} = 2\cos\theta$$

or

$$\cos\theta = \frac{1}{2}\left(z + \frac{1}{z}\right).$$

Similarly,

$$z - \frac{1}{z} = 2i\sin\theta$$

or

$$\sin\theta = \frac{1}{2i}\left(z - \frac{1}{z}\right).$$

Finally, we note that

$$dz = de^{i\theta} = ie^{i\theta}d\theta$$

or

$$d\theta = \frac{dz}{iz}.$$

If the expressions for $\cos\theta$, $\sin\theta$, and $d\theta$ are substituted in the above integral, then (3.48) can be rewritten as

$$I = \frac{1}{2\pi i} \oint_{|z|=1} R\left(\frac{1}{2}(z+z^{-1}), \frac{1}{2i}(z-z^{-1})\right) \frac{dz}{z}. \qquad (3.49)$$

Therefore evaluation of I is reduced to a residue calculation.

Example. Consider

$$J = \frac{1}{2\pi} \int_0^{2\pi} \frac{d\theta}{2+\cos\theta} = \frac{1}{2\pi i} \oint_{|z|=1} \frac{1}{[2+\frac{1}{2}(z+z^{-1})]} \frac{dz}{z}$$

$$= \frac{1}{2\pi i} \oint_{|z|=1} \frac{2\,dz}{z^2+4z+1}.$$

The zeros of the denominator are given by

$$z = \frac{1}{2}\{-4 \pm \sqrt{16-4}\} = -2 \pm \sqrt{3}.$$

Only $-2+\sqrt{3}$ lies inside the unit circle and thus

$$J = \text{Res}\left[\frac{2}{z^2+4z+1}; -2+\sqrt{3}\right] = \frac{2}{(-2+\sqrt{3})-(-2-\sqrt{3})} = \frac{1}{\sqrt{3}}.$$

Case II: We start with an

Example. Consider

$$I = \int_{-\infty}^{\infty} \frac{dx}{x^4+1},$$

which in complex notation is

$$I = \int_{-\infty}^{\infty} \frac{dz}{z^4+1}.$$

To . uate this integral, we first consider the integral

$$\tilde{I} = \oint_C \frac{dz}{1+z^4} = \int_{-R}^{R} \frac{dx}{1+x^4} + \int_{|z|=R, 0\le \arg z \le \pi} \frac{dz}{1+z^4},$$

where as indicated in Figure 3.14 the loop C is along the real axis between $-R$ and R (R is large) and then along the semicircle in the upper half-plane joining these endpoints. Under the limit $R \uparrow \infty$, the first integral tends to I and the second integral clearly vanishes since

$$\left| \int_{|z|=R, 0\le \arg z \le \pi} \frac{dz}{1+z^4} \right| \le \int_{|z|=R, 0\le \arg z \le \pi} \frac{1}{|z|^4} \frac{1}{|1+(1/z^4)|} |dz|$$

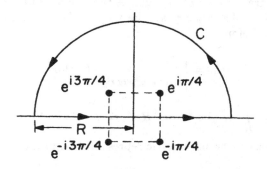

FIGURE 3.14.

$$= \frac{1}{R^3} \int_0^\pi \frac{d\theta}{|1 + (e^{-4i\theta}/R^4)|}.$$

Therefore

$$\lim_{R \uparrow \infty} \tilde{I} = I.$$

$\tilde{I}$ can be evaluated directly by means of residues. As indicated in Figure 3.14, $z^4 + 1$ has simple poles in the upper half-plane at $z = \exp(i\pi/4)$ and at $z = \exp(i\pi 3/4)$. Hence

$$I = 2\pi i \left\{ \text{Res} \left[\frac{1}{1 + z^4}; e^{i\pi/4} \right] + \text{Res} \left[\frac{1}{1 + z^4}; e^{3i\pi/4} \right] \right\}$$

$$= 2\pi i \left\{ \frac{1}{4e^{i\pi 3/4}} + \frac{1}{4e^{9i\pi/4}} \right\} = \frac{\pi i}{2} \{ -e^{-i\pi/4} + e^{-i\pi 4} \}$$

$$= \frac{i\pi}{2} \cdot \left\{ -2i \sin \frac{\pi}{4} \right\} = \frac{\pi}{\sqrt{2}},$$

where we have used the result of Exercise 2(a) in the exercises for this section.

Let us generalize the procedure to integrals of the form

$$I = \int_{-\infty}^{\infty} F(x) dx, \qquad (3.50)$$

for which $F(z)$ is analytic in the upper half-plane except possibly at isolated singularities, say $z_1, \ldots, z_k$. Again consider the contour of Figure 3.14 and the corresponding integral

$$\tilde{I} = \oint_C F(z) dz = \int_{-R}^{R} F(x) dx + \int_{|z|=R, 0 \le \arg z \le \pi} F(z) dz.$$

In the last integral, the notation signifies that we are integrating along the semicircle $|z| = R$, $0 \le \arg z \le \pi$. The magnitude of this last integral can be estimated as before:

$$|\tilde{I}_0| = \left| \int_{|z|=R, 0 \le \arg z \le \pi} F(z)dz \right|$$

$$\le \int_0^\pi |F(Re^{i\theta})| \, |ie^{i\theta} R \, d\theta| = \int_0^\pi |F(Re^{i\theta})| R \, d\theta,$$

where we have used the parametric representation of the circle

$$z = Re^{i\theta}.$$

Therefore, if

$$\lim_{R \uparrow \infty} R|F(Re^{i\theta})| = 0, \tag{3.51}$$

then

$$I = 2\pi i \sum_{n=1}^k \text{Res}[F(z); z_n].$$

If in (3.50) $F = e^{i\alpha z} f(z)$, i.e.,

$$I = \int_{-\infty}^\infty e^{i\alpha x} f(x)dx \tag{3.52}$$

with $\alpha > 0$, and $|f| \downarrow 0$ as $|z|$ condition (3.51) is satisfied. To see this observe

$$|I_0| = \left| \int_{0, |z|=R}^\pi e^{i\alpha(x+iy)} f(Re^{i\theta}) Rie^{i\theta} d\theta \right| \le \int_0^\pi e^{-\alpha y}|f(Re^{i\theta})| R \, d\theta.$$

Take R sufficiently large such that $|f(Re^{i\theta})| < \epsilon$. Then

$$I_0 \le R\epsilon \int_0^\pi e^{-\alpha R \sin \theta} d\theta = 2R\epsilon \int_0^{\pi/2} e^{-\alpha R \sin \theta} d\theta.$$

As a simple figure shows,

$$\frac{2\theta}{\pi} \le \sin \theta, \quad 0 \le \theta \le \pi/2;$$

and from this

$$I_0 \le 2R\epsilon \int_0^{\pi/2} e^{-R2\theta\alpha/\pi} d\theta = \frac{\epsilon(1 - \exp(-\alpha R))}{\alpha\pi} < \frac{\epsilon\pi}{\alpha}.$$

We see therefore that (3.51) is satisfied. This result is known as *Jordan's Lemma*.

Example. Consider

$$J = \int_0^\infty \frac{\cos kx}{1 + x^2} dx$$

with $k > 0$. Since the integrand is an even function of x, we can write

$$J = \frac{1}{2} \int_{-\infty}^\infty \frac{\cos kx}{1 + x^2} dx;$$

and since $\sin kx$ is an odd function of x, we can rewrite this integral as

$$J = \frac{1}{2} \int_{-\infty}^\infty \frac{e^{ikx}}{1 + x^2} dx. \tag{3.53}$$

We then evaluate J by looking at the corresponding integral

$$\begin{aligned}
\tilde{J} &= \frac{1}{2} \oint_C \frac{e^{ikz}}{1 + z^2} dz \\
&= \frac{1}{2} \int_{-R}^R \frac{e^{ikx}}{1 + x^2} dx + \int_{|z|=R, 0 \leq \arg z \leq \pi} \frac{e^{ikz}}{1 + z^2} dz
\end{aligned}$$

with the paths of integration as given in Figure 3.14. Since (3.53) is of the form (3.52) and $f(z) = 1/(1 + z^2) \to 0$ for $|z| \uparrow \infty$, Jordan's Lemma applies. Therefore

$$J = 2\pi i \cdot \frac{1}{2} \cdot \text{Res} \left[\frac{e^{ikz}}{1 + z^2}; i \right] = 2\pi i \cdot \frac{1}{2} \cdot \frac{e^{-k}}{2i} = \frac{\pi}{2} e^{-k}.$$

Note. We could just as well have written

$$J = \frac{1}{2} \int_{-\infty}^\infty \frac{e^{-ikx}}{1 + x^2} dx,$$

in which case the contour cannot be closed in the upper half-plane. In order to evaluate this alternative expression for J, we close the contour in the lower half-plane.

Case III: We start by considering a special case, namely

$$J = \int_0^\infty \frac{\sqrt{x}}{1 + x^2} dx.$$

If we write this in complex form as

$$\int_0^\infty \frac{\sqrt{z}\, dz}{1 + z^2},$$

it requires a statement about the branch of the square root. Since $\sqrt{x}$ is positive in the original integral, we are on the principal branch of the square root.

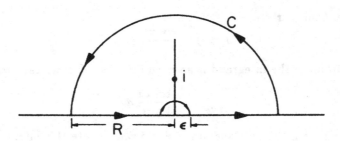

FIGURE 3.15.

Next consider the loop C indicated in Figure 3.15 and the integral

$$\tilde{J} = \tilde{J}(\epsilon; R) = \oint_C \frac{\sqrt{z}}{1+z^2} dz.$$

Since the denominator has a simple zero in the upper half-plane, namely $z = i$, we have (for any small ϵ and for any large R)

$$\tilde{J} = 2\pi i \text{Res}\left[\frac{\sqrt{z}}{1+z^2}; i\right] = 2\pi i \frac{e^{i\pi/4}}{2i} = \pi e^{i\pi/4}.$$

From Figure 3.15 we can separate $\tilde{J}$ into the four parts

$$\tilde{J} = \int_{-R}^{-\epsilon} \frac{\sqrt{x}}{1+x^2} dx + \int_{\epsilon}^{R} \frac{\sqrt{x}}{1+x^2} dx + \int_{\pi,|z|=\epsilon}^{0} \frac{\sqrt{z}}{1+z^2} dz$$
$$+ \int_{0,|z|=R}^{\pi} \frac{\sqrt{z}}{1+z^2} dz,$$

where the notation introduced in the third and fourth integrals is self-explanatory. The last integral vanishes as $R \uparrow \infty$. To see this, we proceed as before and write $z = Re^{i\theta}$ so that

$$\left| \int_{0,|z|=R}^{\pi} \frac{\sqrt{z}\, dz}{1+z^2} \right| \leq \int_0^{\pi} \frac{|R^{1/2}e^{i\theta/2}||R|ie^{i\theta}|d\theta}{|1+R^2e^{i2\theta}|}$$
$$= \frac{1}{R^{1/2}} \int_0^{\pi} \frac{d\theta}{|e^{2i\theta}+(1/R^2)|} \xrightarrow{R\uparrow\infty} 0.$$

The third integral vanishes as $\epsilon \downarrow 0$. To see this we write $z = \epsilon e^{i\theta}$ so that

$$\left| \int_{\pi,|z|=\epsilon}^{0} \frac{\sqrt{z}\, dz}{1+z^2} \right| \leq \int_0^{\pi} \frac{|\epsilon^{1/2}e^{i\theta/2}||e^{i\theta}i|\epsilon\, d\theta}{|1+\epsilon^2e^{2i\theta}|}$$
$$= \epsilon^{3/2} \int_0^{\pi} \frac{d\theta}{|1+\epsilon^2e^{2i\theta}|} \xrightarrow{\epsilon\downarrow 0} 0.$$

The second integral is just J under both of these limits and the first integral is

$$\int_{-\infty}^{0} \frac{\sqrt{x}}{1+x^2} dx,$$

which, if we replace x by $-x$, becomes

$$-\int_{+\infty}^{0} \frac{\sqrt{-x}\, dx}{1+x^2} = \int_{0}^{\infty} \frac{\sqrt{x}\, dx}{1+x^2} e^{i\pi/2} = e^{i\pi/2} J,$$

where $\sqrt{-1} = e^{i\pi/2} = i$ if we are on the principal branch of the square root. Collecting the various results, we have

$$\lim_{\substack{\epsilon \downarrow 0 \\ R \uparrow \infty}} \tilde{J}(\epsilon; R) = \pi e^{i\pi/4} = J + iJ.$$

Therefore, if we solve this for J, we obtain

$$J = \pi \frac{\cos \pi/4 + i \sin \pi/4}{1+i} = \frac{\pi}{\sqrt{2}}.$$

A partial check of such a calculation is the fact that a real integral should yield a real result.

We do not try to present the methods of this case in any generality. Instead it is suggested that they can be used in other problems.

Exercises

1. Suppose that $f(z)$ is analytic in D, C is a loop in D, and $f(z)$ has no zeros or poles on C. Prove

$$\frac{1}{2\pi i} \oint_C \frac{f'}{f} dz = R - P.$$

   R is the total number of zeros inside C and P is the total number of poles inside C—counting multiplicities. [Hint: $f'/f = (d/dz) \ln f$.]

2. (a) If F and G are analytic on D and z_0 is a simple zero of G, show that $\operatorname{Res}[F/G; z_0] = F(z_0)/G'(z_0)$.

   (b) If the same hypothesis as in Part (a) holds here with the exception that z_0 is a zero of order two, show that

$$\operatorname{Res}[F/G; z_0] = \frac{6F'G'' - 2FG'''}{3(G'')^2}.$$

3. If in (3.42) $P(z)$ has only simple zeros, show that (3.43) becomes

$$f(z) = \sum_{n=1}^{k} \frac{Q(z_n)}{(z - z_n)P'(z_n)}.$$

4. Evaluate the following integrals:

(a) $\oint_{|z|=2} \dfrac{e^z}{z^2 - 1} dz,$

(b) $\oint_{|z-1|=2} \dfrac{dz}{z^4 + 1},$

(c) $\oint_{|z|=1} \dfrac{dz}{2z^2 + 3z - 2},$

(d) $\oint_{|z|=2} \dfrac{z^3 + 2z}{z^4 + z^2 + 2} dz,$

(e) $\oint_{|z|=2} \dfrac{dz}{(z-1)^4(z-4)},$

(f) $\oint_{|z|=2} \dfrac{\sin z \, dz}{(z^2 - 1)(z^2 - 9)}.$

[Hint: In Part (d), try finding the residue at infinity.]

5. Evaluate the following integrals:

(a) $\displaystyle\int_{-\infty}^{\infty} \dfrac{dx}{(x^2 + 1)(x^2 - i\sqrt{2} - 1)},$

(b) $\displaystyle\int_{0}^{2\pi} \dfrac{d\theta}{1 + \epsilon \cos \theta}, |\epsilon| < 1,$

(c) $\displaystyle\int_{-\infty}^{\infty} \dfrac{xe^{ix}}{x^2 + 1} dx,$

(d) $\displaystyle\int_{0}^{\pi} \dfrac{a \, d\theta}{a^2 + \sin^2 \theta}$ for $a > 0,$

(e) $\displaystyle\int_{0}^{\infty} \dfrac{x^6 dx}{(1 + x^4)^2},$

(f) $\displaystyle\int_{-\infty}^{\infty} \dfrac{\cos x \, dx}{(x^2 + 1)(x^2 + 4)},$

(g) $\displaystyle\int_{0}^{\infty} \dfrac{\sqrt{x}}{(1 + x^2)^2} dx.$

6. Show that

$$\int_{0}^{\infty} \dfrac{\sin x}{x} dx = \dfrac{\pi}{2}.$$

[Hint: Consider

$$J = \oint_{C} e^{iz} \dfrac{dz}{z},$$

where C is the contour of Figure 3.16. In order to show that the contribution on C_∞ as $R \uparrow \infty$ is vanishingly small, you may find it useful to prove $\sin \theta \geq 2\theta/\pi, 0 \leq \theta \leq \pi/2.$]

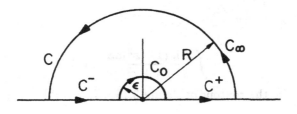

FIGURE 3.16.

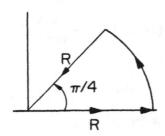

FIGURE 3.17.

7. Regard as true the following:

$$\int_0^\infty \exp(-x^2/2)dx = \frac{\sqrt{2\pi}}{2}.$$

Then show

$$\int_0^\infty \sin(x^2)dx = \int_0^\infty \cos(x^2)dx = \frac{\sqrt{2\pi}}{4}$$

by evaluating

$$\oint_C \exp(-z^2/2)dz$$

on the contour C of Figure 3.17 (let $R \uparrow \infty$).

8. Prove

$$\int_0^\infty \frac{\sin^2 \theta}{\theta^2}d\theta = \frac{\pi}{2}.$$

9. Evaluate

(a) $\displaystyle\int_{-\infty}^\infty \frac{\cos x}{x^2 + a^2}dx,$

(b) $\displaystyle\int_{-\infty}^{\infty} \frac{x \sin x}{x^2 + a^2} dx.$

10. Prove
$$\int_{0}^{\infty} \frac{\ln(1 + x^2) dx}{1 + x^2} = \ln 2^{\pi}.$$

[Hint: Use the contour of Figure 3.16 with $\epsilon = 0$.]

Additional Exercises

1. If $f(z)$ is analytic, show
$$\nabla^2 [f(z)]^2 = \left(\frac{\partial}{\partial x^2} + \frac{\partial}{\partial y^2}\right) |f|^2 = 4 \left|\frac{df}{dz}\right|^2.$$

2. Show
$$\tan^{-1} z = \frac{1}{2i} \ln \frac{1 + iz}{1 - iz}.$$

3. The zeta function is defined by
$$\zeta(z) = \sum_{n=1}^{\infty} \frac{1}{n^z}.$$

Show that this series converges for $\mathrm{Re}\, z > 1 + \epsilon,\ \epsilon > 0$.

4. Find the power series of
$$f(z) = e^z \cos z$$

in the neighborhood of $z = 0$. What is the circle of convergence?

5. What is the domain of convergence of

(a) $\displaystyle\sum_{n=0}^{\infty} \left(\frac{z}{3} - 1\right)^n,$

(b) $\displaystyle\sum_{n=0}^{\infty} \frac{z^n}{n^2 + 1}.$

6. Locate *all* poles and *all* zeros of $(z^3 - 1)/(z^2 + 3z + 2)$.

7. Prove
$$\int_{-\infty}^{\infty} \frac{dx}{(x^2 + a^2)(x^2 + b^2)} = \frac{\pi}{ab(a + b)}, \qquad a, b > 0.$$

8. Find the Laurent expansion of

$$\frac{1}{(z+2)(z-4)} \quad \text{for } 2 < |z| < 4.$$

9. Show

$$\int_0^\infty \frac{dx}{1+x^6} = \frac{\pi}{3}.$$

10. Prove the following: If

$$f(z) = \sum_{n=-\infty}^{\infty} a_n z^n, \quad g(z) = \sum_{n=-\infty}^{\infty} b_n z^n,$$

$$R_1 < |z| < R_2, \quad R_1 < 1 < R_2,$$

then

$$\frac{1}{2\pi} \oint_{|z|=1} f(z)g(z^{-1})dz = \sum_{n=-\infty}^{\infty} a_n b_n.$$

11. Show

(a) $\displaystyle \int_0^{2\pi} \frac{d\theta}{5+3\cos\theta} = \frac{\pi}{2},$

(b) $\displaystyle \int_0^{2\pi} \frac{\cos^2 3\theta \, d\theta}{5-4\cos 2\theta} = \frac{3\pi}{8},$

(c) $\displaystyle \int_0^{\pi} \frac{\cos 2\theta d\theta}{1-2a\cos\theta + a^2} = \frac{\pi a^2}{1-a^2}, \ |a| < 1,$

(d) $\displaystyle \int_0^{2\pi} \frac{d\theta}{a+b\sin\theta} = \frac{2\pi}{\sqrt{a^2-b^2}}, \ a > |b|,$

(e) $\displaystyle \int_0^{2\pi} \frac{d\theta}{(5-3\sin\theta)^2} = \frac{5\pi}{32}.$

12. Show

(a) $\displaystyle \oint_{|z|=5} \frac{e^z \, dz}{\cosh z} = 8\pi i,$

(b) $\displaystyle \oint_C \frac{\cosh z}{z^3} = \pi i$ where C is the square with vertices $\pm 2 \pm 2i,$

(c) $\displaystyle \int_{-\infty}^{\infty} \frac{dx}{x^4+x^2+1} = \frac{\pi\sqrt{3}}{6},$

(d) $\displaystyle \int_0^{\infty} \frac{\cosh x}{(x^2+1)^2} = \frac{\pi e^{-k}(1+k)}{4}, \ k > 0.$

13. Find the Laurent expansion for

(a) $f = 1/(z-3)$, when $|z| < 3$ and when $|z| > 3$;

(b) $f = z/[(1-z)(2-z)]$ when $|z| < 1$, when $1 < |z| < 2$, and when $|z| > 2$.

4

Discrete Linear Systems

4.1 Introduction to Linear Systems

Many systems in diverse fields can be characterized by an input-output analysis. One starts such an analysis by figuratively placing the system in a *black box*. As indicated in Figure 4.1, we denote the system by S, the output by b, and the input by a.

Generally the analysis of such systems follows three different scenarios:

Scenario 1: The system is described by some specific mathematical form, the input is given, and the output is to be determined. For example, the system could be represented by a matrix and the input, by a vector which the matrix (not necessarily square) acts on to give another vector, the output. Or the input could be initial data for a system of differential equations, the solution of which is the output.

Scenario 2: The system and the output are given and the input is to be determined. For example, in the case of linear equations this determination entails the inversion of a matrix, if inversion is possible.

Scenario 3: Both the input and the output are known and the challenge is to characterize the system mathematically. In this case we encounter something unusual. For example, suppose we consider

$$\mathbf{Ma} = \mathbf{b},$$

where $\mathbf{M}$ is a matrix and $\mathbf{a}$ and $\mathbf{b}$ are vectors. It should be clear that a specific $\mathbf{a}$ and $\mathbf{b}$ *will not* determine $\mathbf{M}$. In fact an infinite number of $\mathbf{M}$ can be found which will do the trick. This third scenario therefore requires further consideration, and we return to it later.

For now we consider situations for which the input a and the output b are functions of time. Depending on circumstances and preference, one or both

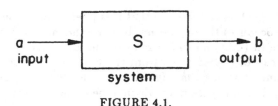

FIGURE 4.1.

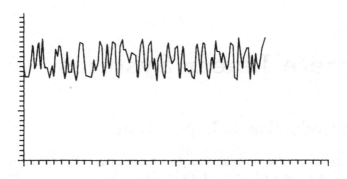

FIGURE 4.2.

of these quantities can be taken as continuous in time or can be defined at discrete (or sampled) times. For example, if the actual output record from the system has the form shown in Figure 4.2, then the corresponding sampled record is shown by dots in Figure 4.3. Suppose the output record is $b(t)$ and the sequence of sampling times, $\{t_n\}$. Then $\{b_n\} = \{b(t_n)\}$ is the sampled data. Customarily, sampling times are uniformly spaced; i.e., $t_n = n\Delta t$. Sampling itself is an example of a Scenario 1 procedure. Formally we can write

$$b_n = b(t_n) = \int_{-\infty}^{\infty} \delta(t - t_n) b(t) dt,$$

where δ, the delta function, characterizes the system. (And the action of the delta function is defined by this equation.)

If the output is as irregular as that illustrated in Figure 4.2 or 4.3, the data may be smoothed or averaged before being considered. For example, in the continuous case we can average over some time slot or *window*, T:

$$\bar{a}(t) = \frac{1}{T} \int_{t-T/2}^{t+T/2} a(s) ds. \tag{4.1}$$

In the dicrete version the analogous form is

$$\bar{a}_n = \frac{1}{2N+1} \sum_{p=-N}^{N} a_{n+p}. \tag{4.2}$$

(We divide by $2N+1$ since there are that many terms in the sum.) Observe that in these procedures the output has become the input for the averaging procedure, so both (4.1) and (4.2) are examples of Scenario 1.

The output can of course come from the black box in smoothed form; i.e., the averaging is done *on the fly*. But then the averaging device cannot be *anticipatory*. It cannot use data which appears in the future. Therefore,

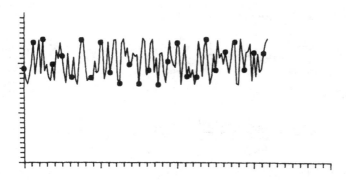

FIGURE 4.3.

in such circumstances, (4.1) and (4.2) should be replaced by

$$\bar{a}(t) = \frac{1}{T} \int_{t-T}^{t} a(s)ds$$

and

$$\bar{a}_n = \frac{1}{N+1} \sum_{p=-N}^{0} a_{n+p},$$

that is, by an average only over some past history of the record.

Since smoothing or averaging is illustrative of a wider range of systems we carry this discussion further. In many situations equal weighting of previous data with current data results in an unwarranted loss of instantaneous information. In such instances we can contemplate an averaging device equipped with *fading memory*. If we denote the weighting at the previous instant s by $W(s)$ (> 0), then the smoothing or averaging of $a(t)$ according to this weighting is defined to be

$$\bar{a}(t) = \frac{1}{W_0} \int_{-\infty}^{0} a(t+s)W(s)ds = \frac{1}{W_0} \int_{-\infty}^{t} a(\tau)W(\tau - t)d\tau, \qquad (4.3)$$

where

$$W_0 = \int_{-\infty}^{0} W(s)ds$$

and where we set $t + s = \tau$ to get the second expression for $\bar{a}(t)$ in (4.3). An example of such a weighting is the exponential $W = \exp(s/T)$, where the constant T is the *fading time*. In this instance (4.3) becomes

$$\bar{a}(t) = \frac{1}{T} \int_{-\infty}^{t} a(\tau)e^{(\tau - t)/T}d\tau. \qquad (4.4)$$

The analogous discrete version of (4.3) is

$$\bar{a}_n = \frac{1}{W_0} \sum_{p=-\infty}^{0} a_{n+p} W_p = \frac{1}{W_0} \sum_{q=-\infty}^{n} W_{q-n} a_q, \quad W_0 = \sum_{p=-\infty}^{0} W_p, \quad (4.5)$$

and of (4.4) is

$$\bar{a}_n = \frac{\sum_{p=-\infty}^{0} a_{n+p} \omega^{-p}}{\sum_{p=-\infty}^{0} \omega^{-p}} = (1-\omega) \sum_{q=-\infty}^{n} a_q \omega^{n-q}, \quad (4.6)$$

where $0 < \omega < 1$.

The operations on a expressed in (4.3) and (4.5) are examples of a *linear, translationally invariant, non-anticipatory* process. Each of these terms will now be discussed.

Linearity. By definition, an operator $\mathcal{L}$ which acts on objects x, y, z, ... is linear if

$$\mathcal{L}(\alpha x + \beta y) = \alpha \mathcal{L}x + \beta \mathcal{L}y,$$

where α, β, ... are constants. For example, if x, y, ... are vectors and $\mathcal{L}$ is a matrix, the above is true. Moreover, it is a consequence of linear algebra that any linear operator on a vector space can be expressed as a matrix. (See Halmos or Nering.) It should therefore not come as a surprise that any linear operator (there is some additional fine print) acting on functions $x(t)$, $y(t)$, ... defined on the real line can be represented by a function of two variables known as a *kernel* $K(s,t)$ in the following way:

$$b(t) = \int_{-\infty}^{\infty} K(s,t) a(s) ds. \quad (4.7)$$

Equation (4.3) is an example of this. The analogy with linear algebra is at least intuitive. If we sample $b(t)$ at discrete times $t_n = n\Delta t$, $n = 1, 2, 3, \ldots$, and as before write $b_n = b(n\Delta t)$, then (4.7) can be approximated by

$$b_n = \sum_{m=-\infty}^{\infty} K_{nm} a_m, \quad (4.8)$$

where K_{nm} is the matrix $K_{nm} = K(n\Delta t, m\Delta t)\Delta t$ and $a_m = a(m\Delta t)$.

Translational Invariance. Many systems which we encounter are invariant in time; i.e., they have no *internal clock* which gives them an absolute time. If the same input is passed through the system at different times, then the same output always appears. This is indicated schematically in Figure 4.4. This property is known as translational invariance because the system S behaves in the same way under *any* translate of time, T.

Since nothing is permanent, translational invariance is bound to be an approximation in any real situation. For example, our visual system under

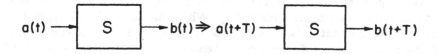

FIGURE 4.4.

identical circumstances at two different times records the same object in the same way. If the time interval between the different times is long, deterioration may set in, and (without the aid of eyeglasses) this invariance may no longer hold. (Of course death marks the ultimate deterioration of any biological system.) Therefore translation invariance, as an approximation, is apt to be appropriate for time scales which are small compared with the long-term *time scale* of the system.

If we apply translational invariance to (4.7), this states

$$b(t + T) = \int_{-\infty}^{\infty} K(s,t)a(s + T)ds$$

for any T. Set $s + T = u$. Then, for any T and any t,

$$b(t + T) = \int_{-\infty}^{\infty} K(u - T, t)a(u)du.$$

In particular, if $t = 0$, then

$$b(T) = \int_{-\infty}^{\infty} K(u - T, 0)a(u)du$$

for any T, which can assume the value t if so desired. Thus we have shown the following:

If a system is translationally invariant, the kernel must take the form

$$K(s,t) = K(t - s), \tag{4.9}$$

and the input-output relation (4.7) takes the form

$$b(t) = \int_{-\infty}^{\infty} K(t - s)a(s)ds \overset{\text{def}}{=} K \star a. \tag{4.10}$$

By analogy the discrete version (4.8) becomes

$$b_n = \sum_{m=-\infty}^{\infty} K_{n-m}a_m \overset{\text{def}}{=} (K \star a)_n. \tag{4.11}$$

Causality. One further ingredient went into the model equation (4.3), namely, the simple idea that the future cannot influence the present. This is known generally as *the principle of causality* and corresponds to our belief that a system cannot be anticipatory. When this constraint is applied to (4.7), it states that $K(s,t) = 0$, $s > t$, while for (4.10), it states that $K(t-s) = 0$ for $s > t$. In the discrete version the analogous statements are that $K_{nm} = 0$, $m > n$, and $K_{n-m} = 0$, $m > n$.

The products defined in (4.10) and (4.11) are called *convolution products*, and each of them is denoted by an asterisk or star. Even though the convolution product is continuous in one instance, (4.10), and discrete in the other, (4.11), no ambiguity should arise. The convolution product is an important form and we will see it in many guises.

The models discussed thus far have been based on the evolution of time-dependent systems. Equally important are models based on events in the *spatial domain.* For example, the gravitational force on an object is the summed forces of matter throughout space. Thus the force per unit mass at a location **x** is given by

$$\mathcal{F} = G \sum_{k} \frac{M_k(\mathbf{x} - \mathbf{x}_k)}{|\mathbf{x} - \mathbf{x}_k|^3},$$

where G is the universal gravitational constant and $\mathbf{x}_k$ is the location of the kth object of mass M_k. In the continuous limit this is written as

$$\mathcal{F} = G \int d\mathbf{y} \, \rho(\mathbf{y}) \frac{(\mathbf{x} - \mathbf{y})}{|\mathbf{x} - \mathbf{y}|^3},$$

where $\rho(\mathbf{y})$ is the mass per unit volume. As this demonstrates, in the spatial domain, unlike the temporal domain, no portion is excluded. All parts of space can influence one another. However, often there is another simplifying phenomenon at work in the spatial domain. This is

Reciprocity Between Cause and Effect. This says that if cause and effect are interchanged, their relation remains the same. As an illustration, consider (4.7), which we write in terms of the independent variables x and y,

$$b(x) = \int_{-\infty}^{\infty} K(x,y)a(y)dy,$$

in order to emphasize that the relation holds in some spatial situation. Broken down into elementary terms, this relation says that a cause (input) at y of strength $a(y)dy$ produces an effect (output) at x of strength

$$K(x,y)a(y)dy,$$

and the above integral is the result of adding all such contributions. The above mentioned *principle* states that the same effect is produced at y if the same cause $(a(y)dy)$ is placed at x. Specifically,

$$K(x,y)a(y)dy = K(y,x)a(y)dy$$

or

$$K(x, y) = K(y, x). \tag{4.12}$$

In this case the kernel is said to be *symmetric*. In the discrete case this results in a symmetric matrix.

The gravitational model expressed above is also an example of the reciprocity of cause and effect. (Note that it also illustrates the principle of translational invariance.)

Fine Print. We observe that real physical systems need not be linear or translationally invariant. Many systems do vary in time or are inhomogeneous in space so that they are not translationally invariant. Finally, few systems are truly linear—although over restricted operating domains they are well approximated by a linear system.

Exercises

1. Prove relation (4.6).

2. Construct the kernels which correspond to the processes specified in (4.1) and (4.2). Prove that the operators are linear and translationally invariant.

3. Suppose a linear system takes a vector of length N and produces a vector of the same length. Furnish a procedure which treats Scenario 3. Next suppose the input is a vector of length $M \neq N$. What then?

4. Prove that the convolution products defined by (4.10) and (4.11) are commutative; i.e., prove

   (a) $f \star g = g \star f$,

   (b) $a \star b = b \star a$.

5. Prove that the following finite convolution products are commutative:

$$f \star g \stackrel{\text{def}}{=} \int_0^t f(t - s)g(s)ds = g \star f,$$

$$(a \star b)_n \stackrel{\text{def}}{=} \sum_{m=0}^n a_{n-m}b_m = (b \star a)_n.$$

4.2 Periodic Sequences

Suppose the sequence $\{g_n\}$ is periodic with period N. By this we mean that

$$g_{n+N} = g_n$$

and from this that

$$g_{n+kN} = g_n, \quad k = \pm 1, \pm 2, \ldots .$$

Since only N terms fully characterize the sequence, it is clearly unnecessary to consider the infinite number of components of $\{g_n\}$. *In fact each such N-periodic sequence $\{g_n\}$ can be represented by a point in N-space.* Instead of following up on this idea we consider what will turn out to be a more interesting and revealing characterization. To arrive at this, we first consider some *standard* periodic sequences.

Recall from Section 1.2, *Roots of a Complex Number,* that an Nth root of unity, i.e., a solution to

$$z^N = 1,$$

is given by

$$\Omega = e^{2\pi i/N}$$

All N roots can then be expressed in terms of Ω by

$$W_k = \Omega^k = e^{(2\pi i/N)k} = \left(\cos \frac{2\pi}{N} + i \sin \frac{2\pi}{N} \right)^k$$

$$= \cos \frac{2\pi k}{N} + i \sin \frac{2\pi k}{N}, \quad k = 1, 2, \ldots, N.$$

Before proceeding further a word about notation is in order. The dependence of Ω and W_k on N has been suppressed in order to avoid a clutter of subscripts and superscripts. The reader should keep in mind this implicit dependence on N. Notation is a general problem in this chapter and such a sacrifice of precision for clarity will occur more than once.

As indicated in Figure 4.5, the W_k form N equally spaced points on the unit circle. Each root W_k, $k = 1, \ldots, N$, itself generates an infinite N-sequence $(W_k)^n$, $n = 0, \pm 1, \ldots$, which is N-periodic. If we substitute the above expression for W_k in this sequence, we obtain N such periodic sequences and they are

$$(W_k)^n = W_{kn} = e^{(2\pi i k n/N)} = \left(\cos \frac{2\pi k n}{N} + i \sin \frac{2\pi k n}{N} \right) = W_{nk},$$

$$n = 0, \pm 1, \pm 2, \ldots \tag{4.13}$$

and $k = 1, \ldots, N$. In terms of Figure 4.5 each W_k corresponds to a dot on the unit circle. Raising W_k to powers of n advances the dot around the circle. If $k = 12$ the dot remains at unity. For $k = 6$, it goes back and forth between 1 and -1. If $k = 1$ the dot advances by $30°$ with each increment of n.

Each of these N-periodic sequences can be regarded as a point in an N-space. (More precisely the real and imaginary parts can each be regarded as

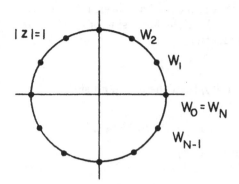

FIGURE 4.5. The roots of unity for $N = 12$.

points in N-space.) Moreover, (4.13) defines N such objects. Thus, if these points span N-space, then we will be able to represent any N-periodic sequence in terms of a linear combination of the sequences drawn from (4.13).

With this approach in mind, consider an arbitrary N-periodic sequence $\{g_n\}$ and provisionally write

$$g_n = \sum_{k=1}^{N} X_k e^{(2\pi ikn/N)} = \sum_{k=1}^{N} X_k W_{kn} \qquad (4.14)$$

where the complex constants X_k $(k = 1, \ldots, N)$ are unknown. The right-hand side of (4.14) is immediately seen to be N-periodic; i.e., if $n + mN$ is substituted for n the sum remains the same. Therefore, if we can satisfy (4.14) for $n = 1, \ldots, N$, we will be finished. All this suggests that we use the *vector* notation

$$\mathbf{g} = (g_1, \ldots, g_N),$$
$$\mathbf{X} = (X_1, \ldots, X_N),$$
$$(\mathbf{W})_{nk} = W_{nk}$$

where $\mathbf{W}$ is a matrix. Then (4.14) may be written in matrix form as

$$\mathbf{g} = \mathbf{WX}. \qquad (4.15)$$

Thus our problem will have a solution if $\mathbf{W}^{-1}$ exists. We now construct this inverse. Multiply both sides of (4.14) by $\exp(-2\pi inm/N)$ and sum on n. The result is

$$\sum_{n=1}^{N} g_n e^{(-2\pi inm/N)} = \sum_{k=1}^{N} X_k \left(\sum_{n=1}^{N} e^{(2\pi i/N)n(k-m)} \right), \qquad (4.16)$$

where th   ders of summation have been interchanged.

The inner sum on the right-hand side of (4.16) is zero if $k \neq m$. To see this, consider the identity

$$(1 - z^N) = (1 - z)(1 + z + \cdots + z^{N-1}).$$

If z is any root of unity, then the left-hand side vanishes; and if in addition $z \neq 1$, we must have

$$1 + z + z^2 + \cdots + z^{N-1} = 0.$$

But the inner sum in (4.16) is

$$\sum_{n=1}^{N} \left(e^{(2\pi i/N)(k-m)} \right)^n = \sum_{n=1}^{N} (W_{k-m})^n = \sum_{n=1}^{N} W_{(k-m)n},$$

where $W_{k-m} = \exp[i(2\pi/N)(k-m)]$ is a root of unity. If $k \neq m$, then the sum vanishes since $W_{k-m} \neq 1$ for $k \neq m$. However, if $k = m$, the sum is just N. So we have just demonstrated that (4.16) has the solution

$$X_m = \frac{1}{N} \sum_{n=1}^{N} g_n e(-2\pi i n m/N) = \frac{1}{N} \sum_{n=1}^{N} g_n \overline{W}_{nm}, \qquad (4.17)$$

where the bar denotes the complex conjugate. In matrix form, we have

$$(\mathbf{W}^{-1})_{mn} = \frac{1}{N} e^{(-2\pi i m n/N)}.$$

We note f   'ure reference that

$$\sum_{n=1}^{N} e^{(2\pi i/N)n(k-m)} = \sum_{n=1}^{N} W_{(k-m)n} = N\delta_{km}, \qquad (4.18)$$

where δ_{km}, known as the *Kronecker delta*, is zero for $k \neq m$ and is unity for $k = m$.

Thus we have proven the following:

Any N-periodic sequence $\{g_n\}$, i.e., *any sequence* $\{g_n\}$ *such that*

$$g_{n+jN} = g_n \qquad (4.19)$$

for all integers j, *has a standard representation*

$$g_n = \frac{1}{N} \sum_{m=1}^{N} \tilde{g}_m W_{mn} \qquad (4.20)$$

known as the (discrete) Fourier series of $\{g_n\}$. $\tilde{g}_n$ *is known as the (discrete) Fourier transform of* $\{g_n\}$ *and is given by*

$$\tilde{g}_n = \sum_{m=1}^{N} g_m \overline{W}_{mn} = N X_n. \qquad (4.21)$$

Equation (4.20) will also be referred to as the *inverse (Fourier) transform*, since it returns g_n after the (Fourier) transform (4.21) is carried out. Thus, *the linear operators expressed by (4.20) and (4.21) are inverse to one another.*

Elementary Properties. (1) Suppose $\{g_n\}$ is even; i.e., suppose $g_{-n} = g_n$. Then

$$g_n = g_{-n} = \frac{1}{N} \sum_{k=1}^{N} \tilde{g}_k e^{-2\pi i(kn/N)},$$

which when added to (4.20) gives

$$g_n = \frac{1}{N} \sum_{k=1}^{N} \tilde{g}_k \cos 2\pi \frac{kn}{N}. \tag{4.22}$$

(2) If $\{g_n\}$ is odd so that $g_{-n} = -g_n$, then

$$g_n = -g_{-n} = -\frac{1}{N} \sum_{k=1}^{N} \tilde{g}_k e^{-2\pi i(kn/N)}.$$

If we add this equation to (4.20), we obtain

$$g_n = i \sum_{k=1}^{N} X_k \sin 2\pi \frac{kn}{N} = \frac{i}{N} \sum_{k=1}^{N} \tilde{g}_k \sin 2\pi \frac{kn}{N}. \tag{4.23}$$

(3) If $\{g_n\}$ is a real sequence, then a real Fourier series should be obtained. In fact, if g_n is real, it is equal to its conjugate:

$$g_n = g_n^*.$$

If (4.20) for g_n is added to (4.20) for g_n^*, then

$$g_n = \frac{1}{2N} \sum_{k=1}^{N} \left(\tilde{g}_k e^{(2\pi i kn/N)} + \tilde{g}_k^* e^{(-2\pi i nk/N)} \right), \tag{4.24}$$

which is manifestly real, since each term of the sum is the sum of a complex number and its conjugate.

In dealing with real sequences, we can write

$$g_n = \sum_{k=1}^{N} a_k \cos \frac{2\pi}{N} nk + \sum_{k=1}^{N} b_k \sin \frac{2\pi}{N} nk \tag{4.25}$$

in place of (4.24) and

$$a_n = \frac{1}{N} \sum_{k=1}^{N} g_k \cos \frac{2\pi}{N} nk, \quad b_n = \frac{1}{N} \sum_{k=1}^{N} g_k \sin \frac{2\pi}{N} nk \tag{4.26}$$

in place of (4.21) for $\tilde{g}_n$. (The proofs of these along with the proofs of other relations are left for the exercises.) If $b_k \equiv 0$, the sequence is said to be even since $g_{-n} = g_n$, while if $a_k \equiv 0$, the sequence is said to be odd ($g_n = -g_{-n}$).

In considering complex periodic sequences $\{g_n\}$, we may split them into their real and imaginary parts. Therefore there is no loss in generality if in the following illustrations we consider real periodic sequences. This being the case we use (4.25) and (4.26) and look at admixtures of the N-periodic sequences

$$\left\{\cos \frac{2\pi}{N} nk\right\}_{n=0,\pm1,\pm2,\dots} \quad , \quad k = 1, 2, \dots, N,$$

$$\left\{\sin \frac{2\pi}{N} nk\right\}_{n=0,\pm1,\pm2,\dots} \quad , \quad k = 1, 2, \dots, N,$$

which are called *Fourier sequences.*

$N = 2$. In this case there are four Fourier sequences and they are represented by

$$\left(\sin \frac{2\pi}{2} nk\right)_{n=1} = \sin \pi k; \quad k = 1, 2, \rightarrow (0,0)$$

$$\left(\sin \frac{2\pi}{2} nk\right)_{n=2} = \sin 2k\pi; \quad k = 1, 2 \rightarrow (0,0)$$

$$\left(\cos \frac{2\pi}{2} nk\right)_{n=1} = \cos \pi k; \quad k = 1, 2 \rightarrow (-1, 1)$$

$$\left(\cos \frac{2\pi}{2} nk\right)_{n=2} = \cos 2\pi k; \quad k = 1, 2 \rightarrow (1, 1).$$

The first two sequences vanish identically. This could have been predicted, since any two-periodic sequence is manifestly even and hence only cosine sequences play a role. The last two vectors are clearly an orthogonal pair and span the space. Note that the second of these two with $n = 2$ (or $n = 0$) corresponds to a constant sequence and that the first of these two corresponds to a sequence which has zero mean.

$N = 4$. In this case there are eight Fourier sequences and the four odd sequences are represented by

$$\left(\sin \frac{2\pi}{4} nk\right)_{n=1} = \sin \frac{\pi k}{2}; \quad k = 1, 2, 3, 4 \rightarrow (1, 0, -1, 0)$$

$$\left(\sin \frac{2\pi}{4} nk\right)_{n=2} = \sin \pi k; \quad k = 1, 2, 3, 4 \rightarrow (0, 0, 0, 0)$$

$$\left(\sin \frac{2\pi}{4} nk\right)_{n=3} = \sin \frac{3\pi}{2} k; \quad k = 1, 2, 3, 4 \rightarrow (-1, 0, 1, 0)$$

$$\left(\sin \frac{2\pi}{4} nk\right)_{n=4} = \sin 2\pi k; \quad k = 1, 2, 3, 4 \rightarrow (0, 0, 0, 0).$$

There is just one independent basis vector! To see that this is reasonable, suppose

$$\{g_n\} = \ldots, g_{-3}, g_{-2}, g_{-1}, g_0, g_1, g_2, g_3, g_4, \ldots$$

is four-periodic and odd. Since it is odd, $g_0 = g_4 = 0$. Since it is four-periodic,

$$g_{-1} = g_3, \quad g_{-2} = g_2, \quad g_{-3} = g_1.$$

Also because it is odd,

$$g_{-2} = -g_2$$

and therefore $g_2 = 0$. Finally,

$$g_{-1} = -g_1, \quad g_{-3} = -g_3.$$

Therefore the sequence has the form

$$\ldots g_{-3} \quad g_{-2} \quad g_1 \quad g_0 \quad g_1 \quad g_2 \quad g_3 \quad \ldots =$$

$$\ldots g_1 \quad 0 \quad -g_1 \quad 0 \quad g_1 \quad 0 \quad -g_1 \ldots,$$

from which we see that there is just one independent component.

Next consider the even sequences:

$$\left(\cos\frac{2\pi}{4}nk\right)_{n=1} = \cos\frac{\pi k}{2}, \quad k = 1, 2, 3, 4 \rightarrow (0, -1, 0, 1)$$

$$\left(\cos\frac{2\pi}{4}nk\right)_{n=2} = \cos\pi k; \quad k = 1, 2, 3, 4 \rightarrow (-1, 1, -1, 1)$$

$$\left(\cos\frac{2\pi}{4}nk\right)_{n=3} = \cos\frac{3\pi}{2}k; \quad k = 1, 2, 3, 4 \rightarrow (0, -1, 0, 1)$$

$$\left(\cos\frac{2\pi}{4}nk\right)_{n=4} = \cos 2\pi k; \quad k = 1, 2, 3, 4 \rightarrow (1, 1, 1, 1).$$

There are just three independent vectors in this orthogonal set, all of which are orthogonal to the above sine series vector (i.e., $(1, 0, -1, 0)$). That the cosine sequences span a three-space could have been anticipated as follows. Any (real) sequence $\{g_n\}$ can be written as the sum

$$\left\{\frac{g_n - g_{-n}}{2}\right\} + \left\{\frac{g_n + g_{-n}}{2}\right\}.$$

Obviously the first sequence in the sum is odd, which from the above discussion can be represented in terms of a linear combination of the basis vectors for a one-dimensional space. Therefore the second sequence, which is even, can be represented in terms of a linear combination of the basis vectors for a $4-1 = 3$-dimensional space.

Examining the four cosine sequences, we see that letting $n = 4$ (or $n = 0$) gives a constant sequence, and that letting $n = 2$ gives a two-periodic sequence (of zero mean). The latter case depicts a *rapid* oscillation in the sequence. On the other hand, letting $n = 1$ (or $n = 3$) gives an irreducible four-periodic sequence (also of zero mean) with one independent component. In this case we have a *slow* oscillation.

As a final example Figure 4.6 gives a pictorial illustration of the decomposition of a six-periodic sequence.

These examples not only illustrate the nature of the Fourier sequences, but in addition underline their naturalness in depicting periodic sequences in general. Such symmetry qualities as evenness and oddness as well as the periodic structure are readily apparent in this framework. The coefficients a_k and b_k also have immediate interpretations in light of the above discussion.

Following the lead of the above discussion, we observe that $X_0 = X_N$ is the arithmetic mean of the sequence $\{g_n\}$; i.e., from (4.17),

$$X_N = X_0 = \frac{1}{N} \sum_{n=1}^{N} g_n.$$

This is referred to as the (complex) amplitude of the constant term. X_1 is the (complex) amplitude of the first harmonic. It corresponds to the sequence of slowest oscillation, namely the sequence with oscillation in the period N. In general, X_k is the (complex) amplitude of the kth harmonic, and it corresponds to a sequence having k oscillations in the period N.

Some additional properties emerge from essentially geometrical arguments. We again consider the N-periodic sequence $\{g_n\}$ and write

$$\mathbf{g} = (g_1, \ldots, g_N)$$

for the sequence in its basic period. Similarly, for another N-periodic sequence $\{f_n\}$, we write

$$\mathbf{f} = (f_1, \ldots, g_N).$$

These two expressions represent vectors in an N-space. From linear algebra the inner product of $\mathbf{f}$ and $\mathbf{g}$ is defined to be

$$(\mathbf{f}, \mathbf{g}) = \sum_{n=1}^{N} \overline{f}_n g_n, \tag{4.27}$$

and the square of the length of, say, $\mathbf{g}$ is given by

$$(\mathbf{g}, \mathbf{g}) = \sum_{n=1}^{N} \overline{g}_n g_n. \tag{4.28}$$

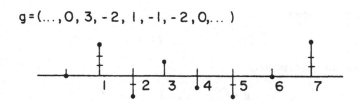

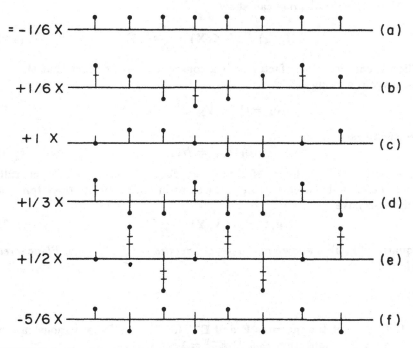

FIGURE 4.6. Graphical representation of the sequence $(\dots, 0, 3, -2, 1, -1, -2, 0, \dots)$. The vertical tick marks represent unit lengths.

We express $\{g_n\}$ in terms of a Fourier series, (4.14), and we express $\{f_n\}$ in a similar manner:

$$f_n = \sum_{k=1}^{N} Y_k W_{kn} = \frac{1}{N} \sum_{k=1}^{N} \tilde{f}_k W_{kn}.$$

Then, if we define

$$\mathbf{Y} = (Y_1, \ldots, Y_N), \quad \tilde{\mathbf{f}} = (\tilde{f}_1, \ldots, \tilde{f}_N),$$

we have the associations

$$\mathbf{g} \leftrightarrow \mathbf{X} \leftrightarrow \tilde{\mathbf{g}}, \quad \mathbf{f} \leftrightarrow \mathbf{Y} \leftrightarrow \tilde{\mathbf{f}}$$

between the N-vectors. But from our earlier discussion, $\mathbf{X}$ and $\tilde{\mathbf{g}}$ can each be regarded as the representation of $\mathbf{g}$ with respect to another orthogonal system. Therefore lengths of vectors should be simply related to one another. Specifically, one can show

$$(\mathbf{g}, \mathbf{g}) = N(\mathbf{X}, \mathbf{X}) = \frac{1}{N}(\tilde{\mathbf{g}}, \tilde{\mathbf{g}}). \tag{4.29}$$

The appearance of the factor N is a consequence of the fact that the new basis vectors, namely,

$$\mathbf{e}_k = (W_k, W_{2k}, \ldots, W_{Nk}) \tag{4.30}$$

are such that

$$(\mathbf{e}_k, \mathbf{e}_m) = N\delta_{km}. \tag{4.31}$$

The vectors $\{\mathbf{e}_k\}$ will be referred to as *Fourier vectors*. More generally, since the *angle* between vectors is the same in both coordinate systems, we must also have

$$(\mathbf{g}, \mathbf{f}) = N(\mathbf{X}, \mathbf{Y}) = \frac{1}{N}(\tilde{\mathbf{g}}, \tilde{\mathbf{f}}). \tag{4.32}$$

Equation (4.29) is known as *Parseval's relation* and (4.32) as *Plancherel's relation*.

Exercises

1. Construct the matrices $\mathbf{E}$ and $\mathbf{E}^{-1}$ for $N = 4$. Then demonstrate by actual computation that $\mathbf{E}\,\mathbf{E}^{-1} = 1$.

2. (a) Prove

$$\sum_{k=1}^{N} \cos \frac{2\pi}{N} kn \cos \frac{2\pi}{N} km = \frac{N}{2}(\delta_{nm} + \delta_{n,N-m}),$$

and from this compute $\tilde{g}_k$ in (4.22) directly.

(b) Prove

$$\sum_{k=1}^{N} \sin \frac{2\pi}{N} kn \sin \frac{2\pi}{N} km = \frac{N}{2}(\delta_{nm} - \delta_{n,N-m}),$$

and from this find $\tilde{g}_k$ in (4.23) directly.

3. Prove

$$\sum_{n=1}^{N} \cos \frac{2\pi}{N} nk \sin \frac{2\pi}{N} nm = 0.$$

Use this with the results of Exercise 2 to demonstrate (4.26).

4. Find the Fourier sequences for $N = 3$ and interpret these in the same way as is done in the text.

5. Prove formulas (4.29), (4.31), (4.32). [Hint: Substitute the Fourier series for f and g in (4.29) and (4.31) and interchange sums.]

6. Find the Fourier series for the following periodic sequences:

   (a) $\ldots, 1, 0, 1, 0, 1, 0, \ldots$;
   (b) $\ldots, 0, 1, -1, 0, 1, -1, 0, \ldots$;
   (c) $\ldots, 3, 1, 2, 1, 3, 1, 2, 1, 3, \ldots$.

   [Hint: Make use of parity (i.e., evenness or oddness) properties inherent in these sequences.]

7. Verify the details given in the pictorial example shown in Figure 4.6. What significance is there to the sequences (a–f) in the figure?

8. For computational purposes it is usually simpler to sum over both negative and positive integers. Verify that

   (a)

$$g_n = \frac{1}{N} \sum_{k=-(N/2)+1}^{N/2} \tilde{g}_k W_{nk}, \quad \tilde{g}_k = \sum_{n=-(N/2)+1}^{N/2} g_n \overline{W}_{nk}$$

   if N is even and

   (b)

$$g_n = \frac{1}{N} \sum_{k=-(N-1)/2}^{(N-1)/2} \tilde{g}_k W_{nk}, \quad \tilde{g}_k = \sum_{n=-(N-1)/2}^{(N-1)/2} g_n \overline{W}_{nk}$$

   if N gives rise to the same transform pairs as in Eqs. (4.20), (4.21).

   (c) What are the consequences of g_n real in (a) and (b)?

4.3   Discrete Periodic Inputs

Consider a discrete, linear, and translationally invariant system. Then from
our earlier remarks, the system is characterized by (4.11), which we write
as

$$f_n = \sum_{m=-\infty}^{\infty} K_{n-m} g_m = (K \star g)_n. \tag{4.33}$$

Causality is not specifically imposed so that (4.33) can be thought of as
existing in the space domain—or if in the time domain, then $K_{n-m} = 0$
when $m > n$. In this section we consider the case of the system being
subjected to periodic inputs $\{g_n\}$.

   If $\{g_n\}$ in (4.33) is N-periodic, then $\{f_n\}$ is also N-periodic. To see this,
observe that for any j (see Exercise 4 at the end of Section 4.1)

$$f_{n+jN} = (K \star g)_{n+jN} = (g \star K)_{n+jN} = \sum_m g_{n+jN-m} K_m$$

$$= \sum_m g_{n-m} K_m = f_n.$$

Since both f_n and g_n are N-periodic, we might hope to reduce (4.33) to an
N-dimensional formulation. Since f_n and g_n are N-periodic, each can be
thought of as a vector in N-space, say $\mathbf{f}$ and $\mathbf{g}$. But since they are linearly
related, there exists a matrix, say $\mathbf{k}$, such that $\mathbf{g} = \mathbf{kf}$. The real problem is
to determine $\mathbf{k}$. With this in mind, we first define

$$k_n = \sum_{j=-\infty}^{\infty} K_{n+jN}, \tag{4.34}$$

and observe that $\{k_n\}$ is a periodic sequence. Also we observe that any
infinite summation can be rewritten as follows:

$$\sum_{n=-\infty}^{\infty} S_n = \sum_{p=-\infty}^{\infty} \sum_{\ell=1}^{N} S_{\ell-pN}$$

(Clearly each term S_n appears once and only once on the right-hand side.)
If this splitting is applied to (4.33), then

$$f_n = \sum_{p=-\infty}^{\infty} \sum_{\ell=1}^{N} K_{n-(\ell-pN)} g_{\ell-pN} = \sum_{p=-\infty}^{\infty} \sum_{\ell=1}^{N} K_{n+pN-\ell} g_\ell$$

$$= \sum_{\ell=1}^{N} g_\ell \sum_{p=-\infty}^{\infty} K_{n+pN-\ell} = \sum_{\ell=1}^{N} k_{n-\ell} g_\ell, \tag{4.35}$$

and the formulation (4.33) does in fact reduce to an N-dimensional one when g_n is N-periodic. Furthermore, if we define the $N \times N$ matrix $\mathcal{K}$ such that

$$(\mathcal{K})_{nm} = k_{n-m}, \tag{4.36}$$

then (4.33) can be said to have been transformed to

$$\mathcal{K}\mathbf{g} = \mathbf{f}, \tag{4.37}$$

a system of N linear equations.

We have arrived at one of the central themes of this course: *the idea of finding some sort of transformation which reduces one problem to another problem in the hope that the second problem is simpler to deal with.* In the present situation we have converted an infinite problem to a finite problem. Indeed, it has been reduces to linear algebra, and elementary methods exist for handling such problems. Actually, it is appropriate for us to be *greedy* in this instance and to seek further transformations. As will be seen, (4.33) can be reduced to scalar equations!

First we remark that matrices for which (4.36) is true are referred to as *circulants.* In more detail we observe that $\mathcal{K}$ and circulants in general have the form

$$\mathcal{K} = \begin{pmatrix} k_N & k_{N-1} & \cdot & \cdot & k_1 \\ k_1 & k_N & \cdot & \cdot & k_2 \\ k_2 & k_1 & k_N & \cdot & k_3 \\ \vdots & \vdots & & & \\ k_{N-1} & k_{N-2} & \cdot & \cdot & k_N \end{pmatrix}. \tag{4.38}$$

Note: *Each row (column) of a circulant matrix is the cyclic permutation of the previous row (column).*

Examples. The unit matrix

$$\mathbf{I} = \begin{pmatrix} 1 & 0 & \cdot & \cdot & \cdot & 0 \\ 0 & 1 & & & & \cdot \\ \cdot & & \cdot & & & \cdot \\ \cdot & & & \cdot & & \cdot \\ 0 & \cdot & \cdot & \cdot & \cdot & 1 \end{pmatrix}$$

is a circulant as is the *shift* or *permutation* operator

$$\mathcal{S} = \begin{pmatrix} 0 & 0 & \cdot & \cdot & 0 & 1 \\ 1 & 0 & \cdot & \cdot & \cdot & \cdot \\ 0 & 1 & 0 & \cdot & \cdot & \cdot \\ \cdot & & & & & \cdot \\ \cdot & & & & & \cdot \\ 0 & & \cdot & \cdot & 1 & 0 \end{pmatrix}$$

other way of formatting (4.37) is through the use of the convolution product, which for N-periodic sequences $\{a_n\}$ and $\{b_n\}$ is defined to be

$$(\mathbf{a} \star \mathbf{b})_n = \sum_{m=1}^{N} a_{n-m} b_m = (\mathbf{b} \star \mathbf{a})_n. \tag{4.39}$$

As with all previously defined convolution products, this too is commutative (Exercise 5 at the end of Section 4.1). [We mention again that no confusion should arise in the diverse ways in which the convolution product has been defined. In each instance the context reveals which particular definition is being used.] With (4.39) defined, we can rewrite (4.37) as

$$\mathbf{f} = \mathbf{k} \star \mathbf{g} \quad \text{or} \quad f_n = (\mathbf{k} \star \mathbf{g})_n. \tag{4.40}$$

Next take the Fourier transform of (4.40). With the use of the vector notation (4.30), we can write the Fourier transform as (see (4.21))

$$\tilde{f}_n = (\mathbf{e}_n, \mathbf{f}). \tag{4.41}$$

In order to deal with the Fourier transform of the right-hand side of (the first equation of) (4.40), we now prove the

Convolution Theorem. *The Fourier transform of a convolution product is the product of the Fourier transforms. In symbols,*

$$\widetilde{(\mathbf{a} \star \mathbf{b})}_n = (\mathbf{e}_n, (\mathbf{a} \star \mathbf{b})) = (\mathbf{e}_n, \mathbf{a})(\mathbf{e}_n, \mathbf{b}) = \tilde{a}_n \tilde{b}_n. \tag{4.42}$$

To see this observe

$$(\mathbf{e}_n, (\mathbf{a} \star \mathbf{b})) = \sum_{m=1}^{N} e^{-2\pi i(nm/N)} \sum_{k=1}^{N} a_{m-k} b_k$$

$$= \sum_{k=1}^{N} b_k e^{-2\pi i(kn/N)} \sum_{m=1}^{N} a_{m-k} e^{-2\pi i(n(m-k)/N)},$$

where orders of summation have been reversed to get the second form. From the periodicity of the terms, we can write

$$\sum_{m=1}^{N} a_{m-k} e^{-2\pi i(n(m-k)/N)} = \sum_{m=1+k}^{N+k} a_{m-k} e^{-2\pi i(n(m-k)/N)}$$

$$= \sum_{s=1}^{N} a_s e^{-2\pi i(ns/N)}.$$

To get the second sum we use periodicity to change the limits of summation. Finally, we set $m - k = s$ to get the last sum. This proves (4.42). Therefore, if we apply the Fourier transform to (4.40), we obtain

$$\tilde{f}_n = \tilde{k}_n \tilde{g}_n. \tag{4.43}$$

This is just a scalar equation; and, as we asserted earlier, the infinite formulation (4.33) with which we started has been reduced to scalar form.

If we return to the introductory section 4.1, then the three scenarios discussed there are easily resolved through the use of (4.43). We recall that in Scenario 1 we are given K_{n-m} and g_m, and the task is to calculate f_n. One way to accomplish this is first to compute $\tilde{k}_n$ and $\tilde{g}_n$ and then to obtain $\tilde{f}_n$ from (4.43). Finally, f_n is computed through

$$f_n = \frac{1}{N} \sum_{m=1}^{N} \tilde{f}_m W_{mn}.$$

[It might be felt, with some justification, that f_n can be more directly calculated from (4.37). In the appendix at the end of this chapter, we show that a computational method known as the fast Fourier transform makes the above seemingly indirect method shorter than the direct method if N is large.] A Scenario 2 calculation is now clearly less time-consuming than a matrix inversion. For from (4.43) we immediately have (provided $\tilde{k}_n \neq 0$ for all n)

$$\tilde{g}_n = \tilde{f}_n / \tilde{k}_n,$$

and the solution for g_n is obtained by substituting this expression for $\tilde{g}_n$ in (4.20). The case of $\tilde{k}_n = 0$ will be considered in Exercise 4 at the end of this section.

The Scenario 3 calculation is perhaps the most interesting. Again, from (4.43) this is formally given by

$$\tilde{k}_n = \tilde{f}_n / \tilde{g}_n.$$

Then, when the right-hand side is substituted for $\tilde{k}_n$ in (4.20), we obtain k_n. It is clear that the input $\{g_n\}$ must be chosen so that all harmonics are present, i.e., so that $\tilde{g}_n \neq 0$ for all n.

Aliasing. It should be recalled that k_n is being determined and not the *system operator* K_n (see (4.34)). In the event that $K_m = 0$ for, say, $|m| > M$, then K_n can be determined from k_n if N is chosen so that $N \geq 2M+1$. If, on the other hand, K_m has an infinite number of components, say

$$K_m = \gamma^{|m|}, \quad 0 < \gamma < 1,$$

then k_n, which is a periodic version of K_n given by (4.34), is said to be an *aliased* version of K_n. The contrast between the two is shown in Figure 4.7. It is important therefore to choose N sufficiently large so that K_m can be determined to good approximation. (See the exercises at the end of this section.)

Circulants. In (4.36) we have given the relationship between an N-periodic sequence and a circulant matrix. To repeat, if $\{c_n\}$ is N-periodic, then the corresponding $N \times N$ circulant matrix C is given by

$$(C)_{nm} = c_{n-m}.$$

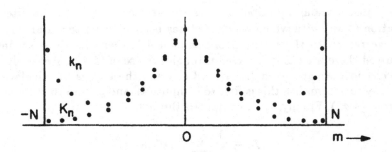

FIGURE 4.7.

The structure of $\mathcal{C}$ is depicted in (4.38).

Next it is interesting to observe that

$$\sum_{1}^{N} c_{n-m} e^{2\pi i(mk/N)} = e^{2\pi i(nk/N)} \sum_{m=1}^{N} c_{n-m} e^{-2\pi i(k(n-m)/N)}$$

$$= \tilde{c}_k e^{2\pi i(nk/N)}.$$

But this is equivalent to

$$\mathcal{C}\mathbf{e}_k = \tilde{c}_k \mathbf{e}_k = \lambda_k \mathbf{e}_k, \tag{4.44}$$

and we have therefore proven the following:

The eigenvectors of a circulant matrix are the Fourier vectors $\mathbf{e}_k$, (4.30), and the corresponding eigenvalues λ_k are the Fourier transforms $\tilde{c}_k$ of the periodic sequence $\{c_n\}$.

Now that we know what the eigenvectors and eigenvalues of a circulant matrix should be, we can avoid repeating the above systematic derivation in order to determine the eigenvectors and eigenvalues of *any* circulant matrix given to us. We offer in its place a simpler method of finding such vectors and scalars based on this hindsight. Consider the circulant matrix

$$\mathbf{C} = \begin{pmatrix} c_0 & c_1 & c_2 & \cdot & \cdot & \cdot & c_{N-1} \\ c_{N-1} & c_0 & c_1 & \cdot & \cdot & \cdot & c_{N-2} \\ \cdot & \cdot & & & & & \\ \cdot & & \cdot & c_0 & & & \\ \cdot & & & \cdot & & & \\ c_1 & c_2 & \cdot & \cdot & \cdot & \cdot & c_0 \end{pmatrix}$$

and z such that

$$z^N = 1.$$

To find the eigenvalues and eigenvectors of $\mathbf{C}$, we let

$$\lambda = c_0 + c_1 z + \cdots + c_{N-1} z^{N-1}.$$

If both sides of this equation are multiplied in turn by z, z^2, ..., we obtain

$$\lambda z = c_{N-1} + c_0 z + \cdots$$
$$\lambda z^2 = c_{N-2} + c_{N-2} z + \cdots$$
$$\vdots \qquad \vdots \qquad \vdots$$

Thus λ is an eigenvalue and $(1, z, z^2, \ldots, z^{N-1})$ is the eigenvector. Since $z^N = 1$ has N roots, this procedure generates the entire system of eigenvalues and eigenvectors.

If we direct our attention to the linear equation

$$C\mathbf{x} = \mathbf{b}, \tag{4.45}$$

then the solution follows from the above discussion and is given by

$$x_n = \frac{1}{N} \sum_{m=1}^{N} \frac{\tilde{b}_m}{\tilde{c}_m} e^{2\pi i(mn/N)} = \frac{1}{N} \sum_{m=1}^{N} \frac{\tilde{b}_m e^{2\pi i(mn/N)}}{\lambda_m}, \tag{4.46}$$

where as before

$$\tilde{b}_m = (\mathbf{e}_m, \mathbf{b}).$$

An interesting interpretation of (4.46) is provided by an elementary corollary to the Convolution Theorem.

The Fourier series constructed from the product of Fourier transforms of two N-periodic sequences is equal to the convolution product of the sequences. Symbolically,

$$(\mathbf{a} \star \mathbf{b})_n = \frac{1}{N} \tilde{a}_m \tilde{b}_m W_{mn}. \tag{4.47}$$

This results immediately from taking the inverse transform of (4.42). (The relation of (4.42) to (4.47) is the same as that of (4.21) to (4.20).)

To apply (4.47) to (4.46), define

$$s_n = \frac{1}{N} \sum_{m=1}^{N} \frac{1}{\tilde{c}_m} W_{mn}. \tag{4.48}$$

Then, instead of (4.46), we can write

$$x_n = (\mathbf{s} \star \mathbf{b})_n = \sum_{m=1}^{N} s_{n-m} b_m. \tag{4.49}$$

For this reason s_n is known as the *fundamental solution* or *solution operator.*

The fundamental solution itself has an interesting interpretation. It gives the solution to (4.45) for that special inhomogeneous term $\mathbf{b}$ whose Fourier transform is

$$\tilde{\mathbf{b}} = (1, 1, \ldots, 1).$$

The vector b itself can be found by inverse transforming $\tilde{b}$. This gives

$$b_n = \frac{1}{N} \sum_{m=1}^{N} W_{mn} = \delta_{nN} \quad (= \delta_{n0})$$

(see (4.30)). Thus s_{n-k} is the solution to (4.45) when $b_n = \delta_{nk}$, and (4.49) with δ_{mk} substituted for b_m takes on intuitively reasonable meaning.

Exercises

1. Apply the Fourier transform (4.21) directly to (4.33) and thereby find (4.43).

2. (a) For the N-periodic sequence $\{g_n\}$ and for

$$K_n = \gamma^{|n|},$$

   where $0 < \gamma < 1$, show that (4.33) can be put in the form (4.40) with

$$k_\ell = \gamma^{|\ell|} + \frac{\gamma^N}{1 - \gamma^N} \left(\gamma^\ell + \frac{1}{\gamma^\ell} \right), \quad |\ell| \le N.$$

   (b) Suppose $N = 20$ and $\gamma = .5$. How bad an approximation is k to K in the fundamental period?

3. What N-periodic sequence has $\ldots, 1, 1, 1, \ldots$ for its transform?

4. (a) Suppose λ_k in (4.46) vanishes for some k. Then the sum (4.46) does not exist. What can be said about solutions of (4.45) in this instance? Can you relate this to a theorem in linear algebra?

   (b) Consider in particular

$$c = \begin{pmatrix} 0 & 1 & 1 & -2 \\ -2 & 0 & 1 & 1 \\ 1 & -2 & 0 & 1 \\ 1 & 1 & -2 & 0 \end{pmatrix}$$

   and

$$b = \begin{bmatrix} 1 \\ q \\ 0 \\ 0 \end{bmatrix}.$$

   Show that c has a zero eigenvalue. Is there a value of q in b which permits (4.45) to be solved? If so, exhibit the solution.

5. Suppose a discrete, linear, and translationally invariant system generates the output

$$\{f_n\} = \ldots, \quad 0, b, a+b, a, 0, b, a+b, a, \ldots$$

when given the input

$$\{g_n\} = \ldots, \quad 1, 1, 0, 0, 1, 1, 0, 0, \ldots.$$

Characterize the system.

6. Suppose that in (4.33)

$$K_0 = \frac{1}{2}, \quad K_{\pm 1} = \frac{1}{4}, \text{ and } \quad K_n = 0 \text{ for } |n| > 1.$$

(This is a weighted average operator.) Then, if

$$\{f_n\} = \ldots, \quad 1, 2, 4, 3, 1, 2, 4, 3, \ldots,$$

what is $\{g_n\}$?

7. Find the inverse of

$$\begin{pmatrix} 0 & 1 & 1 & 1 \\ 1 & 0 & 1 & 1 \\ 1 & 1 & 0 & 1 \\ 1 & 1 & 1 & 0 \end{pmatrix}.$$

8. Show that $\{\exp(2\pi i p n)\}$ for any real p is an eigenvector of the infinite matrix K_{n-m}. What is the corresponding eigenvalue?

9. Show that circulant matrices (a) commute and (b) their product is a circulent.

10. Find the eigenvalues and eigenvectors of

(a)

$$\begin{pmatrix} 2 & 1 & 1 \\ 1 & 2 & 1 \\ 1 & 1 & 2 \end{pmatrix},$$

(b)

$$\begin{bmatrix} 2 & -1 & 0 & -1 \\ -1 & 2 & -1 & 0 \\ 0 & -1 & 2 & -1 \\ -1 & 0 & -1 & 2 \end{bmatrix}.$$

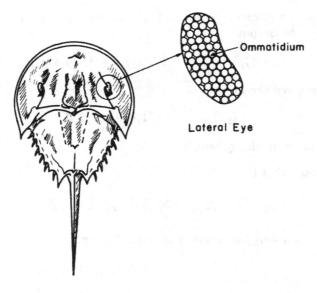

Horseshoe Crab

(Limulus Polyphemus)

FIGURE 4.8.

4.4 Applications

In this section we consider two different biophysical problems which can be formulated and solved by the modeling techniques and methods developed above.

Visual System of the Horseshoe Crab. The horseshoe crab (*Limulus polyphemus*) is familiar to bathers on the East Coast of the United States. It has a compound lateral eye (see Figure 4.8). Each element, referred to as an *ommatidium*, functions as a simple photoreceptor. There are roughly a thousand ommatidia in each eye and from each one there emanates a nerve fiber or axon. These come together in what is the *optic tract*, and carry visual information to higher centers (i.e., the brain?).

Illumination falling on the horseshoe crab eye is transduced by a series of steps into electrical pulses known as *action potentials*, which travel along the axons. All action potentials have essentially identical shapes, so that one must conclude that it is the frequency of these pulses and not their shape which carries visual information. For a single ommatidium that is illuminated—all others receiving no light—we find that the frequency of firing is proportional to the logarithm of the light intensity. To avoid carrying the logarithm, we measure illumination by this excitation; i.e., we

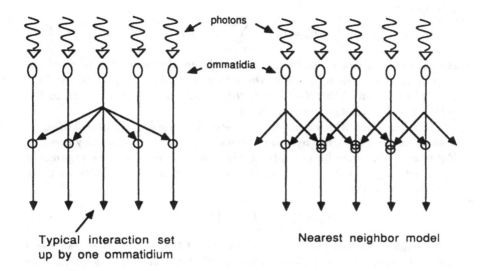

Typical interaction set
up by one ommatidium

Nearest neighbor model

FIGURE 4.9.

measure the intensity of light falling on an ommatidium by the spiking
frequency which such light intensity would produce in the absence of stimulation of other ommatidia. Denoting the response of the nth ommatidium
by r_n (measured in the axon as spiking frequency) and the excitation of
the nth ommatidium by ϵ_n, we have

$$r_n = \epsilon_n$$

in the absence of stimulation of other ommatidia.

Experimentally, it is observed that stimulation of a neighboring ommatidium reduces the response of the test ommatidium. Moreover this phenomenon, known as *lateral inhibition,* is observed to be linear.

To create a simple model which takes into account these effects, we consider one-dimensional stimuli (i.e., illumination patterns only varying in
one dimension) and a one-dimensional system of interactions, as indicated
in Figure 4.9.

The inhibition acting on an ommatidium is proportional to the linearly
summed responses of the neighboring ommatidia. Further, it is reasonable
to assume that as long as we are away from the boundaries of the lateral eye, the inhibitory network is basically the same for all ommatidia.
Therefore, inhibition can be represented by a *linear* and *translationally invariant* operator. Putting the pieces together, the response is governed by

the equation

$$r_n = \epsilon_n - \sum_{m=-\infty}^{\infty} A_{n-m} r_m. \tag{4.50}$$

A_{n-m}, called a *coupling constant*, gives the inhibition at the nth ommatidium due to a unit response in the mth ommatidium. Equation (4.50) is a simplified form of the *Hartline–Ratliff equations*. Before going on to the solution of (4.50), we comment on its form.

The summation which extends from $-\infty$ to $+\infty$ is formal. In real life only a small number (roughly ten) ommatidia enter into the inhibitory process. Equation (4.50) may be put into continuous form by the same arguments which let us pass between (4.10) and (4.11). For future reference we state the continuous version:

$$r(x) = \epsilon(x) - \int_{-\infty}^{\infty} K(x - y) r(y) dy. \tag{4.51}$$

Although we created (4.50) (or (4.51)) as a model of the horseshoe crab visual system, it serves as a model for a variety of other systems. (Naturally, it can be expected to model other visual systems.) Two commonplace instances are xerography and the development process in photography. In the latter case, during development, exposed portions of a film are acted upon chemically by the developer. A number of end products result from the reactions and these retard further film development. In this case, $\epsilon(x)$ in (4.51) represents the level of *exposure* and $r(x)$, the resulting *development*. The integral term then gives the retardation of development due to neighboring developed portions of film. Xerography is quite similar except that light gradations result in a pattern of electric charge. The absence of light corresponds to a positive charge. In the final stage of xerography a black dust falls on the xerography plate and clings to areas of positive charge. It is clear from elementary electrostatics that a charge induces an opposite charge in its neighborhood. Thus, in this case (4.51) can be interpreted as follows: $\epsilon(x)$ represents the illumination pattern and $r(x)$, the resulting positive charge, while the integral term represents the induced negative charge.

For purposes of exposition, we consider a drastically simplified example of (4.50), the so-called *nearest neighbor model* (see Figure 4.9):

$$r_n = \epsilon_n - \alpha r_{n-1} - \alpha r_{n+1}. \tag{4.52}$$

It is further assumed that the coupling is the same on the right and left. This equation, (4.52), says that the response at an ommatidium is due to the excitation falling on it minus a quantity of inhibition that is proportional to the response of its two nearest neighbors. The problem before us is the determination of the response pattern $\{r_n\}$ for a given excitation pattern $\{\epsilon_n\}$.

Let us first consider the solution of (4.52) in the case of a constant illumination, for which we have

$$\epsilon_n = L$$

for all n. It is then reasonable to say that the response is also a constant, say r^0. Then

$$r^0 = L - \alpha(r^0 + r^0)$$

or

$$r^0 = \frac{L}{1 + 2\alpha}. \tag{4.53}$$

Why is the response less than L?

Next consider a periodic illumination pattern superimposed on a constant background:

$$\epsilon_n = L + \frac{1}{N}\sum_{q=1}^{N}\tilde{\epsilon}_q e^{2\pi i(nq/N)} = L + \frac{1}{N}\sum_{q=1}^{N}\tilde{\epsilon}_q W_{qn}, \tag{4.54}$$

where we have made use of the fact that any N-periodic discrete pattern can be expressed in terms of discrete Fourier components. Our governing equation (4.52) is of convolution type so that the solution will also be N-periodic. Express the solution in the form

$$r_n = \frac{L}{1 + 2\alpha} + \frac{1}{N}\sum_{q=1}^{N}\tilde{r}_q W_{qn}, \tag{4.55}$$

where use of (4.53) has been made. (Since (4.52) is linear, a sum of solutions is also a solution.) Substitution of (4.54) and (4.55) for ϵ_n and r_n, respectively, in (4.52) yields

$$\sum_{q=1}^{N}\tilde{r}_q e^{2\pi i(nq/N)} = \sum_{q=1}^{N}\tilde{\epsilon}_q e^{2\pi i(nq/N)}$$

$$- \alpha\sum_{q=1}^{N}\left\{\tilde{r}_q e^{2\pi i((n+1)q/N)} + \tilde{r}_q e^{2\pi i((n-1)q/N)}\right\}$$

$$= \sum_{q=1}^{N}\tilde{\epsilon}_q e^{2\pi i(nq/N)} - \alpha\sum_{q=1}^{N}\tilde{r}_q e^{2\pi i(nq/N)}\left(e^{2\pi iq/N} + e^{-2\pi iq/N}\right).$$

After the dust has settled we have

$$\tilde{r}_q = \tilde{\epsilon}_q - \tilde{r}_q\, 2\alpha \cos\frac{2\pi q}{N}$$

and

$$\tilde{r}_q = \frac{\tilde{\epsilon}_q}{1 + 2\alpha \cos(2\pi q/N)}. \tag{4.56}$$

The solution for r_n is thus given by

$$r_n = \frac{L}{1 + 2\alpha} + \frac{1}{N}\sum_{q=1}^{N}\frac{\tilde{\epsilon}_q W_{qn}}{1 + 2\alpha \cos(2\pi q/N)}. \tag{4.57}$$

On physical grounds we require $0 \leq \alpha < 1/2$; otherwise, the denominator can pass through zero. We see from (4.56) that certain of the Fourier components, $\exp(2\pi i(qn/N))$, are *attenuated* and that others are *amplified*. For example, the Fourier component for $q = 0$ (i.e., for the constant stimulus L) is the most attenuated, by a factor of $1/(1 + 2\alpha)$, while the Fourier component for $q = N/2$ (assuming N is even) is the most amplified, by a factor of $1/(1 - 2\alpha)$. A little thought and reflection on the model depicted in Figure 4.9 reveals why this occurs.

It is unnecessary to consider explicitly the constant stimulus L since we can just as easily carry it in $\tilde{\epsilon}_N$, i.e., the $q = N$ term of the sum (4.57). If we suppress the constant term, we can write

$$\tilde{\epsilon}_q = \sum_{n=1}^{N} \epsilon_n \overline{W}_{nq},$$

which when substituted into the solution yields

$$r_n = \frac{1}{N} \sum_{m=1}^{N} \sum_{q=1}^{N} \frac{\epsilon_m e^{-2\pi i(q(n-m)/N)}}{1 + 2\alpha \cos(2\pi q/N)}.$$

Define

$$s_n = \frac{1}{N} \sum_{q=1}^{N} \frac{e^{-2\pi i(qn/N)}}{1 + 2\alpha \cos(2\pi q/N)}$$

$$= \frac{1}{N} \sum_{q=1}^{N} \frac{\cos(2\pi qn/N)}{1 + 2\alpha \cos(2\pi q/N)}, \tag{4.58}$$

in terms of which we write

$$r_n = \sum_{m=1}^{N} s_{n-m} \epsilon_m = (s \star \epsilon)_n. \tag{4.59}$$

s_n is the solution operator for the problem (see the end of Section 4.3). In this instance it is also referred to as the *point spread* function since it is the solution to (4.52) for $\epsilon_n = \delta_{n0}$ (a *point source*).

We later show that (4.58) can also be written as

$$s_n = C \left\{ \gamma^{|n|} + \left(\gamma^n + \frac{1}{\gamma^n} \right) \frac{\gamma^N}{1 - \gamma^N} \right\}, \tag{4.60}$$

where

$$C = \frac{1}{(1 - 4\alpha^2)^{1/2}}, \quad \gamma = -\frac{1}{2\alpha} + \left(\frac{1}{(2\alpha)^2} - 1 \right)^{1/2} < 0. \tag{4.61}$$

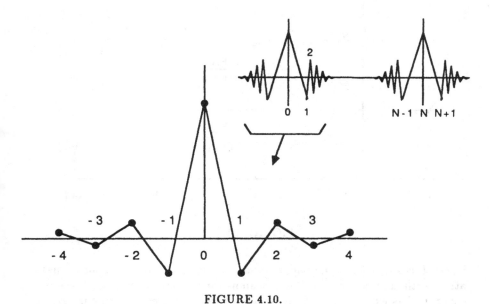

FIGURE 4.10.

It should be noted that as a result of (4.61) and (4.60), s_n alternates in sign. For example, if we have $\alpha = 1/4$, then

$$s_n = (1.15, -.31, .08, -.02, +.005, \ldots).$$

Therefore, if $\epsilon_n = \mathcal{L}\delta_{n0}$ ($\mathcal{L}$ a constant), then

$$r_n = \mathcal{L}s_n.$$

A sketch of the resulting response pattern is given in Figure 4.10.

An intriguing part of this solution is the fact that the response of the zeroth ommatidium is larger than its excitation; for $\alpha = 1/4$ it is fifteen percent above stimulation. The reason for this is as follows. As a result of the stimulation on, say, the zeroth ommatidium, this ommatidium inhibits its nearest neighbors, located at ± 1. This, in turn, results in less inhibition on the zeroth ommatidium as well as on those ommatidia at ± 2—and so forth. This phenomenon is referred to as *disinhibition* and has been observed in many nervous systems. Another point is that we cannot, of course, have a negative response (remember that the response is measured in number of spikes per time). For purposes of our discussion, we imagine that a background illumination is present so that the solution is

$$r_n = \frac{L}{1 + 2\alpha} + \mathcal{L}s_n$$

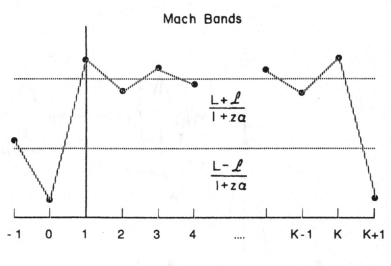

FIGURE 4.11.

where L is the level of the background illumination. Then r_n is never negative. Although our model does not allow for negative responses, there are
other models which do. In these models a threshold effect must be imposed,
in which case the equations become nonlinear.

Mach Bands. Let us consider one other example, namely the response to

$$\epsilon_n = L + \mathcal{L}(\underbrace{1,1,1,\ldots,1}_{k}, \underbrace{-1,-1,\ldots,-1}_{N-k}) = L + \mathcal{L}h_n$$

where as indicated $\{h_n\}$ is the N-periodic sequence such that $h_n = 1$ for
$1 \leq n \leq k$ and $h_n = -1$ for $k+1 \leq n \leq N$. Then, substituting the above
expression for ϵ_n in (4.59), we have as our solution

$$r_n = \sum_{n=1}^{N} s_{n-m}(L + \mathcal{L}h_n) = \sum_{n=1}^{N} s_{n-m}L + \mathcal{L}\sum_{n=1}^{N} s_{n-m}h_n$$

$$= \frac{L}{1+2\alpha} + \mathcal{L}(s \star h)_n.$$

A sketch of the solution is given in Figure 4.11. Actually most of the features
of this sketch can be arrived at by simple considerations.

Well away from an *edge* the local illumination is $(L \pm \mathcal{L})$ (> 0 by assumption), and hence the response is roughly $(L \pm \mathcal{L})/(1 + 2\alpha)$ in these regions.
At an edge an ommatidium is inhibited by one neighbor on the high illumination side and by one on the low illumination side. The result is as
shown in Figure 4.11. Therefore, a visual system of the sort that we believe

Cell Model

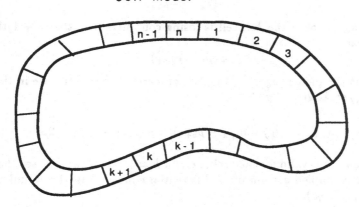

FIGURE 4.12.

describes the horseshoe crab enhances edges. This phenomenon goes by the name of *Mach bands*. If we look carefully at relatively sharp edges, say, of a shadow, we observe that our own visual system also produces Mach bands. Specifically, as the edge is approached from within the shadow, a dark band appears, while if it is approached from the light region, a bright band is perceived. The reason for this is probably the same as that given in the case of the horseshoe crab—lateral inhibition. We mention in passing that Impressionist painters such as Degas, Monet, and others not only noted this effect but overcompensated for it in their paintings.

The above discussion deals with the response to a given illumination pattern. A more complicated problem would be the treatment of the change in time of the response as the illumination pattern changes. The inclusion of temporal change is referred to as *dynamics.* We consider dynamics in the context of another problem. This is the problem of *morphogenesis*, i.e., the evolution of patterns and shapes as, for example, occurs in biology.

Cell Model of Diffusion. In a famous model of morphogenesis, Turing postulates that a ring of cells *communicate* with one another by passing substances, called *morphogens,* across common membranes. For purposes of exposition, we consider just a single morphogen and denote its concentration in the kth cell by ρ_k (see Figure 4.12). The simplest law of diffusion across a membrane states that the flux of concentrate is proportional to the difference in concentrations across the membrane. (The flux is always toward regions of relatively low concentration.) If we denote the proportionality constant by $\mu > 0$, then the flux of concentrate into the kth cell, and hence the rate of increase of ρ_k due to diffusion from the $(k-1)$th cell, is

$$\mu(\rho_{k-1} - \rho_k).$$

By the same token the kth cell is losing concentrate to the $(k + 1)$th cell at the rate

$$\mu(\rho_k - \rho_{k+1}).$$

If we put the pieces together, the rate of increase of concentrate in the kth cell, $d\rho_k/dt$, is given by

$$\frac{d\rho_k}{dt} = \mu(\rho_{k-1} - \rho_k) - \mu(\rho_k - \rho_{k+1}) = \mu(\rho_{k-1} + \rho_{k+1} - 2\rho_k). \quad (4.62)$$

This equation is valid for each cell, i.e., for $k = 1, \ldots, N$, and we see that the result is a coupled system of N differential equations in the N-unknowns $\rho = (\rho_1, \ldots, \rho_N)$.

If we write

$$\mathbf{C} = \mu \begin{pmatrix} -2 & -1 & 0 & \cdot & \cdot & \cdot & \cdot & 1 \\ 1 & -2 & 1 & 0 & \cdot & \cdot & \cdot & 0 \\ 0 & 1 & -2 & 1 & 0 & \cdot & \cdot & \cdot \\ 0 & \cdot & & \cdot & \cdot & & & \\ \vdots & & & & & & & \\ 1 & 0 & \cdot & \cdot & \cdot & \cdot & 1 & -2 \end{pmatrix},$$

then (4.62) can be written in vector form as

$$\frac{d}{dt}\rho = \mathbf{C}\rho = \mathbf{c} \star \rho. \quad (4.63)$$

Observe that $\mathbf{C}$ is a circulant matrix and that $\mathbf{c} = (c_1, c_2, \ldots, c_{N-2}, c_{N-1}, c_N) = \mu(1, 0, \ldots, 0, 1, -2)$ in the convolution is the corresponding vector. Equation (4.63) is to be solved subject to initial conditions giving the initial concentration in each cell, say

$$\rho_k(t = 0) = \rho_k^0, \quad k = 1, 2, \ldots, N.$$

As announced at the outset (4.63) contains an added degree of complexity, namely dynamics (i.e., time variation is now included). However, the right-hand side of (4.63) contains the convolution product, a form with which we have experience.

Because of the circulant matrix $\mathbf{C}$ in (4.63), it is natural to expand the solution in a Fourier series or, equivalently, in the eigenvectors (see Section 4.3, *Circulants*) of $\mathbf{C}$. We write

$$\rho_n = \frac{1}{N} \sum_{k=1}^{N} \tilde{\rho}_k(t) W_{kn} \quad (4.64)$$

or, in vector notation,

$$\rho = \frac{1}{N} \sum_{k=1}^{N} \tilde{\rho}_k(t) e_k.$$

The Fourier vectors e_k are defined in (4.30) and are shown to be the eigenvectors of circulant matrices in (4.44). It is important to note that the coefficients $\tilde{\rho}_k = \tilde{\rho}_k(t)$ are functions of time. The N-periodic sequence $\{\rho_k\}$ changes in time. It therefore has an N-periodic Fourier series at each instant of time and hence its Fourier coefficients change in time. If we substitute (4.64) into (4.63), we obtain

$$\frac{d}{dt} \sum_{k=1}^{N} \tilde{\rho}_k(t) e_k = C \sum_{k=1}^{N} \tilde{\rho}_k e_k = \sum_{k=1}^{N} \lambda_k \tilde{\rho}_k e_k, \qquad (4.65)$$

where the eigenvalue λ_k is given by

$$\lambda_k = \sum_{n=1}^{N} c_n W_{nk}.$$

Then, from the orthogonality condition (4.31),

$$\frac{d}{dt} \tilde{\rho}_k(t) = \lambda_k \tilde{\rho}_k(t), \quad k = 1, \ldots, N. \qquad (4.66)$$

Thus the use of Fourier series has transformed the system of N equations, (4.63), to the relatively simple problem of N scalar equations (4.66).

To solve (4.66), we specify the initial data:

$$\rho_n(t=0) = \rho_n^0 = \frac{1}{N} \sum_{k=1}^{N} \tilde{\rho}_k(0) W_{kn}.$$

Hence

$$\tilde{\rho}_k(0) = \sum_{n=1}^{N} \overline{W}_{kn} \rho_n^0 = \tilde{\rho}_k^0. \qquad (4.67)$$

The solution to (4.66) is immediately given by

$$\tilde{\rho}_k = e^{\lambda_k t} \tilde{\rho}_k^0,$$

and the solution for ρ_n is therefore

$$\rho_n = \frac{1}{N} \sum_{k=1}^{N} e^{\lambda_k t} \tilde{\rho}_k^0 W_{kn}. \qquad (4.68)$$

Note that when t is set to zero the initial data, $\rho_n(t=0)$, is recovered.

An understanding of the time behavior of this solution follows from a consideration of the eigenvalues λ_k. From the above or (4.44), these are given by

$$\lambda_k = \sum_{p=1}^{N} c_p W_{pk}.$$

Then, if we recall that

$$c \doteq \mu(1, 0, 0, \ldots, 0, 1, -2),$$

we see that

$$\lambda_k = \left(e^{(-2\pi i/N)k} + 0 + \cdots + 0 + e^{(-2\pi i/N)(N-1)k} - 2e^{(-2\pi i/N)kN}\right)\mu$$

$$= -2\left(1 - \cos\frac{2\pi k}{N}\right)\mu. \tag{4.69}$$

Therefore λ_k is real and negative for $k \neq N$, and $\lambda_k = 0$ for $k = N$.

The temporal evolution of the cell concentrations is now completely described. Equation (4.69) gives the explicit exponential decay rates and (4.67) gives the constants $\bar{\rho}_k^0$ occurring in (4.68). In fact, if these equations are substituted into (4.68), we find

$$\rho_n = \frac{1}{N}\sum_{k=1}^{N}\sum_{m=1}^{N} e^{\lambda_k t}W_{k(n-m)}\rho_m^0 = \sum_{m=1}^{N} S_{n-m}\rho_m^0 = \mathbf{S} \star \rho^0, \tag{4.70}$$

where

$$S_n = \frac{1}{N}\sum_{k=1}^{N} e^{\lambda_k t}W_{kn}. \tag{4.71}$$

As in previous applications, such an object is called the solution operator or fundamental solution. Again, we note that it corresponds to an initial value problem for which

$$\rho_n^0 = 0, \quad n \neq N,$$

$$\rho_N^0 = 1.$$

We see that the solution (4.68) is represented in terms of a series of harmonics; that each harmonic has its own decay rate; and that the greater the number of oscillations of the harmonic, the faster the decay. Finally, we observe that as $t \uparrow \infty$,

$$\rho_n \rightarrow \frac{1}{N}e^{\lambda_N t}\bar{\rho}_N^0 e^{(2\pi i/N)nN} = \bar{\rho}_N^0/N = \frac{1}{N}(\rho_1^0 + \rho_2^0 + \cdots + \rho_N^0), \tag{4.72}$$

since $\lambda_N = 0$. In final equilibrium each cell contains the same amount of concentrate, namely, the arithmetic mean over the initial concentration distribution.

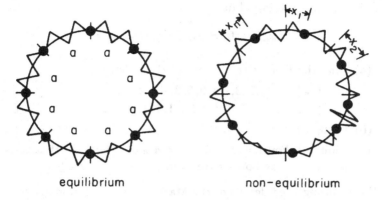

equilibrium   non-equilibrium

FIGURE 4.13.

Exercises

1. Solve
$$r_n = \epsilon_n - \alpha(r_{n-1} + r_{n+1}) - \beta(r_{n-2} + r_{n+2})$$
   for the point spread function s_n.

2. (a) Consider a ring model in which the membrane allows diffusion only in the clockwise direction by an amount proportional to the concentration of the cell. Show that the concentrations are governed by
$$\frac{d}{dt}\rho_n = \mu(\rho_{n-1} - \rho_n).$$

   (b) Solve the equation in Part (a) and obtain an expression for the solution operator by taking
$$\rho_n(t = 0) = \delta_{nN}.$$

3. Show that for the above model as well as for the earlier ring model,
$$\frac{d}{dt}(\rho_1 + \rho_2 + \cdots + \rho_N) = 0.$$
   In other words, show that the total concentration is conserved.

4. Suppose N masses that each have mass m are separated by N springs that each have spring constant k. In equilibrium the masses are equidistant from one another, and when disturbed they have the form shown in Figure 4.13, where the departure from equilibrium is denoted by x_n, $n = 1, 2, \ldots, N$. Show that the governing equation is
$$\frac{d^2 x_n}{dt^2} = \frac{k}{m}(x_{n+1} + x_{n-1} - 2x_n).$$

Solve this for the initial data

$$x_n(t = 0) = \delta_{nN}, \quad \dot{x}_n(t = 0) = 0.$$

5. (a) Find the Fourier transforms of the following:

    i. $\{a_n\} = \ldots, 2, 3, -3, 2, 2, 3, -3, 2, \ldots,$
    ii. $\{b_n\} = \ldots, -1, 1, 1, -1, -1, 1, 1, -1, \ldots.$

    (b) Compute $a \star b$.

    (c) Find the Fourier transform of $a \star b$ and verify that it is the product of the Fourier transforms found in Part (a).

6. Work through the details of the Mach band calculation given in this section. Take $\alpha = 1/4$, $N = 12$, and $k = 6$ and obtain a sketch such as Figure 4.11.

4.5   The Z-Transform and Applications

In this section we solve equations of the form (4.33) when periodicity cannot be assumed; i.e., the problem is *aperiodic*. For example, given an arbitrary pattern of excitation $\{\epsilon_n\}$ in (4.50), find the response pattern $\{r_n\}$. Before going on to the method of solution, we pose other types of problems which yield to the same method of solution.

Coin Tossing, Brownian Motion, Diffusion. The erratic motion of small particles in air or water is easily observed under conditions of strong sunlight. This is just one example of the many diffusion phenomena which are commonly encountered—but is unusual in that it can be analyzed in detail as a result of the relatively large size of the diffusing particles. (Although the diffusion of a dye is observable, individual dye particles are not.) This phenomenon, known as *Brownian motion* (after the English botanist, Robert Brown, who observed it in 1827), is modeled by assuming that a particle moves to the right or left in discrete uniform steps. The motion is random, i.e., determined by chance, which can be in the form of, say, the toss of a coin.

With this background we consider the probability of tossing exactly k heads of a coin in a total of N tosses. This we denote by $P(k; N)$. In the interests of generality we do not assume that we have a *fair coin;* i.e., we assume that the probability, p, of tossing a head is not necessarily equal to the probability, $q = (1 - p)$, of tossing a tail. [A simple combinatorial argument shows $P(k; N) = p^k q^{N-k} N!/((N - k)!k!)$, a result which will be derived by more formal procedures later.]

An equation determining $P(k; N)$ is obtained immediately when it is recognized that the only way to have k heads after N tosses is to have either $k - 1$ heads after $N - 1$ tosses and then a head in the Nth toss or k heads after $N - 1$ tosses and then a tail in the Nth toss. Specifically

$$P(k; N) = pP(k - 1; N - 1) + qP(k; N - 1). \tag{4.73}$$

Predator-Prey Equations. There is a large body of phenomena in diverse fields which can be formulated in terms of *birth and death processes.* (In chemistry the term *mass action* is used.) We now consider a restricted example of this.

Consider a population of hares and wolves living together. The population of each is measured once each year, and the resulting populations are denoted by H_ℓ and W_ℓ, respectively. In setting up the relations governing H_ℓ and W_ℓ, it is appropriate to regard these as the corresponding female populations since such populations measure and control the true populations.

In the absence of wolves we might write

$$H_{\ell+1} = bH_\ell - dH_\ell = \hat{\alpha}H_\ell,$$

where b measures the birth rate of female hares and d, their death rate. For the moment we maintain generality by allowing the coefficients b and d to be variable. The net growth factor $\hat{\alpha} = b - d$ should be greater than or equal to unity since in the absence of wolves the hare population should be nondecreasing. (If $\hat{\alpha}$ is constant, then $H_\ell = \hat{\alpha}^\ell H_0$ satisfies the above equation and we obtain exponential or "Malthusian" growth.) Similarly, in the absence of hares,

$$W_{\ell+1} = \hat{\beta}W_\ell;$$

and $\hat{\beta} < 1$ since the wolves would tend to die out in the absence of their food supply, namely, the hares. The interaction of wolves and hares act to augment the wolf population and diminish the hare population. Thus we write

$$H_{\ell+1} = \hat{\alpha}H_\ell - \hat{\gamma}W_\ell = F(H_\ell, W_\ell),$$
$$W_{\ell+1} = \hat{\beta}W_\ell + \hat{\delta}H_\ell = G(H_\ell, W_\ell). \tag{4.74}$$

In general, $\hat{\alpha}$, $\hat{\beta}$, $\hat{\gamma}$, $\hat{\delta}$ are all functions of H_ℓ and W_ℓ and the problem is *nonlinear.* This is emphasized by the second set of expressions for $H_{\ell+1}$ and $W_{\ell+1}$ in (4.74). The system (4.74) is nevertheless solvable in a step-by-step or iterative manner if we have knowledge of H_0 and W_0 to begin with. We will suppose that (4.74) possesses an equilibrium solution (H, W) such that

$$H = F(H, W), \quad W = G(H, W). \tag{4.75}$$

To consider the problem in the neighborhood of equilibrium, we write

$$H_\ell = H + h_\ell, \quad W_\ell = W + w_\ell,$$

when h_ℓ and w_ℓ are regarded as small. Proceeding in this vein, we approximate F and G by their Taylor expansions. Thus we have

$$F(H + h_\ell, W + w_\ell) \approx F(H, W) + h_\ell \frac{\partial F(H, W)}{\partial H}$$

$$+ w_\ell \frac{\partial F(H, W)}{\partial W} = H + \alpha h_\ell - \gamma W_\ell$$

and an analogous expansion for G. If this is substituted into (4.74) we obtain

$$h_{\ell+1} = \alpha h_\ell - \gamma w_\ell, \quad w_{\ell+1} = \beta w_\ell + \delta h_\ell, \tag{4.76}$$

where β and δ are the differential coefficients which appear in the expansion of G. Equations (4.76) are then to be solved subject to initial conditions

$$h_0 = h^0, \quad w_0 = w^0,$$

which give the initial departure from the equilibrium, $H_\ell = H$, $W_\ell = W$.

An alternative formulation of (4.76) is obtained by writing the second equation of (4.76) as

$$h_\ell = \frac{w_{\ell+1} - \beta w_\ell}{\delta},$$

and then by substituting it into the expressions for h_ℓ and $h_{\ell+1}$ in the first equation to obtain

$$\frac{w_{\ell+2} - \beta w_{\ell+1}}{\delta} = \frac{\alpha}{\delta}(w_{\ell+1} - \beta w_\ell) - \gamma w_\ell$$

or

$$w_{\ell+2} - (\alpha + \beta)w_{\ell+1} + (\gamma\delta + \alpha\beta)w_\ell = 0. \tag{4.77}$$

The initial conditions for this equation are

$$w_0 = w^0, \quad w_1 = \delta h^0 + \beta w^0.$$

Equation (4.77) is an example of a *second order difference equation*—its indices differ by two. Equation (4.76) on the other hand forms a coupled pair of *first order difference equations* (indices differ by one). The reduction of (4.76) to (4.77) is analogous to the reduction of two coupled first order differential equations to a single second order differential equation.

From (4.76) (or (4.77)) and the corresponding initial data, it is clear that we can recursively determine $w_2, w_3, \ldots$ and $h_2, h_3, \ldots$. As we will see in a moment, the populations (w_k, h_k) at any step k can be determined without first computing the prior populations.

Mortgage Retirement. In brief, suppose an amount C_0 is borrowed from a bank at an annual interest rate r, with the stipulation that payments will be made N times per year. What payments P must be made so that the debt is retired at the Mth payment?

We proceed by supposing that P is known. (In some bank situations this is the arrangement that is made.) The debt after the first payment is

$$C_1 = C_0 \left(1 + \frac{r \times 10^{-2}}{N}\right) - P. \tag{4.78}$$

Observe that

$$\theta = r \times 10^{-2}/N$$

appears since interest in percent must be divided by 100 to obtain the decimal form and that annual rates must be divided by the number of payments per year. It is clear that the basic form of (4.78) holds at any stage. For example, the debt after the $(k+1)$th payment is

$$C_{k+1} = (1 + \theta)C_k - P. \tag{4.79}$$

Equation (4.79) is a first order difference equation. Since the posed problem requires us to find P such that $C_M = 0$, recursive solution of (4.79) will not really help in any precise way.

Differential Equations. The analogy between difference equations and differential equations is more than superficial. Consider the simple first order differential equation

$$\frac{dy}{dt} + ay = 0, \tag{4.80}$$

where a is a constant. Of course the solution is simple, but we ignore this and try to solve (4.80) in an approximate way. Suppose that the time-axis is divided into uniform intervals Δ and that we write

$$y(n\Delta) = y_n.$$

Then, from Taylor's Theorem, we can write

$$y_{n+1} = y(n\Delta + \Delta) \approx y(n\Delta) + \Delta y'(n\Delta) = y_n + y'_n \cdot \Delta$$

or

$$y'_n = (y_{n+1} - y_n)/\Delta. \tag{4.81}$$

If y_n and $(y_{n+1} - y_n)/\Delta$ are substituted for $y(n\Delta)$ and $y'(n\Delta)$, respectively, in (4.80), we obtain the first order difference equation

$$y_{n+1} - (1 - a\Delta)y_n = 0. \tag{4.82}$$

In a similar way a second order differential equation can be approximated by a second order difference equation.

Z-Transforms. A general procedure for dealing with difference equations follows from the *Z-transform*. Given a sequence of numbers $\{a_n\}$, $n = 0, 1, \ldots$, its Z-transform is defined by

$$Z[a_n] = \sum_{n=0}^{\infty} a_n z^{-n} = F(z). \qquad (4.83)$$

It will be assumed throughout that for each Z-transform, there exists an R such that (4.83) converges for $|z| > R$.

The use of negative instead of positive powers in (4.83) is little more than a question of taste and convention. If positive powers are used the left-hand side is called the *generating function*, and the series is then supposed to converge in a circle about the origin. An example of the generating function approach is used later in this section and appears in the exercises.

Examples

$$Z[a^n] = \sum_{n=0}^{\infty} \frac{a^n}{z^n} = \frac{1}{1 - a/z} = \frac{z}{z - a}.$$

$$Z\left[\frac{1}{n!}\right] = \sum_{n=0}^{\infty} \frac{1}{z^n n!} = e^{1/z}.$$

As we will see, the Z-transform extends the idea of the discrete Fourier transform. For the present we regard it formally and now obtain properties of (4.83), which is nothing more than a power series in inverse powers of z; i.e., a power series in the neighborhood of ∞.

Inversion. If $F(z)$ is known to be the Z-transform of a sequence $\{a_n\}$, then there exists an R such that the series converges for $|z| > R$. It then follows from residue theory (see in particular (3.44)–(3.46)) that

$$a_n = \frac{1}{2\pi i} \oint_{|z|=\bar{R}>R} z^{n-1} F(z) dz$$

or

$$a_n = \frac{1}{2\pi i} \oint_{\bar{R}} z^{n-1} Z[a_n] dz. \qquad (4.84)$$

For example, if

$$F(z) = Z[a_n] = \frac{z + 3}{z - 2},$$

then F is analytic for $|z| > 2$ and the sequence $\{a_n\}$ can be determined from

$$a_n = \frac{1}{2\pi i} \oint_{|z|>2} z^{n-1} \left(\frac{z + 3}{z - 2}\right) dz = 5 \cdot 2^{n-1}.$$

Of course the same answer is obtained by direct expansion in which

$$F = \left(1 + \frac{3}{z}\right) \frac{1}{1 - 2/z} = \left(1 + \frac{3}{z}\right)\left(1 + \frac{2}{z} + \frac{2^2}{z^2} + \cdots\right),$$

and then identifying the coefficient of z^{-n}. As in this last illustration, often the inversion, i.e., the determination of a_n, is most easily found by a direct expansion of $Z[a_n]$ and inspection of the series.

For the following list of properties of the Z-transform, the sequences $\{a_n\}$ and $\{b_n\}$ are defined such that

$$a_n = b_n = 0, \quad n < 0.$$

Linearity:

$$Z[a_n + b_n] = \sum_{n=0}^{\infty}(a_n + b_n)z^{-n} = Z[a_n] + Z[b_n] \tag{4.85}$$

(with the usual fine print about this being valid in the common domain of convergence of the two series).

Translation: For $N \geq 0$,

$$Z[a_{n-N}] = \sum_{n=0}^{\infty} a_{n-N}z^{-n} = z^{-N}\sum_{n=0}^{\infty} a_{n-N}z^{(N-n)} = z^{-N}Z[a_n], \tag{4.86}$$

$$Z[a_{n+N}] = \sum_{n=0}^{\infty} a_{n+N}z^{-n} = z^{N}\sum_{n=0}^{\infty} a_{n+N}z^{-(n+N)}$$

$$= z^{N}Z[a_n] - z^{N}\sum_{n=0}^{N-1} a_n z^{-n} \tag{4.87}$$

(for (4.86) recall that $a_n = 0$ if $n < 0$).

Multiplication: From (4.83) we have

$$Z[\rho^n a_n] = \sum_{n=0}^{\infty} a_n \left(\frac{\rho}{z}\right)^n = F(z/\rho) \tag{4.88}$$

and

$$Z[na_n] = \sum_{n=0}^{\infty} a_n n z^{-n} = -z\frac{d}{dz}\sum_{n=0}^{\infty} a_n z^{-n} = -zF'(z). \tag{4.89}$$

Convolution: Recall that the convolution product of any two sequences is defined as

$$(a \star b)_n = \sum_{m=-\infty}^{\infty} a_{n-m}b_m,$$

which, if $a_n = b_n = 0$ for $n < 0$, gives

$$(a \star b)_n = \sum_{m=0}^{n} a_{n-m}b_m. \tag{4.90}$$

If we proceed formally, then

$$Z[a \star b] = \sum_{n=-\infty}^{\infty} z^{-n} \sum_{m=-\infty}^{\infty} a_{n-m} b_m,$$

and on interchanging orders of summation

$$Z[a \star b] = \sum_{m=-\infty}^{\infty} b_m \sum_{n=-\infty}^{\infty} z^{-n} a_{n-m}.$$

Finally, if we multiply and divide by z^m we get

$$Z[a \star b] = \sum_{m=-\infty}^{\infty} b_m z^{-m} \sum_{n=-\infty}^{\infty} z^{-(n-m)} a_{n-m} = Z[a_n] Z[b_n]. \qquad (4.91)$$

Application of the Z-Transform to Difference Equations. We have already encountered a second order difference equation in (4.77). Consider the first order linear difference equation

$$x_{n+1} = a x_n + b_n \qquad (4.92)$$

with b_n and x_0 regarded as known. To solve this for x_n we proceed formally and take the Z-transform of (4.92):

$$Z[x_{n+1}] = Z[a x_n + b_n] = a Z[x_n] + Z[b_n].$$

Here, linearity (4.85) has been used, and if we also use (4.87) on the left-hand side, we obtain

$$z Z[x_n] - z x_0 = a Z[x_n] + Z[b_n].$$

Therefore, if we solve for $Z[x_n]$,

$$Z[x_n] = \frac{z x_0}{z - a} + \frac{Z[b_n]}{z - a}.$$

This is such a simple expression that a straightforward expansion of $Z[x_n]$ instead of the use of (4.84) is the method of choice. Thus

$$\frac{z x_0}{z - a} = x_0 \frac{1}{1 - a/z} = x_0 \sum_{n=0}^{\infty} z^{-n} a^n,$$

while

$$\frac{Z[b_n]}{z - a} = \left(\sum_{k=0}^{\infty} \frac{a^k}{z^{k+1}} \right) \left(\sum_{\ell=0}^{\infty} b_\ell z^{-\ell} \right) = \sum_{n=1}^{\infty} \frac{1}{z^n} \left(\sum_{k=0}^{n-1} b_{n-1-k} a^k \right).$$

Hence

$$x_n = a^n x_0 + \sum_{k=0}^{n-1} b_{n-1-k} a_k, \tag{4.93}$$

a result which could have been obtained by repeated direct substitution in (4.92).

To apply this procedure to the mortgage problem (4.79), we set $a = 1+\theta$, $b_n = P$, and $x_n = c_n$. It therefore follows that

$$C_n = (1+\theta)^n C_0 - P \sum_{k=0}^{n-1} (1+\theta)^k$$

$$= (1+\theta)^n C_0 - P \frac{(1+\theta)^n - 1}{\theta}. \tag{4.94}$$

We see that the solution for P, given that $C_M = 0$, is simply

$$P = \theta(1+\theta)^M C_0 / [(1+\theta)^M - 1]. \tag{4.95}$$

It is also of interest to observe that (4.82) has the solution

$$y_n = (1 - a\Delta)^n y_0. \tag{4.96}$$

To see the connection between this and the solution $y = y_0 \exp(-ax)$, of (4.80), observe that if we write $\Delta = x/n$ in (4.96) and let $n \uparrow \infty$ (so that $\Delta \downarrow 0$), then

$$y_n = \left(1 - a\frac{x}{n}\right)^n y_0 \to e^{-ax} y_0, \tag{4.97}$$

which is just the solution of (4.80). If, on the other hand, $1 - a\Delta < 0$, the solution oscillates and there is no analogy with solutions of the ordinary differential equation (4.80).

Finally instead of just treating the predator-prey equation (4.77), we consider the general linear case

$$x_{n+2} + a_1 x_{n+1} + a_2 x_n = b_n, \tag{4.98}$$

with a_1, a_2 constant. With the *initial data*

$$x_0, \quad x_1$$

given, we go ahead as before and define

$$X(z) = Z[x_n] \quad \text{and} \quad B(z) = Z[b_n].$$

Then, if we take the Z-transform of both sides of (4.98) and employ (4.87), we obtain

$$(z^2 + a_1 z + a_2) X(z) = B(z) + z^2 x_0 + z x_1 + a_1 z x_0.$$

We solve for X:

$$X = \frac{B(z)}{z^2 + a_1 z + a_2} + \frac{x_0 z^2 + x_1 z + a_1 z x_0}{z^2 + a_1 z + a_2}.$$

In this case we will assume that for some R, $B(z)$ converges when $|z| > R$. Then we can apply (4.84) to X and obtain

$$x_n = \frac{1}{2\pi i} \oint_{|z|=\overline{R}>R} z^{n-1} X(z)\,dz = \frac{1}{2\pi i} \oint_{|z|=\overline{R}>R} \frac{z^{n-1} B(z)}{p(z)}\,dz$$

$$+ \frac{1}{2\pi i} \oint_{|z|=\overline{R}>R} z^{n-1} \frac{(z^2 x_0 + z x_1 + a_1 z x_0)}{p(z)}\,dz. \qquad (4.99)$$

The circle $|z| = \overline{R}$ represents a path which lies in the domain of convergence of $X(z)$ or in the common domain of convergence of $B(z)/p(z)$ and $(x_0 z^2 + x_1 z + a_1 z x_0)/p(z)$. The radius $\overline{R}$ must therefore be larger than the magnitude of the roots of

$$p(z) = z^2 + a_1 z + a_2 = (z - p_1)(z - p_2),$$

where

$$p_{1,2} = \frac{1}{2}\left\{-a_1 \pm (a_1^2 - 4a_2)^{1/2}\right\}$$

and where it is assumed for simplicity that $p_1 \neq p_2$. Since we may write

$$\frac{1}{p(z)} = \frac{1}{p_1 - p_2}\left\{\frac{1}{z - p_1} - \frac{1}{z - p_2}\right\},$$

(4.99) gives

$$x_n = \sum_{k=0}^{n-1} \frac{p_1^k - p_2^k}{p_1 - p_2} b_{n-1-k} + \frac{1}{p_1 - p_2}[p_1^n(p_1 x_0 + x_1 + a_1 x_0)$$

$$- p_2^n(p_2 x_0 + x_1 + a_1 x_0)].$$

This solution is formally quite close to the solution of a second order ordinary differential equation (with constant coefficients). In spirit the Z-transform lies close to the Laplace transform, which is used to solve differential equations.

Since we know the form of the solution of difference equations, it is not necessary to go through all the details of the Z-transform. To see this consider

$$x_{n+2} + a_1 x_{n+1} + a_2 x_n = 0.$$

If we assume $x_n \propto p^n$, then on substitution we find p must satisfy

$$p^2 + a_1 p + a_2 = 0.$$

Next denote the two solutions of this quadratic by p_1 and p_2. Then the general solution of the difference equation is a sum of these exponentials; i.e.,

$$x_n = Ap_1^n + Bp_2^n,$$

where A and B are to be determined by initial conditions. (By analogy, linear ordinary differential equations with constant coefficients also have exponential solutions.) If the magnitude of either of the roots p_1, p_2 is greater than unity, the solution grows exponentially. If a root is negative or, more generally, complex, then the solution oscillates. (See Exercise 3 at the end of this section for the case of equal roots.)

Coin Tossing Problem. As a change of pace, we use a *generating function* instead of the Z-transform. As explained earlier this involves using positive powers of z. The change is minor, but traditionally the *generating function* appears in problems of probability. The equation to be solved is

$$P(k; N) = pP(k-1; N-1) + qP(k; N-1) \qquad (4.100)$$

with $p + q = 1$ $(p, q > 0)$. This is appropriately called a *partial difference equation*, since two indices are involved. From the origin of the problem we can write

$$P(0; 0) = 1, \quad P(k; N) = 0, \quad k > N.$$

We next introduce the generating function

$$P[z; N] = \sum_{k=0}^{N} z^k P(k; N), \qquad (4.101)$$

since $P(k; N) = 0$ if $k > N$. Applying this, (4.85), and (4.86) to (4.100), we obtain

$$P[z; N] = pz P[z; N-1] + qP[z; N-1]$$

or

$$P[z; N] = (q + pz)P[z; N-1],$$

since $P(z; 0) = 0$. This is a trivial first order difference equation, with initial condition $P[z; 0] = 1$, and has the solution

$$P[z; N] = (q + pz)^N. \qquad (4.102)$$

The right-hand side can be immediately expanded in powers of z using the Binomial Theorem. Therefore we can write

$$P[z; N] = \sum_{k=0}^{N} \binom{N}{k} q^{N-k} p^k z^k, \quad \binom{N}{k} = \frac{N!}{k!(N-k)!}, \qquad (4.103)$$

from which we easily have the solution mentioned earlier, namely

$$P(k; N) = \binom{N}{k} q^{N-k} p^k. \qquad (4.104)$$

Actually the *generating function* is a rich source of additional properties. In dealing with probabilities, one often is interested in their moments:

$$\langle k^n \rangle = \sum_{k=0}^{N} k^n P(k; N). \tag{4.105}$$

For example, if $n = 0$, we obtain the sum of all probabilities and our result should be unity. If $n = 1$, we obtain the average number of heads thrown in N tosses. Another quantity of interest is the variance

$$\langle (k - \langle k \rangle)^2 \rangle,$$

which is easily constructed from (4.105). The actual summation as indicated in (4.105) can be extremely tedious. On the other hand, comparison of (4.105) with (4.102) indicates that

$$\langle k^n \rangle = \left[\left(z\frac{d}{dz} \right)^n (q + pz)^N \right]_{z=1}. \tag{4.106}$$

The right-hand side of this is quite easy to evaluate. For example,

$$\langle k^0 \rangle = (q + p)^N = 1^N = 1$$

and

$$\langle k \rangle = pN(q + p)^{N-1} = pN.$$

Exercises

1. Find the Z-transform of

    (a) $\{\cos \omega n\}$,

    (b) $\{\sin \omega n\}$,

    (c) $\{n^2\}$, $n = 0, 1, \ldots$.

2. Show $Z[1/n] = -\ln(1 - (1/z))$, with the summation starting at $n = 1$. [Hint: Use $Z[n^k a_n] = (-z d/dz)^k Z[a_n]$.]

3. The general form of the second order difference equation having *equal roots* is
    $$x_{n+2} - 2ax_{n+1} + a^2 x_n = 0.$$

    Show by a method other than direct substitution that the two linearly independent solutions are

    $$x_n = (a)^n \quad \text{and} \quad x_n = \frac{1}{a} n(a)^n.$$

4. Solve

   (a) $x_{n+2} + 3x_{n+1} + 2x_n = 0$, $x_0 = 1$, $x_1 = 2$;
   (b) $x_{n+2} - 2x_{n+1} + x_n = 0$, $x_0 = 2$, $x_1 = 0$;
   (c) $x_{n+3} - x_{n+2} - x_{n+1} + x_n = 0$, $x_0 = 1$, $x_1 = x_2 = 0$.

   [Remark: It is not necessary to use the Z-transform once the form of the solutions are known.]

5. Solve the following difference equations without reducing them to second order difference equations:

   $$\text{(a)} \quad y_{n+1} + 3y_n + 2x_n = 0, \quad x_{n+1} - y_n = 0,$$

   $$x_0 = 1, \quad y_0 = 2.$$

   $$\text{(b)} \quad y_{n+1} - 2y_n + x_n = 0, \quad x_{n+1} - y_n = 0,$$

   $$x_0 = 1, \quad y_0 = 0.$$

6. (a) Show that in analogy with (4.81)

   $$y_n'' \approx \frac{y_{n+1} + y_{n-1} - 2y_n}{\Delta^2}.$$

   [Hint: Take the three-term Taylor expansion of y_{n+1} and of y_{n-1}.]

   (b) Use Part (a) to approximate $y'' + ay' + by = 0$ by a second order difference equation.

7. Put the ordinary differential equation

   $$y'' + 3y' + y = 0$$

   in the form of a difference equation. Solve this difference equation and show that under an appropriate limit the resulting solutions tend to those of the differential equation.

8. Furnish the details for obtaining (4.77).

9. For the general second order difference equation

   $$x_{n+2} + 2ax_{n+1} + bx_n = 0$$

   (a) under what conditions will the solutions oscillate?
   (b) under what conditions will the solutions be unstable, i.e., grow?

10. What are the relations analogous to (4.86)–(4.89) for the generating function

    $$G(z) = \sum_{k=0}^{\infty} a_k z^k?$$

4.6　The Double Z-Transform

We return to (4.11), which we write in the form

$$A \star x = \sum_{m=-\infty}^{\infty} A_{n-m} x_m = b_n \qquad (4.107)$$

and which is to be solved for the unknown sequence $\{x_n\}$. Since the subscripts run over the negative as well as the positive integers, the Z-transform is not applicable. For this reason we define the *double Z-transform:*

$$Z_d[x_n] = \sum_{n=-\infty}^{\infty} x_n z^{-n}. \qquad (4.108)$$

This reduces to the Z-transform for sequences in which $x_n = 0$, $n < 0$.

It is clear that the properties associated with linearity, multiplication, and convolution still apply to Z_d. The translation relations however must be modified. In this instance

$$Z_d[a_{n+N}] = \sum_{n=-\infty}^{\infty} a_{n+N} z^{-n} = z^N \sum_{n=-\infty}^{\infty} a_{n+N} z^{-n-N} = z^N Z_d[a_n]$$

$$(4.109)$$

for any integer N.

Unlike the Z-transform, the Z_d-transform is in general a full Laurent expansion and only converges in an *annulus* about the origin in the complex z-plane. *It therefore will be important to examine the various analytic forms that result from taking Z_d-transforms to insure that they have an annulus of convergence.*

In view of our experience with solution operators, we see that there is no loss of generality if we consider instead of (4.107) the equation for the solution operator s_n, namely

$$\sum_{m=-\infty}^{\infty} A_{n-m} s_m = \delta_{n,0}. \qquad (4.110)$$

The solution to the full problem, (4.107), is then given by

$$x_m = \sum_{p=-\infty}^{\infty} s_{m-p} b_p. \qquad (4.111)$$

First let us verify that (4.111) is indeed the solution to (4.107). To show this, convolve both sides of (4.111) with $\{A_n\}$; i.e., consider

$$\sum_{m=-\infty}^{\infty} A_{n-m} x_m = \sum_{m=-\infty}^{\infty} A_{n-m} \sum_{p=-\infty}^{\infty} s_{m-p} b_p$$

$$= \sum_{p=-\infty}^{\infty} b_p \sum_{m=-\infty}^{\infty} A_{n-m} s_{m-p} \tag{4.112}$$

where we have formally reversed the order of summation in the last summation. But if we compute the inner sum using (4.110), we see that

$$\sum_{m=-\infty}^{\infty} A_{n-m} s_{m-p} \xrightarrow{m-p=q} \sum_{q=-\infty}^{\infty} A_{(n-p)-q} s_q = \delta_{(n-p),0} = \delta_{n,p}.$$

Therefore we have verified that (4.111) satisfies (4.107).

Next, in order to solve (4.110), we take its Z_d-transform. Then, from the convolution property (4.91), (4.110) becomes

$$A(z)S(z) = 1, \tag{4.113}$$

where

$$A(z) = Z_d[A_n] \quad \text{and} \quad S(z) = Z_d[s_n].$$

and since $Z_d[\delta_{n,a}] = 1$. The solution for s_n is obtained by solving (4.113) for $S(z)$ and then applying the residue formula, namely

$$s_n = \frac{1}{2\pi i} \oint_R \frac{z^{n-1}}{A(z)} dz$$

where the circle $|z| = \overline{R}$ lies in the annulus of convergence of $1/A(z)$.

Example. Let us return to the Hartline–Ratliff equation (4.50) with coupling coefficients

$$K_n = \begin{cases} \alpha \gamma^{|n|}, & n \neq 0, \\ 0, & n = 0, \end{cases}$$

where $\alpha > 0$ and $0 < \gamma < 1$. (By having $K_{n-m} = 0$ when $n = m$, we have eliminated any *self-inhibition*.) In this model all photoreceptors are coupled (in contrast to the nearest neighbor model (4.52)) and the coupling falls off exponentially.

The equation for the solution operator (which is obtained when we let $\epsilon_n = \delta_{n,0}$) has the form

$$s_n = \delta_{n,0} - \alpha \sum_{m=-\infty}^{\infty\prime} \gamma^{|n-m|} s_m,$$

where the prime on the summation indicates that $m = n$ is deleted. If we take the Z_d-transform of this equation, we obtain

$$S(z) = 1 - K(z)S(z), \tag{4.114}$$

where S is defined as above and where

$$K(z) = \alpha \sum_{m=-\infty}^{\infty\prime} \gamma^{|m|} z^{-m} = \alpha \sum_{m=1}^{\infty} \gamma^{|m|} z^m + \alpha \sum_{m=1}^{\infty} \gamma^{|m|} z^{-m}.$$

$$= \alpha \sum_{m=1}^{\infty} \gamma^m \left(z^m + \frac{1}{z^m} \right) = \alpha \left[\frac{\gamma z}{1 - \gamma z} + \frac{\gamma}{z - \gamma} \right].$$

Observe that $K(z)$ converges in the annulus

$$\gamma < |z| < 1/\gamma.$$

If we substitute the final expression for $K(z)$ in (4.114) and solve for $S(z)$, we obtain

$$S(z) = \frac{1}{1 + K(z)} = \frac{1}{1 + \alpha(\gamma z/(1 - \gamma z) + \gamma/(z - \gamma))}. \qquad (4.115)$$

The remainder of the calculation is left for Exercise 4 below.

Exercises

1. Determine the components s_n of the solution operator from (4.115). Find the solution to (4.50) for $\epsilon_{\pm 1} = \pm 1$ and then for $\epsilon_n = 0$, $|n| \neq 1$.

2. Find the solution operator for the following nearest neighbor model of the Hartline–Ratliff equations:

$$r_n = \epsilon_n - \alpha(r_{n-1} + r_{n+1}), \quad 0 < \alpha < 1/2.$$

3. Consider the cell model of diffusion laid out in a linear fashion. The equation is still

$$\frac{d}{dt} \rho_n = \mu(\rho_{n+1} - \rho_{n-1} - 2\rho_n),$$

but now

$$-\infty < n < +\infty.$$

Solve this infinite system for

$$\rho_n(t = 0) = \delta_{n0}.$$

[Hint: Use the Z_d-transform.]

4. Work through the details to find (4.115) and then complete the solution of

$$s_n = \delta_{n,0} - \alpha \sum_{m=-\infty}^{\infty} {}'\gamma^{|n-m|} s_m$$

for s_n; i.e., find s_n.

4.7 The Wiener–Hopf Method: Discrete Form

One occasionally encounters instead of (4.97) the following problem:

$$b_n = \sum_{m=0}^{\infty} K_{n-m} x_m, \quad n = 0, 1, 2, \ldots.$$

The right-hand side *is not* in the form of a convolution. To solve such problems we will use a procedure known as the Wiener–Hopf method in a somewhat restricted form. This method is less *mechanical* than previous techniques and even with experience still requires some thought when used.

An example of such a problem is encountered in the treatment of the horseshoe crab visual system. A model of the response pattern at the edge of the eye is given by

$$r_n = e_n - \sum_{m=0}^{\infty} K_{n-m} r_m, \quad n = 0, 1, \ldots, \qquad (4.116)$$

where r_0 is the response in the edge ommatidium. In order to treat this problem we proceed as follows. We first define a sequence $\{c_n\}$ by

$$c_n = \begin{cases} -\sum_0^{\infty} K_{n-m} r_m, & n < 0, \\ 0, & n \geq 0, \end{cases}$$

and then also take

$$r_n = 0 = e_n, \quad n < 0.$$

Equation (4.116) may then be *embedded* in the following equation

$$r_n + c_n = e_n - \sum_{m=-\infty}^{\infty} K_{n-m} r_m \qquad (4.117)$$

since for $n \geq 0$ this equation, (4.117), is identical to (4.116). At this point we can take the Z_d-transform of (4.117) to obtain

$$R(z) + C(z) = E(z) - K(z)R(z) \qquad (4.118)$$

where

$$R(z) = \sum_{n=0}^{-\infty} r_n z^{-n}, \quad C(z) = \sum_{n=-1}^{\infty} c_n z^{-n},$$

$$K(z) = \sum_{n=-\infty}^{\infty} K_n z^{-n}, \quad E(z) = -\sum_{n=0}^{\infty} e_n z^{-n}. \qquad (4.119)$$

We arrange equation (4.118) in the form

$$R(z)(1 + K(z)) + E(z) = -C(z). \qquad (4.120)$$

The problem at this point looks insurmountable since (4.120) contains two unknowns, namely $C(z)$ and $R(z)$.

Before proceeding further let us consider the specific case of the nearest neighbor model

$$r_n = e_n - \alpha(r_{n+1} + r_{n-1}).$$

In this particular case

$$K(z) = \alpha \left(z + \frac{1}{z} \right)$$

and we can write

$$R(z) \left(1 + \alpha \left(z + \frac{1}{z} \right) \right) + E(z)$$

$$= R \left\{ \frac{\alpha z^2 + \alpha + z}{z} \right\} + E(z) = -C(z). \qquad (4.121)$$

The quadratic

$$\alpha z^2 + z + \alpha$$

has the property that if $z = \mu$ is a root then $z = 1/\mu$ is also a root. To fix matters suppose $|\mu| < 1$. We then rewrite (4.121) in the form

$$L(z) = \alpha R(z) \left(\frac{(z - \mu)(z - 1/\mu)}{z^2} \right) + \frac{E(z)}{z} = -\frac{C(z)}{z}. \qquad (4.122)$$

For reasons which will be clear in a moment we have divided both sides of the equation by z.

From the definitions (4.119) we see that $C(z)$ and hence $C(z)/z$ are analytic in some circle around the origin. E and R and hence the left-hand side of (4.122), $L(z)$, are analytic in circle enclosing infinity. In particular $L(z)$ is bounded for $|z| \uparrow \infty$, which accounts for the above division by z. We will see after solving the problem, that the right and left sides of (4.122) have a common region of analyticity. Equation (4.122) then says that the right and left are equal on this common region.

We digress to point out *if two analytic functions, say $f(z)$ and $g(z)$, are equal to one another on a line segment, say Γ, then they are identically equal.* To see this consider their difference $F(z) = f(z) - g(z)$, which is zero on Γ. A Taylor series for $F(z)$ about some point of Γ can be constructed by taking derivatives along Γ—which are all just zero. It follows that $F \equiv 0$. When functions defined in different domains are equal on a common curve (or subdomain), they are said to be *analytic continuations* of one another.

To return to the problem it now follows that the right- and left-hand sides of (4.122), being analytic continuations of one another, define a function which is analytic everywhere! Thus by Liouville's Theorem (Section 3.4) the function so defined is at most a constant. Hence

$$L(z) = R(z) \alpha \frac{(z - \mu)(z - 1/\mu)}{z^2} + \frac{E(z)}{z} = k$$

where k is a constant, still to be determined. If we solve for $R(z)$,

$$R(z) = \frac{1}{\alpha} \frac{z}{(z - \mu)(z - 1/\mu)} [kz - E(z)]. \qquad (4.123)$$

It seems that we have satisfactorily solved the problem, but the unknown constant k is still on the loose. A little physical intuition is now required. Consider the determination of the responses r_n:

$$
\begin{aligned}
r_n &= \frac{1}{2\pi i} \oint z^{n-1} R(z) dz \\
&= \frac{1}{2\pi i} \oint \frac{z^n (kz - E(z))}{(z - \mu)(z - 1/\mu)} \alpha,
\end{aligned}
$$

where the path of integration is a sufficiently large circle. If we now shrink the contour we pick residues at $z = 0$, $z = \mu$, and $z = 1/\mu$. The last however contains the coefficient

$$\left(\frac{1}{\mu}\right)^n,$$

which since $|\mu| < 1$ becomes unbounded as $n \uparrow \infty$. This is contrary to physical intuition, which dictates that

$$\lim_{n \uparrow \infty} r_n = 0.$$

To prevent this from happening we force the corresponding pole of (4.123) to disappear with the condition,

$$(kz - E(z))|_{z=1/\mu} = 0$$

or

$$k = \mu E\left(\frac{1}{\mu}\right). \qquad (4.124)$$

The solution to the problem is now complete by use of the integral formula on (4.123).

In the general case, (4.120), one may also always construct the type of splitting just obtained. This we regard as beyond the scope of the course.

Exercises

1. Compute the responses r_n if the excitation is δ_{n0}. Compare this with the "infinite" eye result obtained in Exercise 1 at the end of Section 4.6.

2. Carry out the Wiener–Hopf procedure for the kernel K_n given in Section 4.6.

Appendix: The Fast Fourier Algorithm

A discrete Fourier transform pair is

$$x_q = \sum_{n=1}^{N} a_n e^{2\pi i(nq/N)} \tag{A.1}$$

$$a_n = \frac{1}{N} \sum_{q=1}^{N} x_q e^{-2\pi i(nq/N)}. \tag{A.2}$$

In the terminology of computer science the operation of a multiplication and an addition is referred to as an add-multiply and denoted by AM. Consider for example the quantity a_n. It requires N add-multiplies, and hence N^2 add-multiplies are necessary for the computation of the set $\{a_n\}$. If $2^{10} = 1024$ points are involved, then over 10^6 add-multiplies enter into the calculation of such a transform. We now discuss a procedure which considerably reduces this labor.

Suppose N is factorable, say $N = p_1 p_2$. Then each of the indices n and q can be re-represented as

$$n = n_1 p_2 + n_0, \quad n_0 = 1, 2, \ldots, p_2, \quad n_1 = 0, 1, 2, \ldots, p_1 - 1;$$

$$q = q_1 p_1 + q_0, \quad q_0 = 1, 2, \ldots, p_1, \quad q_1 = 0, 1, 2, \ldots, p_2 - 1.$$

Substitute into the expression for a_n, (A.2):

$$a_n = \frac{1}{N} \sum_{q_1=0}^{p_2-1} \sum_{q_0=1}^{p_1} x_{(p_1 q_1 + q_0)} e^{-2\pi i(n/N)(q_1 p_1 + q_0)}. \tag{A.3}$$

Next note

$$
\begin{aligned}
e^{-2\pi i(n/N)(q_1 p_1 + q_0)} &= e^{-2\pi i(n/N)q_1 p_1} e^{-2\pi i(nq_0/N)} \\
&= e^{-2\pi i(nq_0/N)} e^{-2\pi i(q_1 p_1)(n_1 p_2 + n_0)/N} \\
&= e^{(-2\pi i/N)nq_0} e^{(-2\pi i/N)n_0 q_1 p_1}.
\end{aligned}
$$

If this equivalence is substituted into the expression for a_n in (A.3), we get

$$
\begin{aligned}
a_n &= \frac{1}{p_2} \sum_{q_1=0}^{p_2-1} \frac{1}{p_1} \sum_{q_0=1}^{p_1} x_{(p_1 q_1 + q_0)} e^{(-2\pi i/N)nq_0 - (2\pi i/N)n_0 q_1 p_1} \\
&= \frac{1}{p_1} \sum_{q_0=1}^{p_1} e^{(-2\pi i/N)nq_0} \left(\frac{1}{p_2} \sum_{q_1=0}^{p_2-1} x_{(p_1 q_1 + q_0)} e^{(-2\pi i/N)n_0 q_1 p_1} \right)
\end{aligned}
$$

and

$$a_n = \frac{1}{p_1} \sum_{q_0=1}^{p_1} e^{(-2\pi i/N)nq_0} \tilde{a}_{q_0 n_0}, \tag{A.4}$$

where

$$\tilde{a}_{q_0 n_0} = \frac{1}{p_2} \sum_{q_1=0}^{p_2-1} x_{(p_1 q_1 + q_0)} e^{(-2\pi i/N) n_0 q_1 p_1}.$$

Observe that we have indeed reduced the number of add-multiplies by this procedure. In computing $\tilde{a}_{q_0 n_0}$ there are p_2 add-multiplies and on substituting in (A.4) there are p_1 additional add-multiplies, giving a total of $p_1 + p_2$ to determine a_n. Thus the number of AM to obtain all the a_n is

$$AM = N(p_1 + p_2).$$

If p_2 is factorable the process can be repeated. In fact if

$$N = p_1 p_2 p_3 \ldots p_k,$$

then it is clear that

$$AM = N(p_1 + p_2 + \cdots + p_k)$$

add-multiplies enter. For $p_1 = p_2 = \cdots p_k = p$

$$AM = Nkp = pN \log_p N.$$

In particular, for $p = 2$, which is the most common case,

$$AM = 2N \log_2 N.$$

For the cited case of $1024 = 2^{10}$, $AM = 20 \times 1024$, which is a 50-fold reduction in add-multiplies.

5

Fourier Series and Applications

5.1 Fourier Series – Heuristic Approach

As we pointed out earlier with (3.30) and (3.31), a function $f(z)$ analytic in a domain containing the unit circle can be expressed on the unit circle by a *Fourier series:*

$$f(e^{i\theta}) = F(\theta) = \sum_{n=-\infty}^{\infty} a_n e^{in\theta}, \tag{5.1}$$

where the *Fourier coefficients* a_n are given by

$$a_n = \frac{1}{2\pi} \int_0^{2\pi} F(\theta)e^{-in\theta}d\theta = \tilde{F}_n/2\pi. \tag{5.2}$$

As in the discrete case $\tilde{F}_n$ will be called the *Fourier transform*. This result, shown by Fourier in an entirely different context, was one of the most significant advances in mathematics and has had an immense impact on all areas of applications.

In deriving (5.1) and (5.2), we found that it was sufficient that f be analytic. However, the condition of analyticity is far from necessary. To show this, we take a heuristic approach and start from the observation that formulas (5.1) and (5.2) closely resemble (4.20) and (4.21), respectively, which were found for periodic sequences. These two pairs of relations are close relatives of one another.

Suppose $F(\theta)$ is a function, say continuous, defined on the unit circle; i.e., it is defined for

$$-\pi \leq \theta \leq \pi.$$

On the real line it is a 2π-periodic function. Next we sample $F(\theta)$ at the uniformly spaced points

$$\theta_n = n\left(\frac{2\pi}{N}\right) = n\Delta, \tag{5.3}$$

where N is the number of divisions, of size Δ, of the basic interval 2π (see Figure 5.1). Therefore

$$F_n = F(\theta_n) \tag{5.4}$$

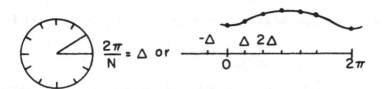

FIGURE 5.1.

is an infinite periodic sequence of period N. It then follows that we can write (for convenience N is taken to be even)

$$F(\theta_n) = F_n = \sum_{m=-(N/2)+1}^{N/2} a_m W_{mn} = \sum_{m=-(N/2)+1}^{N/2} a_m e^{2\pi i(nm/N)}$$

$$= \sum_{m=-(N/2)+1}^{N/2} a_m e^{im\theta_n} \tag{5.5}$$

where (5.3) has been used in the last summation. Also

$$a_m = \frac{1}{N} \sum_{n=1}^{N} F_n \overline{W}_{nm} = \frac{1}{N} \sum_{n=1}^{N} F_n e^{-2\pi i(mn/N)}$$

$$= \frac{1}{2\pi} \sum_{n=1}^{N} F(n\Delta) e^{-imn\Delta} \Delta = \frac{1}{2\pi} \sum_{n=1}^{N} F(\theta_n) e^{-im\theta_n} \Delta. \tag{5.6}$$

In (5.6) we have used $\Delta = 2\pi/N$, which comes from (5.3).

If we proceed to the limit $N \uparrow \infty$, (5.5) becomes (the subscript n can be dropped)

$$F(\theta) = \sum_{m=-\infty}^{\infty} a_m e^{im\theta},$$

which is the same as (5.1). As for (5.6), we recognize that under the same limit (in this case expressed as $\lim_{\Delta \to 0}$) it becomes the Riemann sum of the integral in (5.2). Thus we have obtained, at least heuristically, formulas (5.1) and (5.2) for a function $F(\theta)$ which is only continuous. It should be clear that even continuity was not important to the argument.

Along these same lines we consider (4.31) under the limit $N \uparrow \infty$ by first writing this equation as

$$\delta_{km} = \frac{1}{N}(e_k, e_m) = \frac{1}{N} \sum_{n=1}^{N} e^{-2\pi i(nk/N)} e^{2\pi i(nm/N)}$$

$$= \frac{1}{2\pi} \sum_{n=1}^{N} e^{-ik(n\Delta)} e^{im(n\Delta)} \Delta.$$

Again we recognize the last expression as a Riemann sum and obtain

$$\frac{1}{2\pi} \int_0^{2\pi} e^{ik\theta} e^{im\theta} d\theta = \delta_{km}, \tag{5.7}$$

an expression we have encountered before. The manner in which (5.7) was obtained suggests that we define the inner product of two functions, say $f(\theta)$ and $g(\theta)$, by

$$(f,g) = \int_0^{2\pi} \overline{f}(\theta) g(\theta) d\theta, \tag{5.8}$$

which is the continuous analog of (4.27).

For purposes of exposition, it is convenient to consider functions which have unit period instead of being 2π-periodic. This is easily achieved by writing $\theta = 2\pi x$. Then x varies between 0 and 1 if θ varies between 0 and 2π. In particular, we consider functions $f(x)$ such that

$$f(x + N) = f(x)$$

for all integers N. If this is true we will say that $f(x)$ is one-periodic.

Formally, we assume that we can write

$$f(x) = \sum_{n=-\infty}^{\infty} a_n e^{2\pi i n x} = \sum_{n=-\infty}^{\infty} a_n e_n(x). \tag{5.9}$$

We will use the special notation

$$e_n(x) = \exp(2\pi i n x) \tag{5.10}$$

in order to avoid the repetitious use of the product $2\pi i$. Observe that the functions $\{e_n(x)\}$ form an *orthonormal set* under the inner product over the unit interval, i.e., if

$$(\phi, \psi) = \int_0^1 \overline{\phi}(x) \psi(x) dx$$

then

$$(e_n, e_m) = \delta_{nm}. \tag{5.11}$$

Here and elsewhere we will be casual in specifying the inner product. It may refer to the discrete case (4.27) or the continuous case (5.8). In the latter instance, the interval over which functions are defined may vary. In each situation, the circumstances will dictate which type of inner product is actually being used. Only in the event of ambiguity will special notation be introduced to help distinguish between the different inner products.

If we formally take the inner product of both sides of (5.9) with $\exp(2\pi i m x)$, we find

$$(e_m, f) = a_m = \int_0^1 e^{-2\pi i m x} f(x)dx = \tilde{f}_m. \tag{5.12}$$

Equation (5.9) will be referred to as the Fourier series of f. From (5.12) we see that when the periodic functions are defined over the unit interval, the Fourier coefficient, a_m, and the Fourier transform, $\tilde{f}_m$, are equal. The justification of these formal steps will be given below.

5.2 Riemann–Lebesgue Lemma

A main question now concerns the types of functions $f(x)$ for which the corresponding Fourier series make sense. As a first step we examine the convergence of a Fourier series. This question is intimately connected with the behavior of the coefficients a_n, from (5.12), as $|n| \uparrow \infty$. In this connection suppose $f(x)$ is a piecewise differentiable function on some interval (a, b) (i.e., $f(x)$ and $f'(x)$ are both piecewise continuous in (a, b)). If (a_n, b_n) is a subinterval on which $f(x)$ is differentiable, then we have for the *Fourier integral*

$$\int_{a_n}^{b_n} e^{ikx} f(x)dx = \int_{a_n}^{b_n} f(x)\frac{d}{dx}\frac{e^{ikx}}{ik}dx$$

$$= \left[\frac{f(x)e^{ikx}}{ik}\right]_{a_n}^{b_n} - \frac{1}{ik}\int_{a_n}^{b_n} f'(x)e^{ikx}dx,$$

where k is real. We see from this that the *Fourier integral* vanishes as $|k| \uparrow \infty$.

This is an example of an *asymptotic* or limiting result. In speaking of a function which vanishes asymptotically, it is also useful to indicate the rate at which it vanishes. The Landau symbols $O(\ )$ and $o(\ )$ are useful for this purpose.

Definition. *A function $f(x)$ is said to be $O(x)$ as $x \to 0$ if*

$$\lim_{x \to 0} \left|\frac{f(x)}{x}\right| < \infty.$$

The function is said to be $o(x)$ if

$$\lim_{x \to 0} \left|\frac{f(x)}{x}\right| = 0.$$

In the first case we write

$$f(x) = O(x)$$

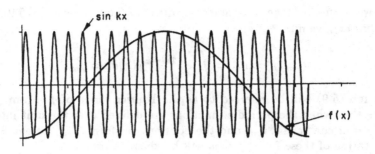

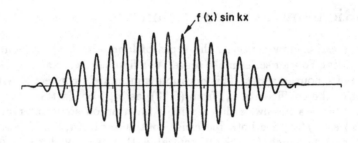

FIGURE 5.2.

and in the second,

$$f(x) = o(x).$$

The additional phrase "as $x \to 0$" is implicit in these relations.

Returning to the above calculation, we can write

$$\int_{a_n}^{b_n} e^{ikx} f(x) dx = O\left(\frac{1}{k}\right).$$

Thus, by appropriate subdivision of the interval of definition (a, b), we have shown the following:

If $f(x)$ is piecewise differentiable on (a, b), then as $|k| \uparrow \infty$

$$\int_a^b e^{ikx} f(x) dx = O\left(\frac{1}{k}\right). \tag{5.13}$$

It is intuitively clear that integrals of this sort vanish for $|k| \uparrow \infty$ under more general hypotheses. The term $\exp(ikx)$ is a rapidly oscillating sinusoid and $f(x)$ (which may be very irregular but does not depend on k) is by comparison slowly varying as $|k| \uparrow \infty$. Their product, $f(x)\exp(ikx)$, results in alternately positive and negative areas that cancel out each other in the integrand—see Figure 5.2.

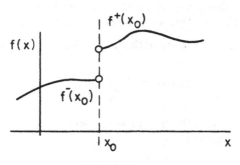

FIGURE 5.3.

The general result on this subject is given by the:

Riemann–Lebesgue Lemma. *If $f(x)$ is absolutely integrable on (a, b), then*

$$\lim_{|k|\uparrow\infty} \int_a^b e^{ikx} f(x) dx = 0$$

or

$$\int_a^b e^{ikx} f(x) dx = o(1) \tag{5.14}$$

as $|k| \uparrow \infty$. This is not proven here. (See Erdelyi for a proof and also Section 6.1 for a proof when $f(x)$ is *square integrable*, i.e., when $|f(x)|^2 = \overline{f}(x)f(x)$ is integrable.) It should be noticed that (5.14) is not as sharp as (5.13), since it tells us nothing of the rate at which the integral vanishes.

5.3 Fourier's Theorem

We consider Fourier series of functions which are piecewise differentiable. At a point of discontinuity x_0, we will write (see Figure 5.3)

$$f^{\pm}(x_0) = \lim_{x \lessgtr x_0,\, x \to x_0} f(x).$$

We now prove

Fourier's Theorem. *If $f(x)$ is one-periodic and piecewise differentiable, then*

$$\lim_{N\uparrow\infty} \sum_{n=-N}^{N} a_n e^{2\pi i n x} = \frac{f^+(x) + f^-(x)}{2}, \tag{5.15}$$

where a_n is given by (5.12). That is, the following is true: (1) the series converges for each x; (2) at a point of continuity the series equals the function; and (3) at a discontinuity point it equals the arithmetic mean of the limits of the function on both sides of the discontinuity.

Proof: To prove this, we start with the partial sum defined by

$$f_N(x) = \sum_{n=-N}^{N} a_n e_n(x), \tag{5.16}$$

with a_n given by (5.12). If we then substitute the expression from (5.12) for a_n in (5.16), we obtain

$$f_N(x) = \sum_{n=-N}^{N} e_n(x)(e_n, f) = \int_0^1 \left(\sum_{n=-N}^{N} e^{2\pi i n(x-y)} \right) f(y) dy.$$

The summation may be put into the compact form

$$D_N(t) = \sum_{n=-N}^{N} e^{2\pi i n t} = 1 + \sum_{n=1}^{N} e^{2\pi i n t} + \sum_{n=1}^{N} e^{-2\pi i n t}$$

$$= 1 + \frac{e^{2\pi i t}(1 - e^{2\pi i N t})}{1 - e^{-2\pi i N t}} + \frac{e^{-2\pi i t}(1 - e^{2\pi i N t})}{1 - e^{-2\pi i t}}, \tag{5.17}$$

where we have again used the identity

$$\frac{1 - x^N}{1 - x} = 1 + x + x^2 + \cdots + x^{N-1}.$$

If we rationalize (5.17), we obtain

$$D_N(t) = \frac{e^{i\pi t} - e^{-i\pi t} - +e^{-i\pi t}e^{-i\pi(2N+1)t} + e^{i\pi(2N+1)t} - e^{i\pi t}}{e^{i\pi t} - e^{-i\pi t}}$$

$$= \frac{\sin(2N+1)\pi t}{\sin \pi t}. \tag{5.18}$$

The function D_N, as expressed in (5.18), is known as the *Dirichlet kernel*.

Three features are noteworthy about the Dirichelt kernel: (1) From (5.17) it follows that $D_N(t)$ is one-periodic. (2) Also from (5.17) it is seen that

$$\int_0^1 D_N(t) dt = 1 = \int_{-1/2}^{1/2} D_N(t) dt = 2 \int_0^{1/2} D_N(t) dt \tag{5.19}$$

where we have used the fact that $D_N(t)$ is one-periodic and even, i.e., $D_N(t) = D_N(-t)$. (3) Finally, from (5.18) it is seen that $D_N(t)$ is sharply peaked at the origin. In fact,

$$D_N(0) = 2N + 1.$$

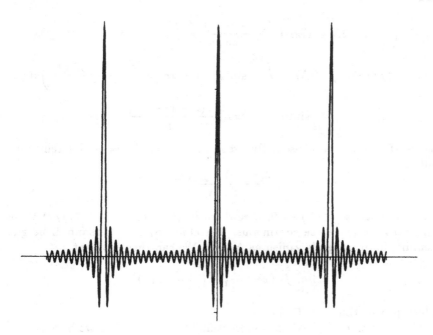

FIGURE 5.4.

A sketch of $D_N(t)$ is shown in Figure 5.4 for $N = 10$.

A strategy for the proof of (5.15) can now be seen. We write (5.16) as

$$f_N(x) = \int_0^1 D_N(x - y)f(y)dy. \tag{5.20}$$

Then, as $N \uparrow \infty$, we obtain the series in (5.9); but from (5.19) and the peaking property shown in Figure 5.4, it is at least intuitively clear that the right-hand side of (5.20) approaches $f(x)$ at a point of continuity.

Since both D_N and f are periodic, the integration in (5.20) can be taken over any unit interval, and, in particular, we take

$$f_N(x) = \int_{-(1/2)+x}^{x+(1/2)} D_N(x - y)f(y)dy.$$

Next we set $x - y = s$, so that

$$f_N(x) = \int_{-1/2}^{1/2} D_N(s)f(x - s)ds.$$

We substitute the expression from (5.18) for $D_N(t)$ and rearrange the integral as follows:

$$f_N(x) = \int_{-1/2}^0 \sin(2N+1)\pi s \left(\frac{f(x - s) - f^+(x)}{\sin s\pi} \right) ds + f^+(x) \int_{-1/2}^0 D_N(s)ds$$

$$+ \int_0^{1/2} \sin(2N+1)\pi s \left(\frac{f(x-s) - f^-(x)}{\sin s\pi} \right) + f^-(x) \int_0^{1/2} D_N(s)ds$$

$$= \frac{1}{2}\{f^+(x) + f^-(x)\} + \int_0^{1/2} \sin(2N+1)\pi s \left(\frac{f(x-s) - f^-(x)}{\sin s\pi} \right) ds$$

$$+ \int_{-1/2}^0 \sin(2N+1)\pi s \left(\frac{f(x-s) - f^+(s)}{\sin s\pi} \right) ds,$$

where (5.19) has been used. By assumption f is piecewise differentiable, and hence

$$\frac{f(x-s) - f^{\pm}(x)}{\sin s\pi}$$

both exist as $(s \gtrless 0) \to 0$. Elsewhere in the interval $(-1/2, 1/2)$ these functions are piecewise continuous. Therefore, by the Riemann–Lebesgue Lemma, both integrals vanish as $N \uparrow \infty$. We can then say that

$$\lim_{N \uparrow \infty} f_N(x) = \frac{1}{2}\{f^+(x) + f^-(x)\},$$

which proves Fourier's Theorem.

The condition of piecewise differentiability under which we have shown convergence of a Fourier series is a *sufficient* condition. The overwhelming majority of cases encountered in applications is covered by this case. In our later discussion we mention weaker conditions under which convergence is obtained.

5.4   Miscellaneous Extensions

Convolution. Equation (5.20) resembles a convolution product, and, in fact, it is under an appropriate definition of convolution product.

If $f(x)$ and $g(x)$ are one-periodic, their convolution is defined as

$$f \star g = \int_0^1 f(x-y)g(y)dy \tag{5.21}$$

(cf. Exercise 5 at the end of Section 4.1 and Equation (4.39)). It is left as an exercise to show that the product is commutative:

$$f \star g = g \star f. \tag{5.22}$$

Since $f(x)$ in (5.21) is one-periodic, it is clear that $f \star g$ is also one-periodic, and hence it too has a Fourier expansion. If we write

$$f = \sum_{n=-\infty}^{\infty} \tilde{f}e^{2\pi i n x}, \quad g = \sum_{n=-\infty}^{\infty} \tilde{g}e^{2\pi i n x},$$

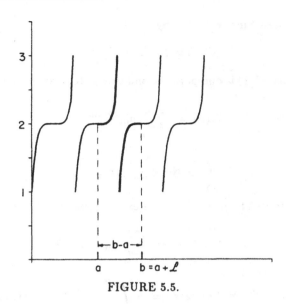

FIGURE 5.5.

with $\tilde{f}$ and $\tilde{g}$ given by (5.12), then from past experience we might suppose

$$f \star g = \sum_{m=-\infty}^{\infty} \tilde{f}_m \tilde{g}_m e^{2\pi i m x}. \qquad (5.23)$$

The proof of this is left for the exercises.

The reverse interpretation of (5.23) should also be noted. Specifically, if we have a Fourier series in which the coefficients can be viewed as products of Fourier coefficients, then the sum may be expressed as a convolution product of the appropriately defined functions.

Nonperiodic Functions. In view of the fact that discontinuous functions are permitted in the above discussion, we can expand an arbitrary nonperiodic function $\mathcal{F}(x)$ in Fourier series. For suppose $\mathcal{F}(x)$ is defined in (a, b). Then it can be extended to the entire real line by periodic repetition; i.e., define $F(x)$ by

$$F(x) = \mathcal{F}(x - N(b - a)),$$

where N is chosen such that $x - N(b-a)\epsilon(a, b)$. The periodic function $F(x)$ then leads to a Fourier series.

Periodic Functions of Arbitrary Period. In general, what we encounter is a periodic function $F(x)$ of arbitrary period $\mathcal{L}$. As Figure 5.5 (heavy curve) indicates, the fundamental interval of definition can be taken as $(a, a + \mathcal{L})$ or, for simplicity, as $(0, \mathcal{L})$. To adapt the above discussion to this case, change variables to

$$\hat{x} = \frac{x}{\mathcal{L}}. \qquad (5.24)$$

Next define a new function $\hat{G}(\hat{x})$ by

$$\hat{G}(\hat{x}) = F(\mathcal{L}\hat{x}) = F(x).$$

By construction, $\hat{G}(\hat{x})$ is one-periodic and can be represented by the Fourier series

$$\hat{G}(\hat{x}) = \sum_{m=-\infty}^{\infty} a_m e^{2\pi i m \hat{x}},$$

with

$$a_m = \int_0^1 e^{-2\pi i m \hat{x}} \hat{G}(\hat{x}) d\hat{x} = (e_m, \hat{G}).$$

If we substitute (5.24) back into these expressions, we obtain

$$\hat{G}(\hat{x}) = F(x) = \sum_{n=-\infty}^{\infty} a_n e_n(x/\mathcal{L}), \qquad (5.25)$$

$$a_n = \frac{1}{\mathcal{L}} \int_0^{\mathcal{L}} e^{-2\pi i n(x/\mathcal{L})} F(x) dx = \frac{1}{\mathcal{L}} \int_{-\mathcal{L}/2}^{\mathcal{L}/2} e^{-2\pi i n(x/\mathcal{L})} F(x) dx \qquad (5.26)$$

which make up the Fourier pair for an $\mathcal{L}$-periodic function $F(x)$.

It follows from this discussion that there is no loss of generality in considering one-periodic functions or, for that matter, functions of any convenient period. For later reference we now obtain some standard forms for Fourier series over the arbitrary period $\mathcal{L}$.

Trigonometric Series. If $\exp(2\pi i n(x/\mathcal{L}))$ is expressed in terms of trigonometric functions, then instead of (5.25) we obtain

$$F(x) = \sum_{n=-\infty}^{\infty} \left(a_n \cos 2\pi n \frac{x}{\mathcal{L}} + i a_n \sin 2\pi n \frac{x}{\mathcal{L}} \right)$$

$$= a_0 + \sum_{n=1}^{\infty} \left[(a_n + a_{-n}) \cos 2\pi n \frac{x}{\mathcal{L}} + i(a_n - a_{-n}) \sin 2\pi n \frac{x}{\mathcal{L}} \right]$$

or

$$F(x) = \frac{A_0}{2} + \sum_{n=1}^{\infty} \left(A_n \cos 2\pi n \frac{x}{\mathcal{L}} + B_n \sin 2\pi n \frac{x}{\mathcal{L}} \right), \qquad (5.27)$$

where from (5.26)

$$A_n = (a_n + a_{-n}) = \frac{1}{\mathcal{L}} \int_0^{\mathcal{L}} F(x) \left(e^{-2\pi i n(x/\mathcal{L})} + e^{+2\pi i n(x/\mathcal{L})} \right) dx$$

$$= \frac{2}{\mathcal{L}} \int_0^{\mathcal{L}} F(x) \cos 2\pi n \frac{x}{\mathcal{L}} dx = \frac{2}{\mathcal{L}} \int_{-\mathcal{L}/2}^{\mathcal{L}/2} F(x) \cos 2\pi n \frac{x}{\mathcal{L}} dx \qquad (5.28)$$

$$B_n = i(a_n - a_{-n}) = \frac{1}{L} \int_0^L F(x)i \left(e^{-2\pi i n(x/L)} - e^{2\pi i n(x/L)} \right) dx$$

$$= \frac{2}{L} \int_0^L F(x) \sin 2\pi n \frac{x}{L} dx = \frac{2}{L} \int_{-L/2}^{L/2} F(x) \sin 2\pi n \frac{x}{L} dx. \qquad (5.29)$$

Equation (5.27) is referred to as the *trigonometric Fourier series*, and A_n and B_n are called the *cosine* and *sine transforms* (or *coefficients*), respectively.

An alternative approach to trigonometric series starts with (5.27) and then uses the relations

$$\frac{2}{L} \int_0^L \left\{ \begin{array}{c} \sin 2\pi m \dfrac{x}{L} \sin 2\pi n \dfrac{x}{L} \\[2mm] \cos 2\pi m \dfrac{x}{L} \cos 2\pi n \dfrac{x}{L} \end{array} \right\} dx = \delta_{nm},$$

$$\frac{2}{L} \int_0^L \cos 2\pi m \frac{x}{L} \sin 2\pi n \frac{x}{L} dx = 0 \qquad (5.30)$$

to obtain the Fourier coefficients (5.28) and (5.29).

Trigonometric expansions are especially convenient to use when a periodic function has symmetry properties.

Even Functions. If $F(x)$ is an even function, then since $\sin 2\pi n(x/L)$ is odd, $B_n = 0$ (see (5.29)) and

$$F(x) = \frac{A_0}{2} + \sum_{n=1}^{\infty} A_n \cos 2\pi n \frac{x}{L}. \qquad (5.31)$$

The coefficients A_n are given by (5.28), which becomes

$$A_n = \frac{4}{L} \int_0^{L/2} F(x) \cos 2\pi n \frac{x}{L} dx$$

since $\cos 2\pi n(x/L)$ is even.

Odd Functions. On the other hand, if $F(x)$ is an odd function, then it follows from (5.28) and (5.29) that $A_n = 0$ and

$$F(x) = \sum_{n=1}^{\infty} B_n \sin 2\pi n \frac{x}{L}, \qquad (5.32)$$

with B_n given by

$$B_n = \frac{4}{L} \int_0^{L/2} F(x) \sin 2\pi n \frac{x}{L} dx.$$

Further symmetries are exploited in the exercises.

Exercises

1. (a) Give a proof of the Riemann–Lebesgue Lemma for piecewise
   continuous functions. [Hint: Split the interval into subintervals
   in which the function is continuous.]

   (b) Prove that $\int_0^{2\pi} e^{ikx} f(x)dx = O(1/k^n)$ if f has n continuous and
   periodic derivatives in $(0, 2\pi)$.

2. (a) Prove (5.22).   (b) Prove (5.23).   (c) Prove (5.30).

3. Show the following:

   (a) If $f(x)$ is $\mathcal{L}$-periodic, even, and such that

   $$f\left(\frac{\mathcal{L}}{4} + x\right) = -f\left(\frac{\mathcal{L}}{4} - x\right),$$

   then

   $$f(x) = \sum_{n=0}^{\infty} A_{2n+1} \cos(2n+1)\pi\frac{2x}{\mathcal{L}}.$$

   (b) If $f(x)$ is 2π-periodic and $f(x - \pi) = f(x)$, then

   $$f(x) = \frac{A_0}{2} + \sum_{n=1}^{\infty}(A_{2n} \cos 2nx + B_{2n} \sin 2nx).$$

   (c) If $f(x)$ is 2π-periodic, odd, and such that $f(x + (\pi/3)) = f(x)$,
   then

   $$f(x) = \sum_{n=1}^{\infty} B_{6n} \sin 6nx.$$

4. Suppose $f(x)$ is one-periodic with

   $$f(x) = \sum_{n=-\infty}^{\infty} a_n e^{2\pi inx}.$$

   Sample $f(x)$ at the equidistant points

   $$x_m = \frac{m}{N}, \quad m = 0, \pm1, \pm2, \ldots.$$

   Then

   $$f\left(\frac{m}{N}\right) = f_m = \sum_{n=-\infty}^{\infty} a_n e^{(2\pi i/N)mn}.$$

   But since f_m is an N-periodic sequence,

   $$f_m = \sum_{n=1}^{N} A_n e^{(2\pi i/N)mn}.$$

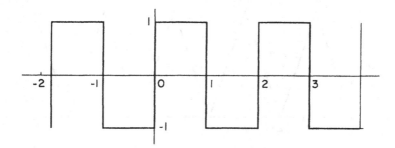

FIGURE 5.6.

Find the relationship between A_n and a_n. Is A_n in any way an approximation to a_n?

5. Formally show that if $F(x)$ is $\mathcal{L}$-periodic, then

$$(F, F) = \frac{1}{\mathcal{L}} \int_{-\mathcal{L}/2}^{\mathcal{L}/2} \overline{F}(x)F(x)dx = \sum_{n=-\infty}^{\infty} \bar{a}_n a_n.$$

What is the result in terms of A_n and in terms of B_n (i.e., the trigonometric coefficients)?

6. Suppose $F(x)$ is one-periodic, we write

$$F(x) = \frac{A_0}{2} + \sum_{n=1}^{\infty} A_n \cos(2n\pi x - \theta_n).$$

Then what are the forms for A_n and θ_n? Is there any loss of generality in this form?

5.5   Examples of Fourier Series

The examples which are given below are real-valued Fourier series. For this reason we use the trigonometric form (5.27) and an interval $\mathcal{L} = 2$. The latter is used since an examination of (5.28) and (5.29) indicates that an interval $\mathcal{L} = 2$ produces the simplest forms.

Square Wave. Consider the function sgn x, $-1 < x < 1$ (see Figure 5.6). The periodic repetition of this function is known as the square wave, and we denote it by $S(x)$. From its definition, $S(x)$ is odd and has period two. Therefore, referring to (5.32), we can write

$$S(x) = \sum_{n=1}^{\infty} B_n \sin n\pi x.$$

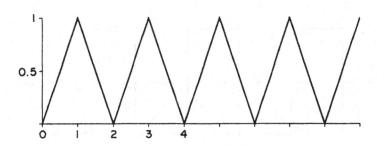

FIGURE 5.7.

Then from (5.29)

$$B_n = \int_{-1}^{1} (\text{sgn } x) \sin n\pi x \, dx = 2 \int_0^1 \sin n\pi x \, dx$$

$$= \frac{2}{\pi} \left. \frac{\cos n\pi x}{n} \right|_1^0 = \begin{cases} 4/n\pi, & n \text{ odd,} \\ 0, & n \text{ even.} \end{cases}$$

Hence

$$S(x) = \frac{4}{\pi} \left(\sin \pi x + \frac{\sin 3\pi x}{3} + \frac{\sin 5\pi x}{5} + \cdots \right). \qquad (5.33)$$

The loss of the even harmonics is a result of the fact that

$$S\left(\frac{1}{2} + x\right) = S\left(\frac{1}{2} - x\right),$$

i.e., that $S(x)$ must become an even function under translation by 1/2. We also note that if we let $x = 1/2$ in (5.33), we obtain

$$\pi = 4 \left(1 - \frac{1}{3} + \frac{1}{5} - \frac{1}{7} \pm \cdots \right),$$

which is a (slowly converging) representation of π.

Triangle Wave. Consider the function $|x|$, $-1 < x < 1$ (see Figure 5.7). When this function is periodically repeated, we create what is known as the triangle wave. It is denoted by $T(x)$ and has the properties of being even and two-periodic by definition. Therefore, from (5.31), we can write

$$T(x) = \frac{A_0}{2} + \sum_{n=1}^{\infty} A_n \cos n\pi x,$$

and from (5.28) we have

$$A_n = \int_{-1}^{1} |x| \cos n\pi x \, dx = 2 \int_0^1 x \cos n\pi x \, dx$$

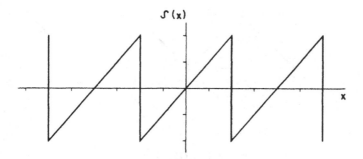

FIGURE 5.8.

$$= 2 \int_0^1 x \frac{d}{dx} \frac{\sin n\pi x}{n\pi} dx = \frac{-2}{n\pi} \int_0^1 \sin n\pi x \, dx$$

$$= \begin{cases} -4/(n\pi)^2, & n \text{ odd}, \\ 0, & n \text{ even}, \end{cases}$$

$$A_0 = 1.$$

Combining these results then gives us

$$T(x) = \frac{1}{2} - \frac{4}{\pi^2} \left\{ \cos \pi x + \frac{\cos 3\pi x}{3^2} + \cdots \right\}. \qquad (5.34)$$

This again leads to a calculation for the value of π.

Sawtooth Wave. As a final evaluation, we consider the sawtooth wave $S(x)$, defined as the periodic repetition of the function x, $-1 < x < 1$ (see Figure 5.8). $S(x)$ is odd and two-periodic and therefore has the expansion

$$S(x) = \sum_{n=1}^{\infty} B_n \sin n\pi x,$$

with

$$B_n = 2 \int_0^1 x \sin n\pi x \, dx = -2 \int_0^1 x \frac{d}{dx} \frac{\cos n\pi x}{n\pi} dx$$

$$= -\frac{2}{n\pi} (-1)^n.$$

We thus have

$$S(x) = \frac{2}{\pi} \left(\sin \pi x - \frac{\sin 2\pi x}{2} + \frac{\sin 3\pi x}{3} \mp \cdots \right). \qquad (5.35)$$

If periodic functions of the same period are added together, their sum is periodic. Moreover, the Fourier series of this sum is obtained by adding the

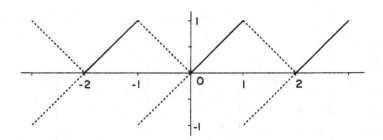

FIGURE 5.9.

Fourier coefficients of the corresponding individual series. As an illustration of this, consider

$$\frac{T(x) + S(x)}{2} = \frac{1}{4} + \frac{1}{\pi}\left(\sin \pi x - \frac{\sin 2\pi x}{2} \pm \cdots\right)$$

$$- \frac{2}{\pi^2}\left(\cos \pi x + \frac{\cos 3\pi x}{3^2} + \cdots\right).$$

On graphically adding the sketch in Figure 5.7 and that in Figure 5.8 corresponding to $T(x)$ and $S(x)$, respectively, we obtain Figure 5.9.

As a final remark, we note that the two functions, $S(x)$ and $S(x)$, with discontinuities have coefficients that are

$$O\left(\frac{1}{n}\right)$$

as $n \uparrow \infty$, while $T(x)$, which is continuous, has coefficients that are

$$O\left(\frac{1}{n^2}\right)$$

as $n \uparrow \infty$. If we compare these results with (5.13) and with Exercise 1(b) of Section 5.4, we find complete agreement.

Exercises

1. Find the Fourier series of the following:

   (a) full-wave rectification of $\cos \pi x$, i.e., $|\cos \pi x|$ (see Figure 5.10);
   (b) half-wave rectification of $\cos \pi x$ as shown in Figure 5.11 (hint: use superposition);
   (c) periodic repetition of

   $$D(x) = \begin{cases} 1, & |x| < \tau < 1, \\ 0, & \text{all other } x \in (-1, 1) \end{cases}$$

   (see Figure 5.12);

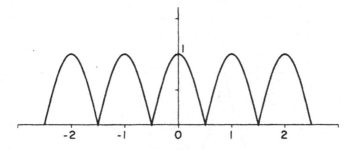

FIGURE 5.10.

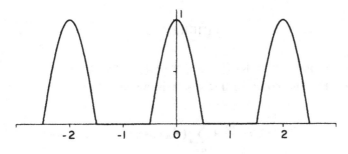

FIGURE 5.11.

(d) periodic repetition of $(x^2 - 1)$, $-1 < x < 1$ (see Figure 5.13).

2. Indicate analytically why neither $T(x)$, given by (5.34), nor $S(x)$, given by (5.33), has even harmonics in its representation.

3. Find the Fourier series for each of the following functions:

   (a) $F(x) = 0$, $-\pi < x < 0$, and $F(x) = 1$, $0 < x < \pi$;

   (b) $F(x) = x^2$, $-1 < x < 1$;

   (c) $F(x) = x^2$, $0 < x < 1$.

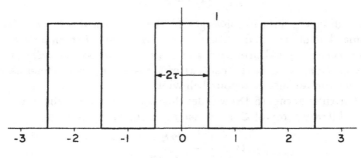

FIGURE 5.12.

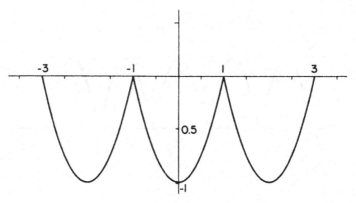

FIGURE 5.13.

4. (a) Expand $\sin x$ for $0 < x < \pi$ in a cosine series.

(b) Expand $\cos x$ for $0 < x < \pi$ in a sine series.

5. If

$$F(x) = \frac{A_0}{2} + \sum_{n=1}^{\infty}(A_n \cos n\pi x + B_n \sin n\pi x),$$

what can be said of the coefficients when

(a) $F(-x) = F(x) = F(\frac{1}{2} - x)$,

(b) $F(x) = F(2x)$,

(c) $F(x + \pi/4) = F(x)$?

5.6   Gibbs' Phenomenon

From our convergence proof in Section 5.3, we know that at a point of discontinuity, say x_0, the Fourier series of a function converges to the arithmetic mean

$$\frac{1}{2}\{f^+(x_0) + f^-(x_0)\},$$

and at ordinary points it converges to the function itself. An interesting and somewhat unexpected feature known as *Gibbs' phenomenon* occurs in the convergence of a Fourier series in the neighborhood of a discontinuity. This is now discussed in the context of the square wave, which as will be seen is generic for discontinuous behavior.

The Dirichlet kernel (5.18) was developed for the case of the unit interval. For an arbitrary interval $\mathcal{L}$, it is easily seen to have the form

$$D_N(t) = \frac{1}{\mathcal{L}} \frac{\sin((2N + 1)\pi t/\mathcal{L})}{\sin(\pi t/\mathcal{L})},$$

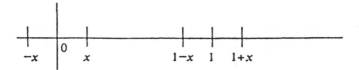

FIGURE 5.14.

and its integral over this interval is

$$\int_{-\mathcal{L}/2}^{\mathcal{L}/2} D_N(t)dt = 1.$$

In particular, if we set $\mathcal{L} = 2$, we find that the Nth order approximation of the square wave $S(x)$ with Fourier series representation (5.33) is given by (see (5.20))

$$
\begin{aligned}
S_N(x) &= D_N * S = \int_{-1}^{1} \frac{\sin(N + (1/2))\pi(x - y)}{2\sin(\pi(x - y)/2)} S(y)dy \\
&= \int_{0}^{1} \frac{\sin(N + (1/2))\pi(x - y)}{2\sin(\pi(x - y)/2)} \\
&\quad - \int_{-1}^{0} \frac{\sin(N + (1/2))\pi(x - y)}{2\sin(\pi(x - y)/2)} dy.
\end{aligned}
$$

If in the first integral we set $y - x = \theta$, and in the second integral, $y - x = -\theta$, then

$$S_N(x) = \int_{-x}^{1-x} \frac{\sin(N + (1/2))\pi\theta}{2\sin(\pi\theta/2)} d\theta - \int_{x}^{1+x} \frac{\sin(N + (1/2))\pi\theta}{2\sin(\pi\theta/2)} d\theta.$$

If we remove the overlap interval $(x, 1 - x)$, then

$$S_N(x) = \int_{-x}^{x} \frac{\sin(N + (1/2))\pi\theta}{2\sin(\pi\theta/2)} d\theta - \int_{1-x}^{1+x} \frac{\sin(N + (1/2))\pi\theta}{2\sin(\pi\theta/2)} d\theta$$

(see Figure 5.14 for the case in which x is a small positive number).

We examine $S_N(x)$ in the neighborhood of the origin ($|x|$ is therefore regarded as small) in the limit of $N \uparrow \infty$. The second integral, by the Riemann–Lebesgue Lemma, is vanishingly small and we can write

$$S_N(x) = \int_{-x}^{x} \frac{\sin(N + (1/2))\pi\theta}{2\sin(\pi\theta/2)} d\theta + o(1)$$

or since the integrand is even

$$S_N(x) = \int_{0}^{x} \frac{\sin(N + (1/2))\pi\theta}{\sin(\pi\theta/2)} d\theta + o(1) = \frac{2}{\pi} \int_{0}^{x} \frac{\sin(N + (1/2))\pi\theta}{\theta} d\theta$$

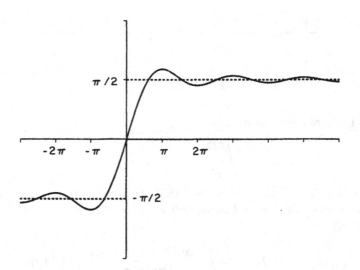

FIGURE 5.15.

$$+ \int_0^x \frac{\sin(N + (1/2))\pi\theta}{\theta} \cdot \left(\frac{\theta}{\sin(\pi\theta/2)} - \frac{2}{\pi} \right) d\theta + o(1)$$

where $2/(\pi\theta)$ has been added and subtracted from the integrand. The function $(1/\sin(\pi\theta/2)) - (1/(\pi\theta/2))$ exists as $\theta \to 0$ and is continuous elsewhere in the interval $(0, x)$. Hence the Riemann–Lebesgue Lemma again applies and in the last expression for $S_N(x)$, the second integral is vanishingly small. Therefore

$$S_N(x) = \frac{2}{\pi} \int_0^x \frac{\sin(N + (1/2))\pi\theta}{\theta} d\theta + o(1)$$

$$= \frac{2}{\pi} \int_0^{(N+(1/2))x\pi} \frac{\sin k}{k} dk + o(1).$$

The resulting integral is related to the tabulated integral

$$S_i(x) = \int_0^x \frac{\sin k}{k} dk,$$

a sketch of which is shown in Figure 5.15. The asymptotic value, $\pi/2$, of $S_i(x)$ as $x \uparrow \infty$ agrees with the asymptotic value obtained from our square wave calculation gotten by keeping x ($x > 0$) fixed and by letting $N \uparrow \infty$, thereby giving $S_N(x) \to 1$ (see Figure 5.6).

What is of interest is the first maximum of $S_i(x)$, which is determined by

$$\frac{d}{dx} S_i(x) = \frac{\sin x}{x} = 0 \Rightarrow x = \pi;$$

and from the tabulation of $S_i(x)$ it is known that

$$S_i(\pi) = 1.8515.$$

Translating this fact into our experience for the square wave, we can say that

$$S_N\left(\frac{1}{N + (1/2)}\right) = \frac{2}{\pi}(1.8515\ldots) + o(1) = (1.179\ldots) + o(1).$$

Therefore, although $S_N(x)$ converges to the square wave $S(x)$, there is always a point $x \approx 1/(N + (1/2))$ for which the overshoot is at the fixed level of 1.179... (see Figure 5.16). Since the square wave has an excursion of two, the percentage overshoot is

$$\frac{(1.179\ldots) - 1}{2} \times 100\% \approx 9\%.$$

(This overshoot is sometimes incorrectly stated to be 18%, and the confusion may exist because there is also an $\approx 9\%$ undershoot.) Gibbs' phenomenon, although just shown for the square wave, appears whenever discontinuities occur.

To show this, consider for simplicity a piecewise differentiable function $f(x)$ having a single discontinuity located at $x = x_0$ in the interval $(-1, 1)$ and having period two. Denote the jump at x_0 by Δ:

$$\Delta = f^+(x_0) - f^-(x_0).$$

Next consider the sawtooth wave $S(x)$, (5.35), shifted by a unit amount. This we denote by $J(x)$ and we have

$$J(x) = \frac{2}{\pi}\left\{\sin \pi(x - 1) - \frac{\sin 2\pi(x - 1)}{2} + \frac{\sin 3\pi(x - 1)}{3} - \cdots\right\}$$

$$= -\frac{2}{\pi}\left\{\sin \pi x + \frac{\sin 2\pi x}{2} + \frac{\sin 3\pi x}{3} + \frac{\sin 4\pi x}{4} + \cdots\right\}. \tag{5.36}$$

The plot of the function $J(x)$ is shown in Figure 5.17. The jump in $J(x)$ is -2 at $x = 0, \pm 2, \ldots$; i.e., $J^+(n) - J^-(n) = -2$, $n = 0, \pm 2, \ldots$.

We return to the function $f(x)$, with a discontinuity at $x = x_0$, and consider

$$F(x) = f(x) + \frac{\Delta}{2}J(x - x_0).$$

Observe that

$$F^+(x_0) - F^-(x_0) = f^+(x_0) - f^-(x_0) + \frac{f^+(x_0) - f^-(x_0)}{2}(-2) = 0.$$

Therefore $F(x)$ is continuous at $x = x_0$ and moveover is piecewise differentiable elsewhere since both $J(x)$ and $f(x)$ are piecewise differentiable.

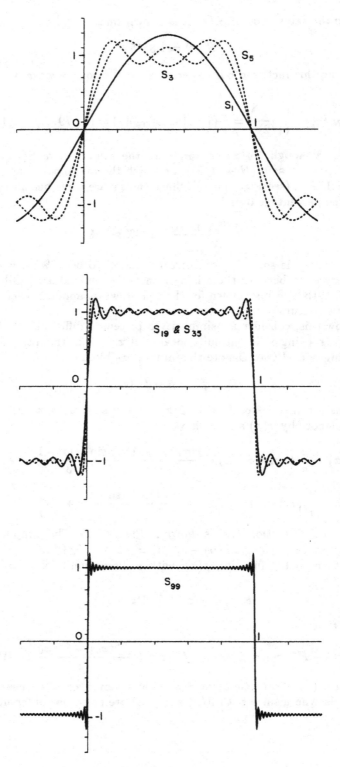

FIGURE 5.16

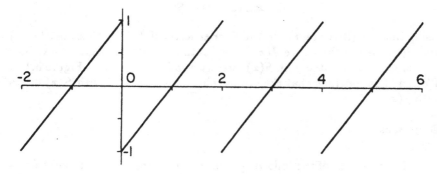

FIGURE 5.17.

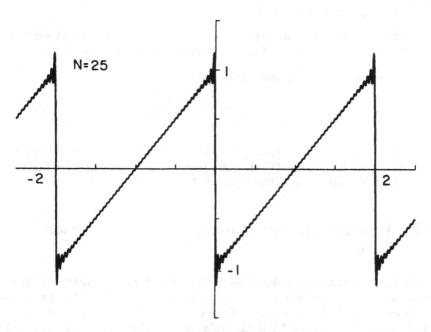

FIGURE 5.18.

Next the Fourier series of $f(x)$ is given by the Fourier series of $F(x)$ minus that of $(\Delta/2)J(x - x_0)$. To see that $J(x)$ gives rise to Gibbs' phenomenon, we simply observe the relation

$$J(x) = J(x - 1) - S(x).$$

Using this identity to represent the Fourier series of $J(x)$ by those of $J(x-1)$ and $S(x)$, we see that since $J(x - 1)$ is continuous at the origin, the Gibbs' phenomenon associated with $S(x)$ directly applies to $J(x)$ (see Figure 5.18). We can thus conclude that the Gibbs' phenomenon found with $S(x)$ arises with $f(x)$.

Exercises

1. Express each of the following periodic functions as a continuous function plus a possible sum of jump functions:

   (a) $f(x) = x^2$, $-\pi < x < \pi$;

   (b) $S(x)$, the square wave;

   (c) $f(x) = x$, $0 < x < 1$.

2. How many terms in $S_N(x)$ must be taken so that $S_N(.1)$ remains within 1% of $S(.1) = 1$ for all $n > N$. Consult tables of $S_i(x)$.

3. Indicate all maxima and minima of

$$S_i(x) = \int_0^x \frac{\sin k}{k}\,dk.$$

4. Evaluate $S_i(\pi)$ by expanding $\sin k$ in a power series and then by retaining five terms. What is the percentage undershoot for the first minimum after the Gibbs overshoot?

5.7   Integration and Differentiation of Fourier Series

An immediate uncertainty develops in discussing the integration of Fourier series. Do we want to integrate the periodic function defined by the series or do we want the Fourier series of the integral of the function defined on the fundamental interval? For example, the constant function $f(x) = 1$ is a periodic function with, say, period one. The integral of this periodic function is x, which of course is not periodic over the entire real line. On the other hand, the Fourier series of the function x restricted to the interval $(0, 1)$ is a one-periodic version of the sawtooth wave function described in

Section 5.5. There is no genuine difficulty in this point and we treat both situations.

Consider a function $f(x)$ which is defined on the interval $(0,1)$ and which has the Fourier series

$$f(x) = \sum_{n=-\infty}^{\infty} e_n(x), \quad a_n = \int_0^1 f(x)e^{-2\pi inx}\,dx.$$

Instead of evaluating the integral of $f(x)$, it will prove convenient to examine the integral

$$F(x) = \int_0^x (f(s) - a_0)ds, \tag{5.37}$$

whose integrand is the periodic function $f(x)$ with its mean value a_0 subtracted off. If, for example, $f(x)$ is piecewise continuous, then $F(x)$ has the piecewise continuous derivative $f(x) - a_0$, and our general treatment tells us that $F(x)$ can be expanded in a Fourier series, with period one. We write this series as

$$F(x) = \sum_{n=-\infty}^{\infty} A_n e^{-2\pi inx}. \tag{5.38}$$

Using parts integration and the fact that $F(1) = F(0) = 0$, we see that if $n \neq 0$,

$$A_n = \int_0^1 F(x)\frac{d}{dx}\left\{\frac{e^{-2\pi inx}}{(-2\pi in)}\right\}dx = \int_0^1 \frac{d}{dx}\left\{\frac{F(x)e^{-2\pi inx}}{(-2\pi in)}\right\}dx$$

$$+ \frac{1}{2n\pi i}\int_0^1 (f(x) - a_0)e^{-2\pi inx}\,dx = \frac{a_n}{2\pi ni}. \tag{5.39}$$

Thus

$$\int_0^x (f(s) - a_0)ds = A_0 + \sum_{n \neq 0} \frac{a_n}{2\pi in}e^{2\pi inx}, \tag{5.40}$$

where

$$A_0 = \int_0^1 \int_0^x (f(s) - a_0)ds\,dx. \tag{5.41}$$

An alternative form of (5.40) is

$$\int_0^x f(s)ds = a_0 x + A_0 + \sum_{n \neq 0} \frac{a_n}{2\pi in}e^{2\pi inx}. \tag{5.42}$$

The ambiguity expressed at the outset of this section can now be completely delineated. If we wish to consider the integral of the periodic function $f(x)$, then it is formed by a periodic function sitting on the "ramp" $a_0 x$ (see Figure 5.19). If, on the other hand, we want the Fourier series of the

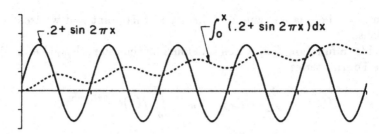

FIGURE 5.19.

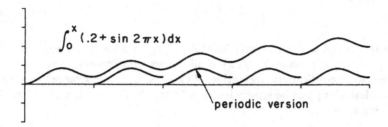

FIGURE 5.20.

integral of $f(x)$ in $(0, 1)$, then we must also expand a_0x as a sawtooth wave function. This is illustrated in Figure 5.20.

From (5.42) we see that the Fourier coefficients of $F(x)$ are given by the term-by-term integration of the formal Fourier series for $f(x)$. Thus we have shown that the integral of a function possesses a convergent Fourier series, *even when the function itself does not*, and that the Fourier series for the integral is given by the integration of the formal Fourier series of the function.

Analytically, the reason behind the increased convergence is easily seen from (5.42) to be that $1/n$ is present as a factor in the Fourier coefficients A_n of the integrated series. The underlying explanation for this is that integration is a smoothing procedure and in particular removes discontinuities. Discontinuities, as the examples have shown, give rise to slowly convergent Fourier series.

Since differentiation is the operation inverse to integration, it is *un-smoothing* in nature; therefore, we cannot expect term-by-term differentiation to result in a convergent series. However, if $f(x)$ possesses a piecewise differentiable derivative, then we can differentiate the Fourier series of $f(x)$ and obtain the Fourier series of $f'(x)$.

5.8 Application to Ordinary Differential Equations

As a simple application of Fourier series, consider the equation of a *sinusoidally driven harmonic oscillator*:

$$\frac{d^2y}{dt^2} + \Omega^2 y = f(t) = A\cos\omega t. \qquad (5.43)$$

A particular solution of this equation is

$$y_p = \frac{A\cos\omega t}{\Omega^2 - \omega^2}$$

(provided $\Omega^2 \neq \omega^2$). More generally, if the harmonic oscillator is driven by an arbitrary $2\pi/\omega$-periodic function $f(t)$, we can expand this driving force in a Fourier series and obtain

$$f(t) = \frac{A_0}{2} + \sum_{n=1}^{\infty}(A_n\cos n\omega t + B_n\sin n\omega t).$$

It then follows that a particular solution is given by

$$y_p = \frac{A_0}{2\Omega^2} + \sum_{n=1}^{\infty}\frac{A_n\cos n\omega t + B_n\sin n\omega t}{\Omega^2 - n^2\omega^2}, \qquad (5.44)$$

which surely converges if the Fourier series for $f(t)$ converges (provided $\Omega^2 - n^2\omega^2 \neq 0$ for all n).

Our main interest, however, lies in another direction. We want to use the idea that a function defined only on an interval can be expanded in a Fourier series (after which it is regarded as defined everywhere by periodic repetition). Thus the solution of an ordinary differential equation defined on some interval is representable by a Fourier series. This approach will be fruitful for our later study of partial differential equations. Although we now limit our discussion to elementary cases, some interesting subtleties occur.

Consider the elementary two-point boundary value problem

$$a^2\frac{d^2}{dx^2}y - y = -1, \quad 0 < x < 1, \quad y(0) = y(1) = 0. \qquad (5.45)$$

Although this problem can be solved by direct means, we wish to solve it by means of Fourier series. A sketch of the expected solution is shown in Figure 5.21 (bold line). From the sketch we ascertain that we should extend our solution $y(x)$ as an odd two-periodic function (solid curve). Otherwise, if we expand $y(x)$ as a one-periodic function, the Fourier series has no interesting properties (such as evenness or oddness); and if we expand $y(x)$

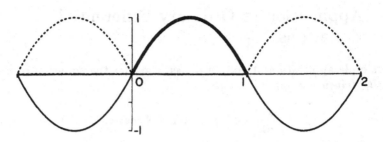

FIGURE 5.21.

as an even two-periodic function (broken curve), $y'(x)$ is discontinuous at the origin (and also at $x = 1$) and the convergence of the Fourier series is lost at the stage of $y''(x)$. The proposed Fourier series solution in the latter case cannot, as a result, be substituted for $y(x)$ in the differential equation (5.45).

We therefore pursue a solution in the form of an odd periodic function (of period two):

$$y(x) = \sum_{n=1}^{\infty} B_n \sin n\pi x.$$

The inhomogeneous term -1 for $0 < x < 1$ is also expanded as an odd two-periodic function, which in fact is given by the square wave (see (5.33)) shifted by a unit amount. Hence we have

$$-S(x) = -\frac{4}{\pi} \sum_{n=0}^{\infty} \frac{\sin(2n+1)\pi x}{2n+1} \quad (= -1,\ 0 < x < 1).$$

If these two Fourier series for $y(x)$ and -1 are substituted in the differential equation (5.45), then we obtain

$$a^2 \frac{d^2}{dx^2} \sum_{n=1}^{\infty} B_n \sin n\pi x - \sum_{n=1}^{\infty} B_n \sin n\pi x$$

$$= -\frac{4}{\pi} \sum_{n=0}^{\infty} \frac{\sin(2n+1)\pi}{2n+1}.$$

If we equate coefficients of $\sin n\pi x$, then

$$B_{2n+1} = \frac{4}{\pi(2n+1)(1 + a^2(2n+1)^2\pi^2)}, \quad B_{2n} = 0, \qquad (5.46)$$

which solves the problem.

Finite Fourier Transform. A subtle distinction now arises in the above method of solution. If we express the solution as an even function (again with period two) and write

$$y(x) = \frac{A_0}{2} + \sum_{n=1}^{\infty} A_n \cos n\pi x, \qquad (5.47)$$

then we cannot, beause of convergence reasons, substitute this series for $y(x)$ in the ordinary differential equation (5.45). On the other hand, since the solution is differentiable, it certainly can be represented by the above form (5.47)—it just cannot as such be substituted into (5.45). A way out of this seeming dilemma is found by multiplying both sides of (5.45) by $\cos n\pi x$ and then integrating both sides over the interval $(0, 1)$:

$$a^2 \int_0^1 \cos n\pi x \frac{d^2y}{dx^2} dx - \int_0^1 (\cos n\pi x) y \, dx$$

$$= - \int_0^1 \cos n\pi x \, dx. \qquad (5.48)$$

This is referred to as *taking a finite Fourier transform*. From (5.47) and (5.28) we have that

$$\int_0^1 y \cos n\pi x \, dx = \frac{A_n}{2}.$$

Also, we know that

$$\int_0^1 \cos n\pi x \, dx = \delta_{n0}.$$

If we use integration by parts, the integral in the first term on the left-hand side of (5.48) becomes

$$\int_0^1 \cos n\pi x \frac{d^2y}{dx^2} dx = \int_0^1 \frac{d}{dx}\left(\cos n\pi x \frac{dy}{dx}\right) dx + n\pi \int_0^1 \sin n\pi x \frac{dy}{dx} dx$$

$$= (-)^n \frac{dy(1)}{dx} - \frac{dy(0)}{dx} - (n\pi)^2 \int_0^1 (\cos n\pi x) y \, dx$$

$$= (-)^n y'(1) - y'(0) - (n\pi)^2 \frac{1}{2} A_n.$$

The appearance of $y'(1)$ and $y'(0)$ is annoying since we have no knowledge of these. To take care of this, we first integrate both sides of (5.45) over the interval $(0, 1)$ and obtain

$$\int_0^1 a^2 \frac{d^2y}{dx^2} dx - \int_0^1 y \, dx = - \int_0^1 dx = -1;$$

or, if the Fourier series (5.47) is substituted for y in the second term on the left-hand side, we come up with

$$a^2(y'(1) - y'(0)) - (A_0/2) = -1.$$

This gives us one relation in the unknowns $y'(0)$ and $y'(1)$. A second relation is obtained if we multiply both sides of (5.45) by dy/dx:

$$a^2 \frac{dy}{dx}\frac{d^2y}{dx^2} = y\frac{dy}{dx} - \frac{dy}{dx},$$

or

$$a^2 \frac{d}{dx}\frac{1}{2}\left(\frac{dy}{dx}\right)^2 = \frac{1}{2}\frac{d}{dx}(y^2 - 2y).$$

If we integrate both sides of this over the interval $(0,1)$, we obtain

$$y'(1)^2 - y'(0)^2 = 0.$$

The appropriate solution is $y'(0) = -y'(1)$, which also follows from symmetry. Combining this with the previous relation, we get

$$-y'(1) = y'(0) = \frac{2 - A_0}{4a^2}. \tag{5.49}$$

We can now substitute these results for $y'(1)$ and $y'(0)$ in the integrated form of the differential equation, (5.48), and we find that

$$A_{2n+1} = 0,$$

$$a^2\left\{-2y'(0) - (2n\pi)^2\frac{A_{2n}}{2}\right\} - \frac{A_{2n}}{2} = -\delta_{n0}.$$

This equation for $n = 0$ is the same as (5.49). For $n \neq 0$,

$$A_{2n} = \frac{-4a^2 y'(0)}{1 + (2n)^2\pi^2} = \frac{A_0 - 2}{1 + (2n)^2\pi^2}, \tag{5.50}$$

which if substituted into (5.47) gives the solution in the form

$$y(x) = \frac{A_0}{2} + (A_0 - 2)\sum_{n=1}^{\infty}\frac{\cos 2n\pi x}{1 + (2n)^2\pi^2}.$$

The solution is completed by imposing the boundary conditions

$$y(0) = y(1) = 0 = \frac{A_0}{2} + (A_0 - 2)\sum_{n=1}^{\infty}\frac{1}{1 + 4n^2\pi^2}$$

so that we then have

$$A_0 = \frac{2\sum_{n=1}^{\infty} 1/(1 + 4n^2\pi^2)}{\frac{1}{2} + \sum_{n=1}^{\infty} 1/(1 + 4n^2\pi^2)}.$$

With the evaluation of the constant A_0, the calculation is complete.

The calculation was tedious and not very useful. It was undertaken mainly to demonstrate that it could be done. It also illustrates an important point. From (5.50) we see that

$$A_{2n} = O(1/n^2),$$

while from (5.46) we have

$$B_n = O(1/n^3).$$

Therefore the first method of solution employing an odd periodic extension is more rapidly convergent. This was anticipated at the outset with our consideration of Figure 5.20—where it was mentioned that (5.47) has a discontinuous derivative when an even periodic function is constructed.

We consider one more example:

$$\frac{d^2}{dx^2}y - y = -1, \quad y(0) = a, \quad y(1) = b. \tag{5.51}$$

One approach in finding the solution is simply to write

$$y(x) = \sum_{n=1}^{\infty} \tilde{B}_n \sin n\pi x \tag{5.52}$$

and to use the finite Fourier transform method discussed above. It is clear that the resulting solution is discontinuous; hence,

$$\tilde{B}_n = O(1/n)$$

so that (5.52) is slowly convergent.

A more rapidly convergent procedure is obtained by first defining a new function $Y(x)$ with the equation

$$y(x) = a + (b - a)x + Y(x). \tag{5.53}$$

Then

$$\frac{d^2Y}{dx^2} - Y = -1 + a + (b - a)x, \quad Y(0) = Y(1) = 0. \tag{5.54}$$

At this point we have homogeneous boundary conditions. Therefore, if we expand $Y(x)$ (and also the right-hand side of (5.54)) by writing

$$Y(x) = \sum_{n=1}^{\infty} B_n \sin n\pi x, \tag{5.55}$$

it is clear from our earlier deliberations that

$$B_n = O(1/n^3).$$

The remainder of the calculation is left for the exercises.

Exercises

1. Find a particular solution of (5.43) if $f(t)$ is an arbitrary $2\pi/\omega$-periodic function and $\Omega^2 = n^2\omega^2$ for some n.

2. Solve (5.51) by assuming that (5.52) holds and by using the finite Fourier transform method.

3. Carry out the details of the solution given by (5.53), (5.54), and (5.55). Compare the results with those obtained by the previous exercise.

4. Use Fourier series to solve the following:

   (a) $\dfrac{d^2}{dx^2}y - y = -x(x-1)$, $y(0) = y(1) = 0$;

   (b) $\dfrac{d^2}{dx^2}y - y = x$, $y(0) = y(1) = 0$;

   (c) $\dfrac{d^2}{dx^2}y - y = \sin \pi x$, $y(0) = y(1) = 0$.

Additional Exercises

1. Show the following:

   (a) $x = \pi - 2 \displaystyle\sum_{n=1}^{\infty} \dfrac{\sin nx}{n}$, if $0 < x < 2\pi$;

   (b) $\dfrac{x^2}{2} = \pi x - \dfrac{\pi^2}{3} + 2 \displaystyle\sum_{n=1}^{\infty} \dfrac{\cos nx}{n^2}$, if $0 \le x \le 2\pi$;

   (c) $x = \dfrac{\pi}{2} - \dfrac{4}{\pi} \displaystyle\sum_{n=1}^{\infty} \dfrac{\cos(2n-1)x}{(2n-1)^2}$, if $0 \le x \le \pi$;

   (d) $x = 2 \displaystyle\sum_{n=1}^{\infty} \dfrac{(-1)^{n-1}\sin nx}{n}$, if $-\pi < x < \pi$;

   (e) $x^2 = \dfrac{\pi^2}{3} + 4 \displaystyle\sum_{n=1}^{\infty} \dfrac{(-1)^n \cos nx}{n^2}$, if $-\pi \le x \le \pi$.

2. Verify the following:

   (a) $x^2 = \dfrac{4}{3}\pi^2 + 4 \displaystyle\sum_{n=1}^{\infty} \left(\dfrac{\cos nx}{n^2} - \dfrac{\pi \sin nx}{n} \right)$, if $0 < x < 2\pi$;

   (b) $\cos x = \dfrac{8}{\pi} \displaystyle\sum_{n=1}^{\infty} \dfrac{n \sin 2nx}{4n^2 - 1}$, if $0 < x < \pi$;

(c) $\sin x = \dfrac{2}{\pi} - \dfrac{4}{\pi} \displaystyle\sum_{n=1}^{\infty} \dfrac{\cos 2nx}{4n^2 - 1}$, if $0 < x < \pi$;

(d) $x \cos x = -\dfrac{1}{2} \sin x + 2 \displaystyle\sum_{n=2}^{\infty} \dfrac{(-1)^n n \sin nx}{n^2 - 1}$, if $-\pi < x < \pi$;

(e) $x \sin x = 1 - \dfrac{1}{2} \cos x - 2 \displaystyle\sum_{n=2}^{\infty} \dfrac{(-1)^n \cos nx}{n^2 - 1}$, if $-\pi \leq x \leq \pi$.

3. Consider the square wave

$$S(x) = \frac{4}{\pi} \sum_{n=1}^{\infty} \frac{\sin(2n-1)x}{2n-1}, \quad -\pi < x < \pi.$$

If

$$S_N(x) = \frac{4}{\pi} \sum_{n=1}^{N} \frac{\sin(2n-1)x}{2n-1},$$

then show

(a) $S_N = \dfrac{2}{\pi} \displaystyle\int_0^x \dfrac{\sin 2ns}{\sin s} \, ds$;

(b) the local maxima and minima are located at $x_m = m\pi/2N$, $m = 1, \ldots, 2N$;

(c) $S_N(x_1)$ is the maximum.

4. Prove

(a) $\log \left| \sin \dfrac{x}{2} \right| = -\log 2 - \displaystyle\sum_{n=1}^{\infty} \dfrac{\cos nx}{n}$, if $x \neq 2k\pi$ (k an integer);

(b) $\log \left| \cos \dfrac{x}{2} \right| = -\log 2 - \displaystyle\sum_{n=1}^{\infty} \dfrac{(-1)^n \cos nx}{n}$, if $x \neq (2k+1)\pi$;

(c) $\log \left| \tan \dfrac{x}{2} \right| = -2 \displaystyle\sum_{n=1}^{\infty} \dfrac{\cos(2n-1)}{2n-1}$, if $x \neq k\pi$.

6

Spaces of Functions

6.0 Introduction

The contents of Chapters 4 and 5 bear a close resemblance to one another. In fact the case of periodic functions considered in Chapter 5 was developed heuristically from its discrete counterpart, periodic sequences, given in Chapter 4. In the latter instance each periodic sequence could be identified with a point in finite dimensional space, while in the former case a Fourier representation required an *infinite dimensional* specification. Both the discrete and continuous cases had a geometrical flavor. It is this that we now develop further. As an experiment in presentation we do this in parallel.

6.1 Discrete and Continuous Fourier Expansions — Geometrical Extensions

Discrete

Let $\{f_n\}$ represent an N-periodic sequence or, equivalently, an N-vector $\mathbf{f} = (f_1, \ldots, f_N)$ which has been periodically repeated. This leads to the discrete Fourier pair

$$f_n = \sum_{k=1}^{N} a_k W_{kn}, \qquad (6.1)$$

$$a_k = \frac{1}{N} \sum_{n=1}^{N} f_n W_{nk} = \frac{1}{N} \tilde{f}_n. \qquad (6.2)$$

$\tilde{f}_k$ is referred to as the Fourier transform.

A symmetrical version is obtained by setting

$$a_n = \frac{1}{\sqrt{N}} \hat{f}_n : \qquad (6.3)$$

$$\begin{cases} f_n = \frac{1}{\sqrt{N}} \sum_{k=1}^{N} \hat{f}_k W_{kn}, \\ \hat{f}_k = \frac{1}{\sqrt{N}} \sum_{n=1}^{N} f_n \overline{W}_{nk}. \end{cases}$$

The transform relations are a direct consequence of the orthogonality relations

$$\frac{1}{N} \sum_{n=1}^{N} \overline{W}_{nk} W_{n\ell} = \delta_{k\ell}, \tag{6.4}$$

or, if the complex inner product

$$(\mathbf{a}, \mathbf{b}) = \sum_{n=1}^{N} \bar{a}_n b_n \tag{6.5}$$

is used,

$$\frac{1}{N} (\mathbf{e}_k, \mathbf{e}_\ell) = \delta_{k\ell}. \tag{6.6}$$

Continuous

Let $f(t)$ represent a T-periodic function or, equivalently, the periodic extension of a function defined for $t \in (0, T)$. The corresponding Fourier transform pair is then given by

$$f(t) = \sum_{k=-\infty}^{\infty} \alpha_k e_k(t/T) \tag{6.7}$$

$$\alpha_k = \frac{1}{T} \int_0^T \bar{e}_k(t/T) f(t) dt = \frac{1}{T} \tilde{f}_k. \tag{6.8}$$

$\tilde{f}_k$ is referred to as the Fourier transform.

A symmetrical version is obtained by setting $\alpha_k = \tilde{\alpha}_k / \sqrt{T}$:

$$f(t) = \frac{1}{\sqrt{T}} \sum_{k=-\infty}^{\infty} \hat{\alpha}_k e_k(t/T),$$

$$\hat{\alpha}_k = \frac{1}{\sqrt{T}} \int_0^T \bar{e}_k(t/T) f(t) dt. \tag{6.9}$$

The transform relations are a direct consequence of the orthogonality relations

$$\frac{1}{T} \int_0^T \bar{e}_k(t/T) e_\ell(t/T) dt = \delta_{k\ell}, \tag{6.10}$$

or, if the inner product

$$(f, g) = \int_0^T \bar{f} g \, dt \tag{6.11}$$

is used,

$$\frac{1}{T} (e_k(t/T), e_\ell(t/T)) = \delta_{k\ell}. \tag{6.12}$$

Formal passage from the discrete to the continuous version was shown in Section 5.1. We recall this by taking the formal process in reverse order. First we sample a continuous function at N equally distant, by Δt, points in the interval such that

$$\frac{T}{\Delta t} = N.$$

Thus we write

$$f_n = f(n\Delta t).$$

If this is substituted into (6.7), we obtain (6.1) with α_n and a_n related by

$$a_n = \sum_{q=-\infty}^{\infty} \alpha_{n+qN} \tag{6.13}$$

(see Section 4.3), which is exact. Alternatively, if we multiply numerator and denominator of the right-hand side of (6.2) by Δt, it can be viewed as the Riemann sum approximation to (6.8).

The inner products (6.5) and (6.11) can be identified with (6.11) and (6.12) with the use of a Riemann sum approximation for the integral in (6.11). In this manner

$$\int_0^T \overline{f}g\, dt \approx \Delta t \sum_{n=1}^{N} \overline{f}(n\Delta t)g(n\Delta t)$$

or

$$\frac{1}{T}\int_0^T \overline{f}g\, dt = \frac{1}{T}(f,g) \approx \frac{1}{N}\sum_{n=1}^{N}\overline{f}_n g_n = \frac{1}{N}(\mathbf{f},\mathbf{g}).$$

This comparison is already foreshadowed by (6.6) and (6.12).

The idea of an inner product can be more abstractly formalized, with (6.6) and (6.12) as two specific examples. For a general set of *points* or *elements* f, g, h,... and complex numbers c, d,..., an inner product by definition is a bilinear form that satisfies the following four rules:

$$(f,g) = \overline{(g,f)},$$

$$(f,g+h) = (f,g) + (f,h), \tag{6.14}$$

$$(f,cg) = c(f,g),$$

$$(f,f) \geq 0 \quad \text{with} \quad (f,f) = 0 \Leftrightarrow f = 0.$$

There is one piece of fine print, namely that the *points* or *elements* belong to a linear space. This means that $cf + dg$ belongs to the space if f and g do. (We refer to this property as closure under addition and multiplication by complex numbers.)

The idea of distance follows from that used in N-space; namely, the distance between points of the space, f and g, is defined to be

$$\|f - g\| = (f - g, f - g)^{1/2}. \tag{6.15}$$

The distance symbol $\|\ \|$ is also referred to as the *norm*. The norm of f,

$$\|f\| = (f, f)^{1/2},$$

in effect gives the distance of the point f from the origin of the space. These rules, (6.14), give rise to a number of purely geometric relations:

$$\text{Schwarz Inequality} \quad |(f, g)| \leq \|f\| \|g\|, \tag{6.16}$$

$$\text{Triangle Inequality} \quad \|f \pm g\| \leq \|f\| + \|g\|, \tag{6.17}$$

From (6.16) we can define an angle θ between f and g by the relation

$$\cos\theta = \frac{\text{Re}(f, g)}{\|f\| \|g\|},$$

from which, in turn, we obtain the *law of cosines*

$$\|f - g\|^2 = \|f\|^2 + \|g\|^2 - 2\|f\| \|g\| \cos\theta. \tag{6.18}$$

Discrete Case

The transformation (6.2) can be regarded as a mapping from one N-space to another N-space. In vector notation we can write $\mathbf{f}$ for $\{f_n\}$, $n = 1, 2, \ldots, N$, in the fundamental period, and $\tilde{\mathbf{f}}$ for $\{\tilde{f}_n\}$, $n = 1, \ldots, N$, to represent its Fourier transform. To relate these two spaces further we denote by $\mathbf{g}$ and $\tilde{\mathbf{g}}$ another N-periodic sequence (in its fundamental period) and its transform. It is then easily seen (Exercise 3) that

$$(\mathbf{f}, \mathbf{g}) = (\tilde{\mathbf{f}}, \tilde{\mathbf{g}})/N. \tag{6.19}$$

A symmetrical form is obtained if (6.3) is used instead. A convenient way of doing this is through

$$\hat{\mathbf{e}}_k = \mathbf{e}_k / \sqrt{N}, \tag{6.20}$$

which are orthonormal with

$$(\hat{\mathbf{e}}_k, \hat{\mathbf{e}}_\ell) = \delta_{k\ell}. \tag{6.21}$$

From these it follows that

$$\begin{cases} \mathbf{f} = \sum_{k=1}^{N} \hat{f}_k \hat{\mathbf{e}}_k, \\ \hat{f}_k = (\hat{\mathbf{e}}_k, \mathbf{f}). \end{cases} \tag{6.22}$$

Then instead of (6.19) we have

$$(\mathbf{f}, \mathbf{g}) = (\hat{\mathbf{f}}, \hat{\mathbf{g}}) \qquad (6.23)$$

and, in particular,

$$\|\mathbf{f}\|^2 = (\mathbf{f}, \mathbf{f}) = (\hat{\mathbf{f}}, \hat{\mathbf{f}}) = \|\hat{\mathbf{f}}\|^2. \qquad (6.24)$$

Equation (6.23) is referred to as *Parseval's Relation* and states that the *angle* between two vectors is the same when viewed either in the original orthonormal basis or with respect to the new orthonormal basis (6.20). Equation (6.24) is referred to as *Plancherel's Relation* and simply states that the distances from the origin of the vector and its transform are the same in the two spaces. Finally we point out the relatively trivial relation

$$\|\mathbf{f}\|^2 \geq \sum_{n=1}^{M} \hat{f}_n^* \hat{f}_n, \quad M \leq N, \qquad (6.25)$$

known as *Bessel's Inequality*. The continuous analogue of (6.25) is quite important. In fact from it all features of the finite case carry over to the infinite dimensional function space analogue.

Continuous Case

For the case of functions we introduce the set

$$\hat{e}_k(t/T) = \frac{e_k(t/T)}{\sqrt{T}} \qquad (6.26)$$

of orthonormal functions,

$$(\hat{e}_k, \hat{e}_\ell) = \delta_{k\ell}, \qquad (6.27)$$

on the interval $(0, T)$.

For a function $F(t)$ defined on $(0, T)$ the Fourier coefficients are

$$\hat{a}_k = (\hat{e}_k, f). \qquad (6.28)$$

Consider the finite Fourier approximation to the function f,

$$f_N = \sum_{k=-N}^{N} \hat{a}_k \hat{e}_k. \qquad (6.29)$$

Then

$$0 \leq \|f - f_N\|^2 = (f - f_N, f - f_N)$$

$$= (f,f) - (f_N,f) - (f,f_N) + (f_N,f_N).$$

But

$$(f_N,f) = (f,f_N) = (f_N,f_N), \tag{6.30}$$

and we therefore have

$$\|f\|^2 \geq \|f_N\|^2 = \sum_{k=-N}^{N} \hat{a}_k^* \hat{a}_k \tag{6.31}$$

which is called *Bessel's Inequality*.

Inspection of Bessel's Inequality, (6.31), yields two remarkable results. Since $\|f_N\|^2$ is a series of positive terms, it follows that $\|f_N\|^2$ is bounded from above if $\|f\|^2$ exists. (If the latters exists, we say that f is *square integrable*.) Thus, if f is square integrable and the $\hat{a}_n$ denote the Fourier coefficients, then

$$\sum_{n=-\infty}^{\infty} \hat{a}_n^* \hat{a}_n < \infty. \tag{6.32}$$

[Since

$$\hat{a}_n = \int_0^T \hat{e}_n(t/T) f(t) dt,$$

this gives us a *cheap* proof of the Riemann–Lebesgue Lemma (see Section 5.2) where (6.32) implies $|\hat{a}_n| \to 0$ as $n \uparrow \infty$.]

An immediate question concerns the relation between the formal Fourier series

$$\sum_{n=-\infty}^{\infty} \hat{a}_n \hat{e}_n(t/T)$$

and the function f which gives rise to it. This is settled by the *Riesz–Fischer Theorem: To each sequence $\{\hat{a}_n\}$ such that $\sum_{n=-\infty}^{\infty} \hat{a}_n^* \hat{a}_n < \infty$, there corresponds a function f such that $\|f\|^2 < \infty$, and vice versa. Moreover,*

$$\lim_{N \uparrow \infty} \|f - f_N\|^2 = 0;$$

i.e., the formal Fourier series converges to the function according to the distance function of the space. Going to the limit, we get

$$\|f\|^2 = \sum_{n=-\infty}^{\infty} \hat{a}_n^* \hat{a}_n, \tag{6.33}$$

which is known as *Plancherel's Relation*. Observe that it says that the distance from the origin is the same in the two spaces.

It is left to the exercises to show that the inner product is preserved, namely that *Parseval's Relation* holds:

$$\int_0^T f^* g \, dt = (f, g) = (\hat{a}, \hat{b}) = \sum_{n=-\infty}^{\infty} \hat{a}_n^* \hat{b}_n, \qquad (6.34)$$

where $\hat{b}_n$ is a Fourier coefficient of the square integrable function g.

The proof of the Riesz–Fischer Theorem (see Courant & Hilbert) is beyond the s   is course, but in this connection there is one important piece of fine pi...   . ...ust be mentioned. For this theorem, in speaking of integration a more general integral known as the *Lebesgue integral* is meant and not the Riemann integral (see Naylor & Sell). The former is a more general and hence a more forgiving definition. It allows for functions which, from the ordinary viewpoint, seem quite pathological.

We continue by examining some geometrical properties of N-space and carry these over to function space.

Discrete Case

Suppose $u_1, u_2, \ldots, u_M$, $M \leq N$, is a set of orthonormal vectors in N-space. These establish a *subspace* or *hyperplane*, say, H_M.

Given an arbitrary vector f in N-space, we ask what vector, f_M, of H_M is *nearest* to f? A reasonable criterion for *nearest* is that f_M be minimally distant, according to (6.15), from f. Since f_M belongs to H_M, it can be written as

$$f_M = \sum_{n=1}^{M} a_n u_n. \qquad (6.35)$$

The solution is possibly evident from geometrical considerations. We simply drop a perpendicular from f to H_M and this locates f_M. Said in other terms, $f - f_M$ is perpendicular to H_M, or

$$(u_i, f - f_M) = 0 \qquad (6.36)$$

for $i = 1, \ldots, M$. This yields

$$a_n = (u_n, f). \qquad (6.37)$$

To prove this we attempt a solution of the problem by minimizing the distance $d^2 = \|f - \hat{f}_M\|^2$, where $\hat{f}_M$ is some arbitrary vector in H_M.

Then, if we use (6.35) with coefficients (6.37),

$$\|\mathbf{f} - \hat{\mathbf{f}}_M\|^2 = \|\mathbf{f} - \mathbf{f}_M + \mathbf{f}_M - \hat{\mathbf{f}}_M\|^2 = \|\mathbf{f} - \mathbf{f}_M\|^2 + \|\mathbf{f}_M - \hat{\mathbf{f}}_M\|^2$$

(why?), and clearly d^2 is minimal if $\hat{\mathbf{f}}_M = \mathbf{f}_M$. For obvious reasons $\mathbf{f}_M$ is referred to as the *projection of* $\mathbf{f}_M$ *on* H_M.

One final point concerns the *enlargment* of H_M. Suppose we add the orthonormal vector $\mathbf{u}_{M+1}$ to H_M, thus generating H_{M+1}, and again ask for the nearest vector of H_{M+1} to f. The answer is clearly

$$\mathbf{f}_{M+1} = \sum_{n=1}^{M+1} a_n \mathbf{u}_n,$$

with a_n given by (6.37). The main point to note is that all previously computed coefficients remain unchanged in this procedure!

Continuous Case

Suppose $\{u_n(t)\}$ is a set of orthonormal functions on the interval $(0, T)$. Specifically we can consider

$$u_n = \hat{e}_n(t/T) = e^{2\pi i n t/T}/\sqrt{T}.$$

A finite collection of these can be regarded as spanning a *subspace* in which an element is given by

$$\hat{f}_M(t) = \sum_{n=-M}^{M} \hat{a}_n u_n. \tag{6.38}$$

Call this space H_M. For an arbitrary T-periodic function $f(t)$, we can ask for the best approximation to $f(t)$ in the subspace H_M. The criterion for *best* is that

$$D^2 = \|f - \hat{f}_M\|^2 \tag{6.39}$$

be minimal. Denote by f_M the element of H_M having the Fourier coefficients

$$a_n = (u_n, f) \tag{6.40}$$

in the sum (6.38) such that

$$f_M = \sum_{n=-M}^{M} a_n u_n. \tag{6.41}$$

We write (6.39) in the form

$$D^2 = \|f - f_M + f_M - \hat{f}_M\|^2.$$

By simple manipulation

$$D^2 = \|f - f_M\|^2 + \|f_M - \hat{f}_M\|^2,$$

which proves that D^2 is minimized by choosing Fourier coefficients (6.40) in the sum (6.38) (since this makes the second term vanish). We will say that

$$f_M = \sum_{n=-M}^{M} a_n u_n = \sum_{n=-M}^{M} (u_n, f) u_n \qquad (6.42)$$

is the projection of f onto the subspace H_M.

If we now consider a larger subspace, say H_p, such that $H_p \supset H_M$ (i.e., H_p contains H_M), then again the nearest element of H_p is given by the projection, in the above sense, of f onto H_p. Remarkable or not, it also means that the coefficients in (6.42) do not change as the number of elements $(u_1, u_2, \ldots)$ increases. The coefficients are the Fourier coefficients, and these are unaltered as the approximating space is increased in size.

The present section presents a geometrical picture of finite and infinite dimensional spaces. The latter arises from considering functions and is an example of a *function space*. Although the discussion followed from the expansion of functions in trigonometric series, the actual formalism did not depend on this feature. In the remainder of this chapter we consider functions other than sines and cosines for the purpose of representing functions.

Exercises

1. Prove (6.13) where α_k is given by (6.8) and a_n by (6.2) and where $f_n = f(n\Delta t)$ and $N\Delta t = T$. [Hint: Use the Dirichlet kernel (5.17) and the fact that it tends to a delta "function."]

2. Assume f and g are real and prove the following:

   (a) Schwarz Inequality (6.16). [Hint: $\|f + xg\|^2 = \|f\|^2 + 2x(f, g) + x^2\|g\|^2 \; (> 0)$ has no real roots if $f \neq g$.]

   (b) Triangle Inequality (6.17). [Hint: First show

   $$\|f \mp g\|^2 = (\|f\| \mp \|g\|)^2 \pm 2(\|f\| \, \|g\| \mp (f, g))$$

   and use the Schwarz Inequality.]

(c) The law of cosines (6.18).

3. Demonstrate Parseval's Relation (6.23) and also (6.19).

4. Transform $\{\sin(2\pi nt/T)\}$, $\{\cos(2\pi nt/T)\}$ into an orthonormal system.

5. Given the real-valued function

$$f = \frac{a_0}{2} + \sum_{n=1}^{\infty}(a_n \cos n\theta + b_n \sin n\theta),$$

what is $\int_0^{2\pi} f^2(\theta)d\theta$ in terms of the Fourier coefficients?

6. Evaluate the integral in Exercise 5 for a square wave and for a triangle wave.

6.2   Orthogonal Functions

The main goal of the previous section was to show that many of the ideas which enter into the discussion of an N-dimensional vector space extend to *function spaces*. Although the latter was illustrated by periodic functions with trigonometric functions as basis elements, the formalism used was almost entirely free of the specific features of trigonometric functions. In this section we extend these ideas to other *function spaces*.

Tchebycheff Polynomials

A famous theorem of Weierstrass [see Courant & Hilbert] states that *a function continuous on the interval* $[-1, 1]$ *(or for that matter any closed interval) can be uniformly approximated to within arbitrary accuracy by polynomials.* It is important to make the distinction between such an approximation and the Taylor expansion of a function, which also approximates a function by a polynomial. For example, $\sqrt{|x|}$ does not possess a continuous derivative at the origin and therefore *does not* possess a Taylor expansion in the neighborhood of the origin; but according to Weierstrass's theorem, since it is continuous, it can be approximated arbitrarily closely by a polynomial. In another connection, the function $1/(4x^2 + 1)$ is infinitely differentiable on $[-1, 1]$ and, for example, has an infinite Taylor expansion about the origin. However, we also know from complex variable theory that this function has poles at $\pm i/2$ and therefore that the Taylor series only converges for $|x| < 1/2$. On the other hand, according to the Weierstrass Approximation Theorem there is a polynomial approximation which is *uniformly* valid on the entire interval.

We first demonstrate this theorem in a less than general framework. Consider functions which are defined on $-1 \leq x \leq 1$ and which are piecewise differentiable on the interval. If we introduce the transformation

$$x = \cos \theta, \tag{6.43}$$

then the function $f(x)$ defined on this interval becomes

$$f(x) = f(\cos \theta) = F(\theta), \tag{6.44}$$

and from (6.43) the interval $[-1, 1]$ maps into $[-\pi, 0]$. Furthermore, from (6.44) the function $F(\theta)$ can be redefined as an even function for $-\pi \leq \theta \leq \pi$. Since f is continuous it follows that $F(-\pi) = F(\pi)$. From these properties we see that the function $F(\theta)$ possesses a Fourier cosine series:

$$F(\theta) = \frac{A_0}{2} + \sum_{n=1}^{\infty} A_n \cos n\theta \tag{6.45}$$

with

$$A_n = \frac{2}{\pi} \int_0^{\pi} F(\theta) \cos n\theta d\theta. \tag{6.46}$$

If we now return to the variable x, then

$$F(\theta(x)) = f(x) = \frac{A_0}{2} + \sum_{n=1}^{\infty} A_n \cos n\theta(x) = \frac{A_0}{2} + \sum_{n=1}^{\infty} A_n T_n(x). \tag{6.47}$$

The functions T_n, known as *Tchebycheff functions*, are given by

$$T_n(x) = \cos n\theta(x) = \frac{1}{2}(e^{in\theta(x)} + e^{-in\theta(x)})$$

$$= \frac{1}{2}\{(\cos \theta + i \sin \theta)^n + (\cos \theta - i \sin \theta)^n\},$$

which from (6.43) and the relation

$$\sin \theta = -\sqrt{1 - x^2}$$

can also be written as

$$T_n(x) = \frac{1}{2}\{(x - i\sqrt{1 - x^2})^n + (x + i\sqrt{1 - x^2})^n\}. \tag{6.48}$$

Since $T_n(x)$ is real, all odd powers of $(1 - x^2)^{1/2}$ should cancel and this is easily seen to be true. Clearly $T_n(x)$ is a polynomial of degree n. It is unnecessary to enter into a discussion of the convergence of the Tchebycheff expansion (6.47) to $f(x)$ since such questions reduce to questions already considered for Fourier series. In particular, we have therefore proven Weierstrass's theorem on polynomial approximations for the class of functions which are piecewise differentiable. Moreover, it follows from our results on Fourier series that at a discontinuity x_0 of $f(x)$,

$$\lim_{N \uparrow \infty} \left\{ \frac{A_0}{2} + \sum_{n=1}^{N} A_n T_n(x_0) \right\} = \frac{f^+(x_0) + f^-(x_0)}{2}. \qquad (6.49)$$

We can also obtain the coefficients A_n directly in terms of the Tchebycheff polynomials. From (6.46)

$$A_n = \frac{2}{\pi} \int_{-\pi}^{0} F(\theta) \cos n\theta d\theta = \frac{2}{\pi} \int_{-1}^{1} f(x) T_n(x) \frac{dx}{(1 - x^2)^{1/2}}, \qquad (6.50)$$

where in the second integral we have used

$$\frac{d\theta}{dx} = \frac{-1}{\sin \theta} = \frac{1}{(1 - x^2)^{1/2}},$$

since $\sin \theta = -\sqrt{1 - x^2}$ when $-\pi < \theta < 0$. In addition we note that for $n \neq 0$

$$\delta_{nm} = \frac{1}{\pi} \int_{-\pi}^{\pi} \cos n\theta \cos m\theta \, d\theta = \frac{2}{\pi} \int_{0}^{\pi} \cos n\theta \cos m\theta \, d\theta$$

$$= \frac{2}{\pi} \int_{-1}^{1} T_n(x) T_m(x) \frac{dx}{(1 - x^2)^{1/2}}. \qquad (6.51)$$

This last form suggests that we introduce the inner product

$$(f, g) = \int_{-1}^{1} f(x) g(x) \frac{dx}{(1 - x^2)^{1/2}}. \qquad (6.52)$$

(In the present case, it is sufficient to consider a real inner product.) This differs from earlier forms in that a *weight function* $1/(1 - x^2)^{1/2}$ appears. The demonstration that (6.52) satisfies the requirements of an inner product (see (6.14)) is left to the exercises.

In terms of the inner product (6.52), the orthogonality relation (6.51) can be written as

$$(T_n, T_m) = \frac{\pi}{2} \delta_{mn}, \quad n \neq 0, \quad (T_0, T_m) = \pi \delta_{0m}, \qquad (6.53)$$

and the expansion (6.47) as

$$f(x) = \frac{(T_0, f)}{\pi} + \frac{2}{\pi} \sum_{n=1}^{\infty} T_n (T_n, f). \qquad (6.54)$$

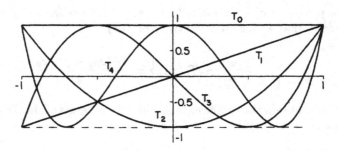

FIGURE 6.1.

Properties of the Tchebycheff Polynomials

(1) Since $T_n(\cos\theta) = \cos n\theta$ and since $\cos n\theta$ has n zeros in $(-\pi, 0)$, it follows that $T_n(x)$ has n zeros in $(-1, 1)$.

(2) These zeros are located at $(n\theta = -(2j-1)\pi/2 = n\cos^{-1} x_j)$

$$x_j = \cos\left(-\frac{\pi}{2}(2j-1)\right), \quad j = 1, \ldots, n.$$

(3) $T_n(1) = 1$ and $T_n(-1) = (-1)^n$.

(4) It follows from (6.48) that $T_n(x)$ is an odd or even function depending on the parity of the index n.

(5) The Tchebycheff polynomials, $T_n(x)$, can be constructed from (6.48). A short list of these is:

$$T_0(x) = 1, \quad T_1(x) = x, \quad T_2(x) = 2x^2 - 1,$$

$$T_3(x) = 4x^3 - 3x, \quad T_4(x) = 8x^4 - 8x^2 + 1,$$

$$T_5(x) = 16x^5 - 20x^3 + 5x.$$

(6) A sketch of several $T_n(x)$ is shown in Figure 6.1. Note that the zeros are skewed toward the endpoints and that T_n varies between $+1$ and -1.

(7) *Recurrence relations:* From the trigonometric identity

$$\cos(n+1)\theta + \cos(n-1)\theta = 2\cos\theta\cos n\theta,$$

it follows that

$$T_{n+1}(x) = 2xT_n(x) - T_{n-1}(x). \tag{6.55}$$

Thus starting with $T_0 = 1$ and $T_1 = x$ the Tchebycheff polynomials can be generated recursively from (6.55).

(8) *Generating Function:* This last relation, (6.55), is a second order difference relation (in which x is regarded as a parameter) of the sort we considered in Chapter 4 with the initial data given by

$$T_0 = 1, \quad T_1 = x.$$

Equation (6.55) can be solved by means of the Z-transform (Section 4.5). To solve (6.55), write it in the form

$$T_{n+2} = 2xT_{n+1} - T_n,$$

multiply by z^{-n}, sum from $n = 0$ to ∞, and use (4.87) to obtain

$$z^2 Z[T_n] - z^2 \sum_{n=0}^{1} T_n z^{-n} = 2x \left(z\, Z[T_n] - z \sum_{n=0}^{0} T_n z^{-n} \right) - Z[T_n]$$

or

$$(z^2 - 2xz + 1)Z[T_n] = z^2 \left(1 + \frac{x}{z} \right) - 2xz$$

and hence

$$Z[T_n] = \frac{z^2 - xz}{z^2 - 2xz + 1}. \tag{6.56}$$

Actually it is customary to consider the generating function

$$G(x; z) = \sum_{n=0}^{\infty} T_n z^n, \tag{6.57}$$

instead of $Z[T_n]$. The form of this follows directly from (6.56) and is given by

$$G(x; z) = \frac{(1/z^2) - (x/z)}{(1/z^2) - (2x/z) + 1} = \frac{1 - xz}{1 - 2xz + z^2}. \tag{6.58}$$

(9) *Integral representation:* The inversion formula for T_n now follows from (6.57) and (6.58) and Cauchy's Integral Formula. Applying this to (6.58) yields

$$T_n = \frac{1}{2\pi i} \oint_C \frac{1}{z^{n+1}} \frac{1 - xz}{1 - 2xz + z^2} dz, \tag{6.59}$$

where the path of integration C is a sufficiently small circle about the origin—with the roots of $1 - 2xz + z^2$ lying outside it. Further properties of $T_n(x)$ are given in the exercises.

(10) *Representation of monomials:* It is clear that any monomial, x^n, can be represented by a linear combination of the T_k for $k \le n$. To see this, observe that if $c \ne 0$ represents the leading coefficient of T_n, then $x^n - T_n(x)/c$ represents a polynomial of degree $n - 1$. After a finite number of such operations, we obtain the desired representation. Equivalently, if we use the expansion (6.54), it then follows that

$$x^n = \frac{(T_0, x^n)}{\pi} + \frac{2}{\pi} \sum_{k=1}^{n} (T_k, x^n) T_k. \tag{6.60}$$

Geometrical Considerations

Following the discussion of Section 6.1, we ask for what constants, B_n, in the series

$$f_N = \sum_{n=0}^{N} B_n T_n(x) \tag{6.61}$$

will the squared distance

$$(f - f_N, f - f_N) = \int_{-1}^{1} (f - f_N)^2 \frac{dx}{(1 - x)^{1/2}}$$

be minimized. Then, using arguments similar to those of Section 6.1, it follows that these are given by

$$\begin{cases} B_n = 2(T_n, f)/\pi, & n \geq 1, \\ B_0 = (T_0, f)/\pi, \end{cases} \tag{6.62}$$

i.e., the *Fourier coefficients*. (The proof is left to the exercises.)

This also proves that a partial sum of (6.54) is the best polynomial approximation with respect to the weight $(1 - x^2)^{-1/2}$ on the interval $[-1, 1]$. To see this, suppose we have a polynomial of degree N,

$$\sum_{n=0}^{N} a_n x^n,$$

with undetermined coefficients a_n. Then we may ask for what choice of a_n will

$$\int_{-1}^{1} \left(f(x) - \sum_{n=0}^{N} a_n x^n \right)^2 \frac{dx}{(1 - x^2)^{1/2}}$$

be minimal? Since each monomial x^n is representable in terms of Tchebycheff polynomials, the above sum is representable in terms of Tchebycheff polynomials and hence the minimization or the best polynomial fit to $f(x)$ is obtained through the *Fourier coefficients* (6.62).

Continuing with the geometrical viewpoint we observe that the functions $\{T_n\}$ represent orthogonal *vectors* (with respect to the weight $1/(1 - x^2)^{1/2}$). The best approximation to a function f in terms of a finite collection $\{T_n\}$ is given by its projection on the space spanned by this set. This projection is given by the (Fourier) series (6.61) with coefficients (6.62).

Examples

In view of the connection between $\cos n\theta$ and $T_n(\cos \theta)$, a number of illustrations of Tchebycheff expansions follow from our earlier deliberation. We first show that

$$|x| = \frac{2}{\pi} - \frac{4}{\pi} \sum_{n=1}^{\infty} (-)^n \frac{T_{2n}(x)}{4n^2 - 1}. \tag{6.63}$$

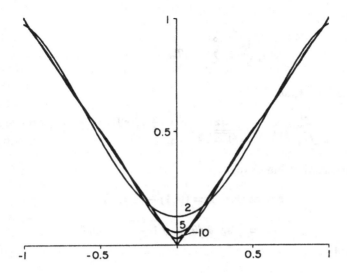

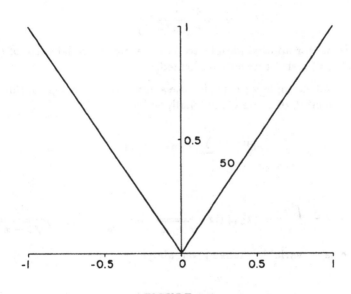

FIGURE 6.2.

Formally write

$$|x| = \frac{A_0}{2} + \sum_{n=1}^{\infty} A_n T_n(x), \quad -1 \le x \le 1.$$

It then follows that

$$A_n = \frac{2}{\pi} \int_{-1}^{1} |x| T_n(x) \frac{dx}{\sqrt{1-x^2}} = \begin{cases} \frac{4}{\pi} \int_0^1 x T_n(x) \frac{dx}{\sqrt{1-x^2}}, & n \text{ even,} \\ 0, & n \text{ odd.} \end{cases}$$

Consider the variable change

$$x = \cos\theta, \quad x = (0,1) \to \theta = (0, \pi/2),$$

$$dx = -\sin\theta\, d\theta, \quad \sqrt{1-x^2} = \sin\theta.$$

This yields

$$A_{2n} = \frac{4}{\pi} \int_0^1 x T_{2n}(x) \frac{dx}{\sqrt{1-x^2}} = \frac{4}{\pi} \int_0^{\pi/2} \cos\theta \cos 2n\theta\, d\theta$$

$$= -\frac{4}{\pi} \frac{(-)^n}{(4n^2 - 1)}.$$

See Figure 6.2 for approximations to $|x|$ in terms of partial sums of (6.63). The number of terms carried is indicated.

As a second example we consider the signum function, sgn x. Since this is an odd function of x, we can formally write

$$\text{sgn } x = \sum_{n=0}^{\infty} A_{2n+1} T_{2n+1}(x)$$

with

$$A_{2n+1} = \frac{2}{\pi} \int_{-1}^{1} \text{sgn } x\, T_{2n+1}(x) \frac{dx}{\sqrt{1-x^2}} = \frac{4}{\pi} \int_0^1 T_{2n+1}(x) \frac{dx}{\sqrt{1-x^2}}.$$

Under the same variable change given above, we have

$$A_{2n+1} = \frac{4}{\pi} \int_0^{\pi/2} \cos(2n+1)\theta\, d\theta = \frac{4}{\pi} \frac{(-)^n}{2n+1}.$$

This therefore leads to

$$\text{sgn } x = \frac{4}{\pi} \sum_{n=0}^{\infty} (-)^n \frac{T_{2n+1}(x)}{2n+1}. \tag{6.64}$$

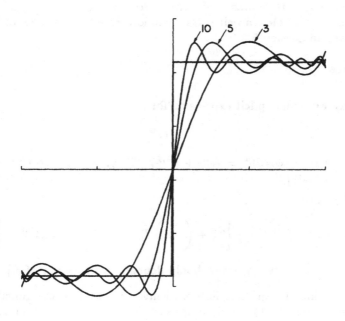

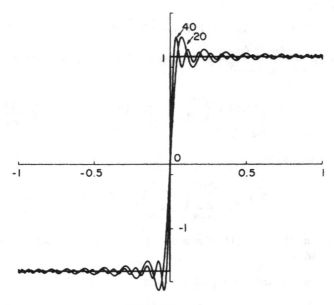

FIGURE 6.3.

Approximations to sgn x in terms of partial sums of (6.64) are shown in Figure 6.3, where the number of terms carried is indicated.

The above examples as well as others follow from the examples of Fourier series given in Section 5.5.

Exercises

1. (a) Find an explicit expression for

    $$(x^n, T_k).$$

    [Hint: $(\cos \theta)^n = (e^{i\theta} + e^{-i\theta})^n/2^n$ yields the explicit form of (6.60).]

    (b) Show

    $$x^n = \frac{1}{2^{n-1}} \left[T_n + \binom{n}{1} T_{n-2} + \binom{n}{2} T_{n-4} + \cdots \right].$$

    [Hint: The last term should depend on the parity of n.]

2. In certain instances it is more convenient to consider functions in the interval $(0,1)$. To accomplish this, set $x = 2y - 1$ in $T_n(x) = T_n(2y-1) = T_n^*(y)$, $0 < y < 1$. Show the following properties of $\{T_n^*(y)\}$:

    (a) $T_{n+1}^*(y) = 2(2y-1)T_n^*(y) - T_{n-1}^*(y)$.

    (b) $\displaystyle\int_0^1 \frac{T_n^*(y)T_m^*(y)dy}{\sqrt{y(1-y)}} = 0$, $m \neq n$. What does the integral equal if $m = n$?

    (c) $y^k = \dfrac{2}{2^{2k}} \left[T_k^*(y) + \binom{2k}{1} T_{k-1}^*(y) + \cdots + \dfrac{1}{2}\binom{2n}{k} \right]$. [Hint: See Exercise 1.]

3. Demonstrate that

    $$T_n(x) = \sum_{m=0}^{[n/2]} \binom{n}{2m} x^{n-2m}(x^2 - 1)^m,$$

    where $[n/2] = (n-1)/2$ if n is odd. [Hint: Use the binomial expansion on (6.48).]

4. Demonstrate the *Rodrigues formula*

    $$T_n(x) = \frac{(-)^n(1-x^2)^{1/2}}{2^{n+2}(n-1/2)(n-3/2)\cdots(1/2)} \frac{d^n}{dx^n} \left(\frac{1}{(1-x^2)^{1/2}} \right)^n.$$

5. Show the following:

(a) $T_m(x)T_n(x) = \frac{1}{2}[T_{m+n}(x) + T_{m-n}(x)]$. [Hint: This follows from trigonometry.]

(b) $T_m(T_n(x)) = T_{mn}(x)$. [Hint: Use (6.48).]

6. Demonstrate the *Dirichlet* relation

$$\frac{1}{2} + T_1(x) + \cdots + \frac{1}{2}T_n(x) = \frac{T_n'(x)(1+x)}{2n}.$$

7. Demonstrate that (6.52) satisfies the requirements of an inner product.

8. Furnish the details in the derivation of the Tchebycheff expansions (6.63) and (6.64).

9. Prove the minimization property of the coefficients (6.62).

10. Tchebycheff polynomials of the *second kind* are defined by

$$U_n(x) = \frac{\sin(n+1)\theta}{\sin\theta}, \quad x = \cos\theta, \quad n = 0, 1, \ldots.$$

Demonstrate

(a) $U_n(x)$ is a polynomial of degree n,

(b) $\displaystyle\int_{-1}^{1} U_m(x)U_n(x)\sqrt{1-x^2}\,dx = \frac{\pi}{2}\delta_{mn}$,

(c) $U_n(x) = 2xU_{n-1}(x) - U_{n-2}(x)$.

Legendre Polynomials

As we saw in the above, appearance of the weight function $1/(1-x^2)^{1/2}$ arose for the special reason that we transformed from the trigonometric case. It is also natural to discuss the approximation of functions by polynomials with respect to a *unit weight*. One way to pose this problem is to ask for what values of the constants a_n in

$$f_N = \sum_{n=0}^{N} a_n x^n \tag{6.65}$$

will

$$D = \int_{-1}^{1} \left(f(x) - \sum_{n=0}^{N} a_n x^n \right)^2 dx \tag{6.66}$$

be a minimum? This is known as the *method of least squares*.

For t'ʰ purpose it is useful to introduce the real inner product

$$(f, g) = \int_{-1}^{1} f(x)g(x)\,dx, \qquad (6.67)$$

which evidently satisfies the rules laid out in (6.14). A direct (but mindless) approach is then to solve the system of $N + 1$ equations

$$\frac{\partial D}{\partial a_k} = 0, \quad k = 0, 1, \ldots, N. \qquad (6.68)$$

This constitutes a necessary condition for D to have a minimum and, after a little manipulation, leads to the matrix equation

$$\sum_{n=0}^{N} (x^k, x^n)a_n = (x^k, f), \quad k = 0, \ldots, N. \qquad (6.69)$$

There are several shortcomings to such an approach. Most notable is the following: (1) it involves a matrix inversion; and (2) if N is changed (say, made larger to improve the calculation), all previously computed $\{a_n\}$ must be recomputed. The Fourier and Tchebycheff developments avoided both these shortcomings through the use of orthogonal functions. To accomplish a similar procedure, we must therefore construct an appropriate system of orthogonal polynomials. We start with a recollection of the analogous procedure in an N-dimensional vector space.

Given a collection of linearly independent vectors $\{\mathbf{u}_n\}$, $n = 1, \ldots, M$ ($M \leq N$), in N-space, we wish to construct a set $\{\mathbf{v}_n\}$ of orthogonal vectors from these. To accomplish this, set

$$\begin{cases} \mathbf{v}_1 = \mathbf{u}_1, \\ \mathbf{v}_k = \mathbf{u}_k - \sum_{n=1}^{k-1} a_n \mathbf{v}_n, \quad k = 2, 3, \ldots, M, \end{cases} \qquad (6.70)$$

which, since $(\mathbf{v}_k, \mathbf{v}_n) = 0$ for $k > n$, gives

$$a_n = \frac{(\mathbf{v}_n, \mathbf{u}_k)}{(\mathbf{v}_n, \mathbf{v}_n)}, \quad n = 1, 2, \ldots, k - 1. \qquad (6.71)$$

It should be clear that in a successive way this determines an orthogonal set of vectors.

Gram–Schmidt Procedure

This simple recipe, known as the *Gram–Schmidt* orthogonalization procedure, is adaptable virtually without change to the case of function spaces. For a given set of linearly independent functions $\{f_n(x)\}$ and an inner product (f, g), we can construct an orthogonal set $\{v_n(x)\}$ as follows:

$$\begin{cases} v_1 = f_1, \\ v_k = f_k - \sum_{n=1}^{k-1} a_n v_n, \end{cases} \qquad (6.72)$$

which, from the orthogonality of the $\{v_n\}$, gives

$$a_n = (v_n, f_k)/(v_n, v_n), \quad n = 1, 2, \ldots, k - 1. \tag{6.73}$$

The $\{v_n\}$ can in turn be made into an orthonormal set $\{w_n\}$ by setting

$$w_n = \frac{v_n}{\|v_n\|}.$$

Neither the length of the interval, the details of $\{f_n\}$, nor the nature of the inner product entered into the construction of (6.72). (Note: Since the number of independent variables is unspecified as well, the Gram–Schmidt procedure also applies in higher dimensional function spaces.)

For the set of linear independent functions, we take the monomials $\{x^n\}$, and denote the orthogonal set generated from (6.72) by $\{p_n\}$. It then follows that

$$p_0 = 1, \quad p_1 = x - (1, x)/(1, 1) = x,$$

where $(1, x) = 0$ (the indices of (6.72) have been *downshifted* by one). In fact, it is clear from the Gram–Schmidt construction that the proposed set $\{p_n\}$ is composed of alternately even and odd polynomials. The next member is

$$p_2 = x^2 - (1, x^2)/(1, 1) = x^2 - \frac{1}{3}$$

since

$$(1, x^2) = \int_{-1}^{1} x^2 dx = \frac{2}{3}, \quad (1, 1) = \int_{-1}^{1} 1 \, dx = 2.$$

If we proceed further, we find

$$p_3 = x^3 - \frac{3x}{5}, \quad p_4 = x^4 - \frac{6}{7}x^2 + \frac{3}{35},$$

$$p_5 = x^5 - \frac{10}{9}x^3 + \frac{5}{21}x.$$

Rather than continue in this way, we assert that the above polynomials are given by the *Rodrigues formula*

$$p_n(x) = \frac{n!}{(2n)!} \frac{d^n}{dx^n}(x^2 - 1)^n. \tag{6.74}$$

Clearly p_0 and p_1 agree with the above. The assertion then follows from a few simple observations. First we note that $(x^2 - 1)^n$ is an even polynomial of degree $2n$. Therefore, on differentiating n times, we obtain a polynomial of degree n which is odd or even depending on the parity of n. The factor $n!/(2n)!$ in (6.74) was introduced so that the coefficient of x^n is unity. All this is in agreement with the Gram–Schmidt construction given above. Then, since the Gram–Schmidt procedure leads to a unique construction, we now only have to show that the $p_n(x)$ given by (6.74) are orthogonal on the interval $(-1, 1)$. To see this consider

$$\int_{-1}^{1} p_n(x)p_m(x)dx,$$

with p_n and p_m defined by (6.74). To be specific we take $n > m$. If (6.74) is substituted, then after repeated parts integration we find

$$\int_{-1}^{1} p_n(x)p_m(x)dx = \frac{n!m!}{(2n)!(2m)!} \int_{-1}^{1} (x^2 - 1)^n (-)^n \frac{d^{n+m}}{dx^{n+m}}(x^2 - 1)^m dx, \tag{6.75}$$

which is zero since $n + m > 2m$.

In actual practice a normalization other than that given in (6.74) is used. The standard form, known as *Legendre polynomials*, is given by the Rodrigues formula

$$P_n(x) = \frac{1}{2^n n!} \frac{d^n}{dx^n}(x^2 - 1)^n. \tag{6.76}$$

The first few Legendre polynomials are given by

$$P_0 = 1, \quad P_1 = x, \quad P_2 = \frac{3}{2}x^2 - \frac{1}{2}, \quad P_3 = \frac{5}{2}x^3 - \frac{3}{2}x,$$

$$P_4 = \frac{35}{8}x^4 - \frac{15}{4}x^2 + \frac{3}{8}.$$

As we will later verify,

$$\int_{-1}^{1} P_n(x)P_m(x)dx = \frac{2}{2n + 1}\delta_{nm}. \tag{6.77}$$

(See also Exercise 2 at the end of Section 6.3.)

The form (6.76) renormalizes (6.74) so that $P_n(1) = 1$ (and hence $P_n(-1) = (-1)^n$)—see Exercise 5 at the end of Section 6.3. A sketch of several Legendre polynomials is given in Figure 6.4. Note that in contrast to the Tchebycheff polynomials, the Legendre polynomials decrease in amplitude as we approach the origin. We mention (to be demonstrated later) that P_n has exactly n zeros in $(-1, 1)$ (as does $T_n(x)$).

To obtain the recurrence relation for Legendre polynomials, we start with an identity from elementary calculus:

$$x\frac{d^n}{dx^n}F = \frac{d^n}{dx^n}xF - n\frac{d^{n-1}}{dx^{n-1}}F, \tag{6.78}$$

where F is an arbitrary sufficiently smooth function. If we take $F = (x^2 - 1)^n/2^n n!$ in (6.78), then from (6.76)

$$xP_n = \frac{1}{2^n n!}\frac{d^n}{dx^n}x(x^2 - 1)^n - \frac{n}{2^n n!}\frac{d^{n-1}}{dx^{n-1}}(x^2 - 1)^n.$$

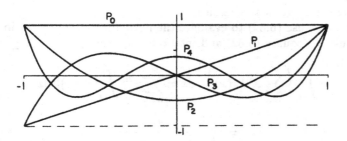

FIGURE 6.4.

But

$$\frac{1}{2^n n!}\frac{d^n}{dx^n}x(x^2-1)^n = \frac{1}{2^n n!}\frac{1}{2(n+1)}\frac{d^{n+1}}{dx^{n+1}}(x^2-1)^{n+1} = P_{n+1}(x),$$

and hence

$$xP_n(x) = P_{n+1} - \frac{n}{2^n n!}\frac{d^{n-1}}{dx^{n-1}}(x^2-1)^n. \tag{6.79}$$

A similar relation is obtained by considering

$$xP_n = \frac{x}{2^{n-1}(n-1)!}\frac{d^{n-1}}{dx^{n-1}}x(x^2-1)^{n-1},$$

which is gotten by carrying out one differentiation on the right-hand side of (6.76). If we apply (6.78) to the right-hand side of this expression, we obtain

$$xP_n = \frac{1}{2^{n-1}(n-1)!}\frac{d^{n-1}}{dx^{n-1}}x^2(x^2-1)^{n-1} - \frac{(n-1)}{2^n n!}\frac{d^{n-1}}{dx^{n-1}}(x^2-1)^n$$

$$= \frac{1}{2^{n-1}(n-1)!}\frac{d^{n-1}}{dx^{n-1}}(x^2-1)^n + \frac{1}{2^{n-1}(n-1)!}\frac{d^{n-1}}{dx^{n-1}}(x^2-1)^{n-1}$$

$$- \frac{(n-1)}{2^n n!}\frac{d^{n-1}}{dx^{n-1}}(x^2-1)^n = \frac{n+1}{2^n n!}\frac{d^{n-1}}{dx^{n-1}}(x^2-1)^n + P_{n-1}. \tag{6.80}$$

If (6.79) is divided by n, (6.80) by $(n+1)$, and the resulting two equations are added, then

$$(2n+1)xP_n = (n+1)P_{n+1} + nP_{n-1}, \tag{6.81}$$

which is the desired recurrence relation.

With the use of the Z-transform, it can be shown that the generating function is given by

$$\sum_{n=0}^{\infty} z^n P_n(x) = \frac{1}{(1-2xz+z^2)^{1/2}}. \tag{6.82}$$

The details are left as an exercise.

We can now use (6.82) to evaluate the normalization constant in (6.77). To see this, we square (6.82) and integrate over the interval $-1 \leq x \leq 1$:

$$\int_{-1}^{1} \left(\sum_{n=0}^{\infty} z^n P_n(x) \right)^2 dx = \sum_{n=0}^{\infty} z^{2n} \int_{-1}^{1} [P_n(x)]^2 dx$$

$$= \int_{-1}^{1} \frac{dx}{1 - 2xz + z^2}.$$

The second expression follows directly from the orthogonality of the Legendre polynomials. Integration of the final expression is elementary, and we have

$$\int_{-1}^{1} \frac{dx}{1 - 2xz + z^2} = -\frac{1}{2z} \int_{-1}^{1} \frac{d}{dx} \ln(1 - 2xz + z^2) dx = \frac{1}{z} \ln \left(\frac{1+z}{1-z} \right).$$

Thus

$$\sum_{n=0}^{\infty} z^{2n} \int_{-1}^{1} [P_n(x)]^2 dx = \frac{1}{z} \ln \left(\frac{1+z}{z-1} \right) = 2 \sum_{n=0}^{\infty} \frac{z^{2n}}{2n+1}, \qquad (6.83)$$

where a straightforward power series expansion has been used. The required relation, (6.77), is now immediate.

Let us now return to the problem posed at the outset of our discussion; viz., what is the best polynomial fit to a function if the weight function is unity? Past experience suggests that we rewrite f_N, (6.65), in the form

$$f_N = \sum_{n=0}^{N} \alpha_n P_n(x). \qquad (6.84)$$

Equation (6.66) is replace by

$$D = \int_{-1}^{1} \left[f(x) - \sum_{n=0}^{N} \alpha_n P_n(x) \right]^2 dx, \qquad (6.85)$$

and the minimization condition (6.68) is replaced by

$$\frac{\partial D}{\partial \alpha_n} = 0, \quad n = 0, 1, \ldots, N,$$

which is easily seen to lead to

$$\alpha_n = \frac{(P_n, f)}{(P_n, P_n)} = \frac{2n+1}{2} (P_n, f).$$

As before, we will refer to

$$\beta_n = (P_n, f) \qquad (6.86)$$

as a *Fourier coefficient* and to the *Fourier series*

$$\tilde{f}_N = \sum_{n=0}^{N} \frac{2n+1}{2} \beta_n P_n \qquad (6.87)$$

as the projection of f on the subspace spanned by $\{P_n\}$, $n = 0, 1, \ldots, N$. If we use the inner product (6.67) to define a norm, we have the *Pythagorean Theorem*

$$\|f - f_N\|^2 = \|f - \tilde{f}_N\|^2 + \|f_N - \tilde{f}_N\|^2 \qquad (6.88)$$

for arbitrary f_N, (6.84). This demonstrates that $\tilde{f}_N$, (6.87), is the true minimum.

Example. As a concrete example of expansion in Legendre polynomials, we consider $|x|$, i.e.,

$$|x| = \sum_{n=0}^{\infty} a_{2n} P_{2n}(x). \qquad (6.89)$$

Only even indices appear since $|x|$ is even, and

$$a_{2n} = \frac{4n+1}{2}(P_{2n}, |x|) = \frac{4n+1}{2} \int_{-1}^{1} |x| P_{2n} dx = (4n+1) \int_{0}^{1} x P_{2n} dx$$

$$= (4n+1) \int_{0}^{1} x \frac{d^{2n}}{dx^{2n}} \frac{(x^2-1)^{2n}}{2^{2n}(2n)!} dx$$

$$= \frac{(4n+1)}{2^{2n}(2n)!} \int_{0}^{1} \frac{d}{dx} \left[x \frac{d^{2n-1}}{dx^{2n-1}}(x^2-1)^{2n} \right] dx$$

$$- \frac{(4n+1)}{2^{2n}(2n)!} \int_{0}^{1} \frac{d^{2n-1}}{dx^{2n-1}}(x^2-1)^{2n} dx.$$

The first integral vanishes at the lower endpoint because of the factor x and at the upper endpoint because $(x^2-1)^{2n}$ has a zero of order $2n$ at $x = 1$. The second integrand is a perfect differential which vanishes, for the same reason just given, at $x = 1$. Hence

$$a_{2n} = \frac{4n+1}{2^{2n}(2n)!} \left[\frac{d^{2n-2}}{dx^{2n-2}}(x^2-1)^{2n} \right]_{x=0}.$$

From the Binomial Theorem

$$(x^2-1)^{2n} = \sum_{k=0}^{2n} (x^2)^{2n-k}(-)^k \begin{pmatrix} 2n \\ k \end{pmatrix}.$$

The only term of the sum which contributes to a_{2n} is that for which $k = n+1$, so

$$a_{2n} = \frac{(4n+1)(-)^{n+1}}{2^{2n}} \frac{(2n-2)!}{(n-1)!(n+1)!}, \quad n \geq 1, \quad a_0 = \frac{1}{2}. \qquad (6.90)$$

6.3 Comparison of Tchebycheff and Legendre Expansions

We now have two different methods for the polynomial approximation of functions on the interval $(-1, 1)$ (and, by renormalization, on an arbitrary interval). A comparison is therefore in order.

We first consider $|x|$ as represented by the two different polynomial systems. From (6.63) we have

$$|x| = \frac{2}{\pi} - \frac{4}{\pi}\left(-\frac{T_2(x)}{3} + \frac{T_4(x)}{15} - \frac{T_6(x)}{35} + \cdots\right)$$

$$= \frac{4}{\pi}\left(\frac{1}{10} + \frac{6x^2}{5} - \frac{8x^4}{15} + O\left(\frac{T_6}{35}\right)\right)$$

$$= .127 + 1.53x^2 - .68x^4 + O\left(\frac{T_6}{35}\right), \tag{6.91}$$

while from (6.89) and (6.90),

$$|x| = \frac{P_0(x)}{2} + \frac{5}{8}P_2(x) - \frac{3}{16}P_4(x) + \frac{13}{128}P_6(x) + \cdots$$

$$= \frac{P_0(x)}{2} + \frac{5}{8}P_2(x) - \frac{3}{16}P_4(x) + \frac{13}{128}P_6(x) + \cdots$$

$$= \frac{15}{64}\left(\frac{1}{2} + 7x^2 - \frac{7}{2}x^4\right) + O\left(\frac{13}{128}P_6\right)$$

$$= .117 + 1.64x^2 - .82x^4 + O\left(\frac{13P_6}{128}\right). \tag{6.92}$$

In both (6.91) and (6.92) we follow the time-honored practice of estimating the error by the first neglected term.

If we denote by $|x|_N$ the Nth degree polynomial approximation to $|x|$, we find the following numerical values at the origin and at $x = 1$:

	$N = 0$	1	2	3
$|x|_N(0) =$	.6366	.2122	.1273	.091
$|x|_N(1) =$	.6366	1.061	.9762	1.0125

Tchebycheff Approximation

	$N = 0$	1	2	3
$|x|_N(0) =$	.5	.1875	.1171	.0854
$|x|_N(1) =$	.5	1.125	.9375	1.0391

Legendre Approximation

Four—Term Tchebycheff Expansion.

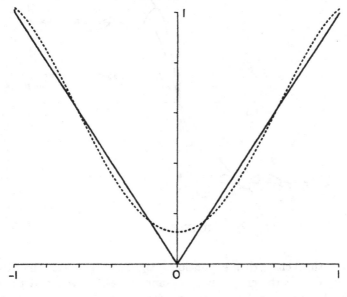

FIGURE 6.5.

Four—Term Legendre Expansion.

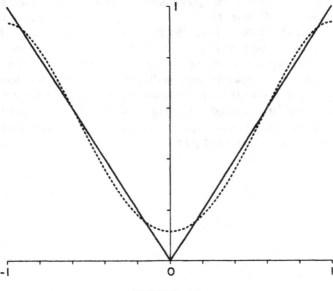

FIGURE 6.6.

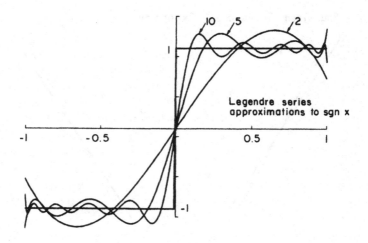

FIGURE 6.7.

We see that the Tchebycheff expansion is significantly better than the Legendre expansion at the endpoint $x = 1$ and somewhat worse at the origin. This is further borne out by Figures 6.5 and 6.6. As another comparison we consider the signum function sgn x. Figures 6.7 and 6.8 again illustrate the fact that the Tchebycheff expansion shows less error at the endpoints. Elsewhere the errors are comparable.

All in all, we see that Tchebycheff approximations are better at $x = 1$ but worse at $x = 0$ than the corresponding Legendre approximations. To some extent this follows from (6.52). Since it has the weight function $1/(1-x^2)^{1/2}$, the Tchebycheff approximation strives to reduce errors in the neighborhood of $|x| \approx 1$. Also, we we have noted earlier, $T_n(x)$ varies between ± 1 with the maxima and minima distributed throughout the interval. Thus, as is implicit from the error estimate in (6.91), the error committed is more or less uniform throughout the interval. The Legendre expansion is based on minimizing the mean square error, (6.66). Thus it can tolerate a relatively large error at isolated points.

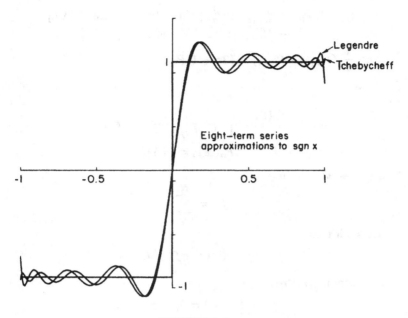

FIGURE 6.8.

Exercises

1. Verify the forms given for p_3, p_4, p_5 in Section 6.2 by the Gram–Schmidt process.

2. Show (6.77), namely

$$\int_{-1}^{1} P_n(x)P_m(x)dx = \frac{2}{2n+1}\delta_{nm}.$$

[Hint: First show

$$\int_{-1}^{1} P_n^2 dx = \frac{(2n)!}{2^{4n}(n!)^2} \int_{-1}^{1} (1-x^2)^n dx,$$

then make the substitution $x = \cos\theta$, and finally apply repeated parts integration.]

3. Prove

$$x\frac{d^n}{dx^n} = \frac{d^k}{dx^k}x\frac{d^{n-k}}{dx^{n-k}} - k\frac{d^{n-1}}{dx^{n-1}},$$

which is used in proving the recurrence relation (6.78).

4. Show the following:

(a) $(n+1)P_n = P'_{n+1} - xP'_n$. [Hint: Differentiate (6.79).]

(b) $xP'_n = nP_n + P'_{n-1}$. [Hint: Differentiate (6.80).]

(c) $\dfrac{d}{dx}(P_{n+1} - P_{n-1}) = (2n+1)P_n$.

5. Prove

$$P_n(0) = \begin{cases} 0, & n \text{ odd}, \\ \dfrac{(-)^{n/2}n!}{2^n((n/2)!)^2}, & n \text{ even}; \end{cases}$$

$$P_n(\pm 1) = (\pm 1)^n.$$

6. Suppose the a_n in

$$f_N = \sum_{n=0}^{N} a_n x^n$$

are su.h that

$$\int_{-1}^{1} (f - f_N)^2 dx$$

is a minimum. Prove

$$(P_m, f_N) = 0, \quad m > N.$$

7. Show

$$P_n(x) = \frac{1}{2^n} \sum_{k=0}^{[n/2]} \frac{(-)^k (2n-2k)!}{k!(n-k)!(n-2k)!} x^{n-2k}.$$

8. Demonstrate (6.82).

9. Find the N-term error of the Tchebycheff and Legendre expansions of $|x|$ at $x = 0$ and $x = 1$. [Hint: Use Exercise 5 above in (6.89) and Exercise 3 of Section 6.2 in (6.63).]

10. Find the Legendre polynomial expansion for sgn x.

11. The minimization problem posed by (6.66) is known as the method of least squares. Show that (6.69) results from this approach.

12. Use the method of least squares to fit an arbitrary function on $(-1, 1)$ by a cubic polynomial. Prove that this least-squares cubic polynomial fit is identical to a third order Legendre polynomial fit.

13. Use the Gram–Schmidt procedure to obtain an orthogonal basis for

(a) the set $\{x^n\}$, $n = 0, 1, 2, \ldots$, the interval $-1 < x < 1$, and the weight function $|x|^{1/2}$;

(b) the set $\{e^{-nx}\}$, $n = 1, 2, \ldots$, the interval $0 < x < \infty$, and the unit weight function;

(c) the set $\{x^n\}$, $n = 0, 1, 2, \ldots$, the interval $-\infty < x < \infty$, and the weight function $e^{-|x|}$.

6.4 Orthogonal Functions—Continued

Thus far we have encountered three systems of orthogonal functions: trignometric functions, Tchebycheff polynomials, and Legendre polynomials. As may be imagined, there are a limitless number of such systems. A certain small number of these are referred to as the *classical special functions*. In this section we consider two more such *standard* orthogonal systems.

Hermite Polynomials

The Hermite polynomials, denoted by $He_n(x)$, are defined on the real line $(-\infty, \infty)$ and can be generated from the monomials $\{x^n\}$ and the inner product

$$(f, g) = \int_{-\infty}^{\infty} \frac{e^{-x^2/2}}{\sqrt{2\pi}} f(x)g(x)dx = \int_{-\infty}^{\infty} \omega(x)f(x)g(x)dx, \qquad (6.93)$$

where again only a real inner product is necessary. The weight function appearing in (6.93) is the Gaussian,

$$\omega(x) = \frac{e^{-x^2/2}}{\sqrt{2\pi}}. \qquad (6.94)$$

As can be shown (Exercise 1 at the end of this subsection)

$$\int_{-\infty}^{\infty} \omega(x)dx = 1. \qquad (6.95)$$

Then it can be seen that from mathematical induction (Exercise 1)

$$\int_{-\infty}^{\infty} x^{2n}\omega(x)dx = \frac{(2n)!}{n!2^n} \qquad (6.96)$$

and from parity considerations

$$\int_{-\infty}^{\infty} x^{2n+1}\omega(x)dx = 0.$$

The orthogonal set of Hermite polynomials, $He_n(x)$, is easily generated from the Gram–Schmidt process and from the use of (6.96).

Rather than follow this approach we assert that the Hermite polynomials, $He_n(x)$, are given by the Rodrigues formula

$$He_n(x) = \frac{(-)^n}{\omega(x)} \frac{d^n}{dx^n} \omega(x). \qquad (6.97)$$

From (6.97) we have that

$$He_0(x) = 1, \quad He_1(x) = x, \quad He_2(x) = x^2 - 1, \quad He_3(x) = x^3 - 3x,$$

$$He_4(x) = x^4 - 6x^2 + 3, \quad He_5(x) = x^5 - 10x^3 + 15x. \qquad (6.98)$$

A number of properties derive directly from (6.97); e.g., it is seen immediately that the polynomials, He_n, are either odd or even depending on the index n. Therefore

$$He_{2n+1}(0) = 0$$

and, after a little work,

$$He_{2n}(0) = \frac{(-)^n (2n)!}{n! 2^n}. \qquad (6.99)$$

Also

$$He_n(x) = (-)^n \frac{1}{\omega} \frac{d}{dx} \frac{d^{n-1}}{dx^{n-1}} \omega$$

$$= \frac{(-)^1}{\omega(x)} \frac{d}{dx} [\omega(x) He_{n-1}] = -He'_{n-1} + x He_{n-1}. \qquad (6.100)$$

If just one differentiation is carried out in (6.97), we get

$$He_n(x) = (-)^{n+1} \frac{1}{\omega} \frac{d^{n-1}}{dx^{n-1}} x\omega,$$

which from (6.78) gives the recurrence relation

$$He_n(x) = (-)^{n+1} \frac{1}{\omega} \left[x \frac{d^{n-1}}{dx^{n-1}} \omega + (n-1) \frac{d^{n-2}}{dx^{n-2}} \omega \right]$$

$$= x He_{n-1}(x) - (n-1) H_{n-2}(x). \qquad (6.101)$$

If we subtract this from (6.100), we obtain

$$\frac{d}{dx} He_n(x) = n He_{n-1}(x). \qquad (6.102)$$

To show orthogonality, consider

$$\int_{-\infty}^{\infty} \omega He_n He_m \, dx = \int_{-\infty}^{\infty} He_n(x)(-)^m \frac{d^m}{dx^m} \omega(x) \, dx,$$

where we have substituted the expression in (6.97) for He_m. To be specific we take $m \geq n$. If we parts integrate m times, we obtain

$$\int_{-\infty}^{\infty} \omega He_n(x) He_m(x) \, dx = \int_{-\infty}^{\infty} \omega \frac{d^m}{dx^m} He_n(x) \, dx.$$

Since $He_n(x)$ is of degree n, it thus vanishes unless $m = n$. Also, from (6.102),

$$\frac{d^n}{dx^n} He_n(x) = n!$$

since $He_0 = 1$. We have therefore shown

$$\int_{-\infty}^{\infty} \omega He_n He_m \, dx = n!\delta_{mn}. \qquad (6.103)$$

It is tempting to apply the Z-transform to the recurrence relation (6.101) in order to obtain a generating function; but $He_n \approx x^n$ for $|x| \uparrow \infty$, and from (6.99) we know that at $x = 0$ neither $\sum_{n=0}^{\infty} He_n z^n$ nor $\sum_{n=0}^{\infty} He_n z^{-n}$ exists. Therefore such an approach is doomed from the start. Instead we can show (Exercise 4) that

$$e^{zx-(z^2/2)} = \sum_{n=0}^{\infty} \frac{He_n(x)z^n}{n!} = G(x;z). \qquad (6.104)$$

Note that $n!$ in the denominator gives the required convergence.

It follows from these deliberations that the functions

$$\{he_n(x)\} = \{\omega^{1/2} He_n(x)\} \qquad (6.105)$$

represent an orthogonal set of functions on the real line $(-\infty, \infty)$, i.e.,

$$\int_{-\infty}^{\infty} he_n(x) he_m(x) \, dx = n!\delta_{mn}, \qquad (6.106)$$

which is just a reformatted version of (6.103). It can be shown (but is beyond the scope of this text) that the set of functions (6.105) is complete for the space of square integrable functions (but see Section 6.5),

$$\int_{-\infty}^{\infty} f^2(x) \, dx < \infty. \qquad (6.107)$$

Thus, for any function satisfying (6.107), there exist coefficients $\{a_n\}$ such that

$$f_N = \sum_{n=0}^{N} a_n he_n(x) = \omega^{1/2} \sum_{n=0}^{N} a_n He_n(x) \qquad (6.108)$$

converges to f in the sense that

$$\lim_{N \uparrow \infty} \int_{-\infty}^{\infty} [f(x) - f_N(x)]^2 dx = 0. \qquad (6.109)$$

In this event, as in previous cases we will write simply

$$f(x) = \omega^{1/2} \sum_{n=0}^{\infty} a_n He_n(x). \qquad (6.110)$$

The determination of the constants a_n is straightforward. If we multiply (6.110) by $\omega^{1/2} He_n(x)$ and integrate, then

$$a_n = \frac{1}{n!} \int_{-\infty}^{\infty} \omega^{1/2} f(x) He_n(x) \, dx \qquad (6.111)$$

or

$$f(x) = \omega^{1/2} \sum_{n=0}^{\infty} \frac{He_n(x)}{n!} \int_{-\infty}^{\infty} \omega^{1/2} f \, He_n \, dx. \tag{6.112}$$

It is important to note that other expansions in hermite polynomials are also possible and sometimes necessary. For example, if f is square integrable, $\omega^{1/2} f$ is also square integrable and the above development applies. When this holds we can write

$$f = \sum_{n=0}^{\infty} \frac{He_n(x)}{n!} \int_{-\infty}^{\infty} \omega(x) f(x) He_n(x) dx. \tag{6.113}$$

Of more practical importance is the case when f is not square integrable but $\omega^{1/2} f$ is—in which case (6.113) is valid. At the other extreme suppose f vanishes quickly enough as $|x| \uparrow \infty$ so that $f/\omega^{1/2}$ is square integrable. Then the above considerations imply a development in the form

$$f = \omega \sum_{n=0}^{\infty} \frac{He_n(x)}{n!} \int_{-\infty}^{\infty} He_n(x) f(x) dx. \tag{6.114}$$

Example. The function e^x is certainly not square integrable and an expansion in the form (6.112) would be unjustified. To overcome this, consider $\omega^{1/2} e^x$, which is certainly square integrable so that we can write

$$e^x = \sum_{n=0}^{\infty} \frac{He_n(x)}{n!} \int_{-\infty}^{\infty} \omega e^x He_n(x) dx.$$

In fact, comparison with (6.104) shows that

$$e^x = \sqrt{e} \sum_{n=0}^{\infty} \frac{He_n(x)}{n!}. \tag{6.115}$$

Other Hermite polynomial expansions are indicated graphically in Figures 6.9 and 6.10.

Exercises

1. Verify (6.95) and (6.96). [Hint: $\int_{-\infty}^{\infty} x^{2n+2} \omega \, dx = - \int_{-\infty}^{\infty} x^{2n+1} \frac{d}{dx} \omega \, dx$.]

2. Obtain the first four Hermite polynomials through the use of the Gram–Schmidt procedure.

3. Verify (6.99).

4. Verify (6.104). [Hint: Form the sum $G(x; z)$ of (6.104) in (6.101). This should yield a first order ordinary differential equation for G. To obtain the initial data for G use (6.99).]

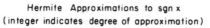

Hermite Approximations to sgn x
(integer indicates degree of approximation)

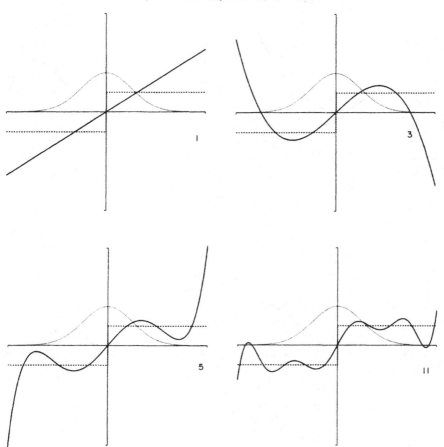

FIGURE 6.9.

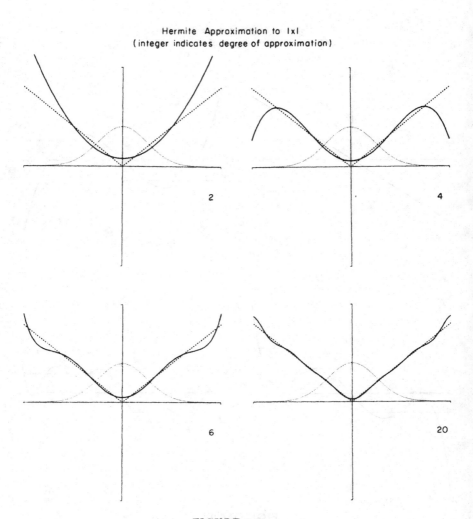

Hermite Approximation to |x|
(integer indicates degree of approximation)

FIGURE 6.10.

5. Find Hermite polynomial expansions for

   (a) sgn x,
   (b) $|x|$.

   [Hint: Neither of these are square integrable. See Figures 6.9 and 6.10.]

6. Show

$$\left(2x - \frac{d}{dx}\right)^n 1 = H_n(x).$$

   [Hint: Use mathematical induction.]

7. (a) Show

$$x^{2r} = \frac{(2r)!}{2^{2r}} \sum_{n=0}^{r} \frac{H_{2n}(x)}{(2n)!(r-n)!}.$$

   (b) What is the expansion of x^{2r+1}?

Laguerre Polynomials

In many instances we are interested in polynomial expansions on the positive real axis. Although Hermite expansions still apply, it is useful to have a separate treatment for the interval $(0, \infty)$. To this end we consider the monomials $\{(-x)^n\}$ and the weight function e^{-x}. The inner product of choice is

$$(f, g) = \int_0^{\infty} e^{-x} f(x)g(x)dx.$$

The resulting polynomials are called *Laguerre polynomials* and are denoted by $L_n(x)$. The Gram–Schmidt process is especially simple since only the simple integral

$$\int_0^{\infty} e^{-x} x^n dx = n! \tag{6.116}$$

appears. As indicated this is the factorial. If we follow the usual procedure we find

$$\begin{cases} \tilde{L}_0(x) = 1, \quad \tilde{L}_1(x) = -x + 1, \quad \tilde{L}_2(x) = x^2 - 4x + 2, \\ \tilde{L}_3(x) = -x^3 + 9x^2 - 18x + 6, \\ \tilde{L}_4(x) = x^4 - 16x^3 + 72x^2 - 96x + 24. \end{cases} \tag{6.117}$$

The tilda has been introduced since these do not have the standard normalization.

For this purpose and as a change of pace we develop this case by starting with the recurrence relation for Laguerre polynomials,

$$(n + 1)L_{n+1}(x) - (2n + 1 - x)L_n(x) + nL_{n-1}(x) = 0, \tag{6.118}$$

with $L_0 = 1$. It can then be seen that

$$n! L_n(x) = \tilde{L}_n(x).$$

To *solve* (6.118) for the Laguerre polynomials, we introduce the generating form

$$G(x; z) = \sum_{n=0}^{\infty} z^n L_n(x), \qquad (6.119)$$

which for $z = 0$ is

$$G(x; 0) = 1 \qquad (6.120)$$

since we take $L_0(x) = 1$. To carry out summation in (6.118), we rewrite it with index incremented by one:

$$(n + 2) L_{n+2}(x) - (2n + 3 - x) L_{n+1}(x) + (n + 1) L_n(x) = 0.$$

This is multiplied by z^{n+2} and summed for $0 \le n \le \infty$. The first term is seen to give

$$\sum_{n=0}^{\infty} (n + 2) L_{n+2}(x) z^{n+2} = z \frac{dG}{dz} - z(1 - x),$$

and similar findings exist for the other terms. The result is

$$(1 - 2z + z^2) \frac{dG}{dz} + (z - 1 + x) G = 0, \qquad (6.121)$$

and the integral of (6.121) under the initial condition (6.120) is given by

$$G = \frac{e^{-xz/(1-z)}}{1 - z}. \qquad (6.122)$$

This is convergent for $|z| < 1$, and we can construct the Laguerre polynomials using Cauchy's Integral Formula:

$$L_n(x) = \frac{1}{2\pi i} \oint_{|z|=\epsilon} \frac{G(x; z)}{z^{n+1}} dz = \frac{1}{2\pi i} \oint_{|z|=\epsilon} \frac{e^{-xz/(1-z)}}{(1 - z) z^{n+1}} dz, \qquad (6.123)$$

where as usual the path of integration is a small circle about the origin. The integration is facilitated by the transformation

$$\frac{xz}{1 - z} = t - x \rightarrow z = \frac{t - x}{t}.$$

Thus

$$L_n(x) = \frac{e^x}{2\pi i} \oint_{|t-x|=\epsilon} \frac{t^n e^{-t}}{(t - x)^{n+1}} dt,$$

and from the formula for a derivative (3.17) we get the Rodrigues formula

$$L_n(x) = \frac{e^x}{n!}\frac{d^n}{dx^n}(x^n e^{-x}).\tag{6.124}$$

As in previous cases orthogonality follows directly from (6.124). To obtain the constant in

$$\int_0^\infty e^{-x}L_n(x)L_m(x)dx = C(n)\delta_{nm},$$

observe that

$$\int_0^\infty \left(\sum_{n=0}^\infty L_n(x)z^n\right)^2 e^{-x}dx = \int_0^\infty \sum_{n=0}^\infty (L_n(x))^2 z^{2n} e^{-x}dx$$

$$= \int_0^\infty e^{-x}(G(x;z))^2\,dx = \int_0^\infty e^{-x}\frac{e^{-2xz/(1-z)}}{(1-z)^2}dx = \frac{1}{1-z^2}.$$

Hence the constant $C(n)$ is equal to unity, or

$$\int_0^\infty e^{-x}L_n(x)L_m(x)dx = \delta_{nm}.\tag{6.125}$$

Finally, if $f(x)$ is square integrable on $(0,\infty)$, then

$$f_N = e^{-x/2}\sum_{n=0}^N L_n(x)\int_0^\infty e^{-x/2}f(x)L_n(x)dx\tag{6.126}$$

approaches f in the sense that

$$\lim_{N\uparrow\infty}\int_0^\infty (f - f_N)^2 dx = 0.$$

When this is the case, we write

$$f = e^{-x/2}\sum_{n=0}^\infty a_n L_n(x),$$

$$a_n = \int_0^\infty e^{-x/2}L_n(x)f(x)dx.$$

Exercises

1. Equation (6.116) defines the factorial for noninteger values of n. Customarily this is defined through the *gamma* function

$$\Gamma(z) = \int_0^\infty e^{-t}t^{z-1}dt.$$

(a) Show that this function is analytic for Re $z > 0$.

(b) Show the factorial property, viz., $\Gamma(z+1) = z\Gamma(z)$.

(c) Show that $\Gamma(z)$ has no other singularities but simple poles at $z = 0, -1, -2, -3, \ldots$. [Hint: Since $\Gamma(z) = (1/z)\Gamma(z+1)$ and $\Gamma(z+1)$ is analytic for Re $z > -1$ from (a), $\Gamma(z)$ is analytic for Re $z > -1$ except for a pole at $z = 0$.]

2. Actually use the Gram–Schmidt procedure to obtain (6.117).

3. Complete the derivation of (6.121).

4. Show

$$e^{-ax} = \frac{1}{1+a} \sum_{n=0}^{\infty} \left(\frac{a}{1+a}\right)^n L_n(x).$$

5. Show that for integer r

$$x^r = (r!)^2 \sum_{n=0}^{r} \frac{(-)^n L_n(x)}{n!(r-n)!}$$

[Hint: Use the Rodrigues formula (6.124).]

6. The associated Laguerre polynomials are defined by

$$L_n^k(x) = (-)^k \frac{d^k}{dx^k} L_{n+1}(x).$$

(a) Find L_0^k, L_1^k, L_2^k.

(b) Prove

$$\frac{e^{-x} x^{-k}}{n!} \frac{d^n}{dx^n} (e^{-x} x^{n+k}) = L_n^k.$$

(c) Prove

$$\int_0^\infty e^{-x} x^k L_n^k(x) L_m^k(x) dx = \frac{(n+k)!}{n!} \delta_{m,n}.$$

6.5   Sturm–Liouville Theory

We have encountered five systems of orthogonal functions, viz., trigonometric, Tchebycheff, Legendre, Hermite, and Laguerre. In each instance a certain standard framework appeared. Included in this framework was orthogonality, recurrence relations, Rodrigues relations, properties of zeros, generating functions, and so forth. A general unified development of this framework would be beyond the scope of this course, but a limited version can be presented. For this purpose it is useful to start with the various differential equations satisfied by the orthgonal functions. These are as follows:

Trigonometric:

$$\frac{d^2}{dx^2}(\sin nx, \cos nx) = -n^2(\sin nx, \cos nx).$$

Tchebycheff:

$$(1-x^2)\frac{d^2}{dx^2}T_n - x\frac{d}{dx}T_n = -n^2T_n \quad \text{or}$$

$$\frac{d}{dx}\left((1-x^2)^{1/2}\frac{d}{dx}T_n\right) = \frac{-n^2T_n}{(1-x^2)^{1/2}}. \tag{6.127}$$

Legendre:

$$(1-x^2)\frac{d^2P_n}{dx^2} - 2x\frac{dP_n}{dx} = -n(n+1)P_n \quad \text{or}$$

$$\frac{d}{dx}\left((1-x^2)\frac{dP_n}{dx}\right) = -n(n+1)P_n. \tag{6.128}$$

Hermite:

$$\frac{d^2He_n}{dx^2} - x\frac{dHe_n}{dx} = -nHe_n \quad \text{or}$$

$$\frac{d}{dx}\left(e^{-x^2/2}\frac{dHe_n}{dx}\right) = -ne^{-x^2/2}He_n. \tag{6.129}$$

Laguerre:

$$x\frac{d^2L_n}{dx^2} + (1-x)\frac{dL_n}{dx} = -nL_n \quad \text{or}$$

$$\frac{d}{dx}\left(xe^{-x}\frac{dL_n}{dx}\right) = -ne^{-x}L_n. \tag{6.130}$$

The differential equation for trigonometric functions is well known. To obtain the Tchebycheff differential equation, we first refer back to its definition, which from (6.47) and (6.43) is

$$T_n(x) = \cos n\theta(x) = \cos n(\cos^{-1} x).$$

Next we observe

$$\frac{dT_n}{dx} = -n\frac{d\theta}{dx}\sin n\theta$$

and

$$\frac{d^2T_n}{dx^2} = -n^2\cos n\theta\left(\frac{d\theta}{dx}\right)^2 - n\sin n\theta\frac{d^2\theta}{dx^2}.$$

If we differentiate $x = \cos\theta$, we obtain

$$\frac{d\theta}{dx} = -\frac{1}{\sin\theta}$$

and on a second differentiation

$$\frac{d^2\theta}{dx^2} = \frac{\cos\theta}{\sin^2\theta}\frac{d\theta}{dx}.$$

Substitution then yields

$$\frac{d^2T_n}{dx^2} = -n^2\frac{T_n(x)}{\sin^2\theta} + \frac{\cos\theta}{\sin^2\theta}\left(-n\sin n\theta\frac{d\theta}{dx}\right).$$

The term in the brackets is simply dT_n/dx, and on substitution of $x = \cos\theta$ and $\sin^2\theta = 1 - x^2$ we obtain the required form, (6.127).

An even simpler demonstration occurs in the case of Hermite polynomials. In this case we note from (6.102) that

$$He_{n-1} = \frac{1}{n}\frac{dHe_n}{dx}.$$

Iterating on this expression gives

$$He_{n-2} = \frac{1}{(n-1)n}\frac{d^2He_n}{dx^2},$$

which on substitution into (6.101) yields

$$He_n = \frac{x}{n}\frac{dHe_n}{dx} - \frac{(n-1)}{(n-1)n}\frac{d^2He_n}{dx^2} = \frac{x}{n}\frac{dHe_n}{dx} - \frac{1}{n}\frac{d^2He_n}{dx^2}$$

and hence the above form of the Hermite differential equation, (6.129).

Each of the above forms is a special case of the *Sturm–Liouville* (S–L) *differential equation*

$$Lu = \frac{d}{dx}\left(p(x)\frac{du}{dx}\right) - q(x)u = -\lambda r(x)u. \qquad (6.131)$$

The left-hand side of (6.131), L, will be referred to as the *S–L operator*.

Associated with the S–L operator is the *S–L boundary value problem*, namely that (6.131) be solved in an interval (a, b) subject to the zero or homogeneous boundary conditions

$$A_1u(a) + B_1u'(a) = 0, \qquad A_2u(b) + B_2u'(b) = 0. \qquad (6.132)$$

It should be noted at the outset that the S–L boundary value problem might seem unusual in that $u = 0$ is immediately a solution of the problem. It may be conceptually useful to point out that the same can be said for the matrix problem $\mathbf{Ax} = \lambda\mathbf{x}$, i.e., that the zero vector $\mathbf{x} = 0$ is a solution. Yet as we know that for special values of λ, the eigenvalues, this equation has nonzero solutions. Thus we refer to a value of λ for which the S–L problem (6.131), (6.132) has a solution as an *eigenvalue* and a corresponding solution as its *eigenfunction*.

Definition. An S–L problem is said to be *regular* if the following conditions hold:

(1) $r(x) > 0$ and $p(x) > 0$ for $x \in [a, b]$.

(2) $q(x)$ is continuous for $x \in [a, b]$.

(3) (a, b) is finite.

If (1) and (2) are valid, then the differential equation itself, (6.131), is said to be regular. If any of the conditions fail, then the S–L problem is said to be *singular*.

Of the list given at the outset of this section, only the trigonometric case is regular. For all others, $p(x)$ vanishes at an endpoint. Thus only the trigonometric case of our list is a possible regular case. (We come to the question of boundary conditions later.)

Example 1. As an example consider the problem posed by

$$\frac{d^2u}{dx^2} = -\lambda u, \quad u(0) = u(1) = 0. \tag{6.133}$$

Since $p = 1$, $r = 1$, $q = 0$, the interval is finite, and the boundary conditions satisfy (6.132), this is a regular S–L problem. If $\lambda < 0$, this equation is solved in terms of $\sinh\sqrt{\lambda}x$ and $\cosh\sqrt{\lambda}x$. The boundary condition at the origin forces the solution to be $\sinh\sqrt{\lambda}x$; but the boundary condition at $x = 1$ then cannot be satisfied, and we must conclude that only $\lambda > 0$ is possible. The solution then is

$$u = \sin\sqrt{\lambda}x,$$

which, to satisfy the boundary condition at $x = 1$, requires that $\sqrt{\lambda} = n\pi$, n an integer, or that

$$\lambda_n = n^2\pi^2 \tag{6.134}$$

is the eigenvalue that corresponds to the eigenfunction

$$u_n = \sin n\pi x. \tag{6.135}$$

The S–L problem thus leads us to an infinite number of eigenvalues (6.134) and corresponding eigenfunctions (6.135).

If instead in (6.133) we take as a boundary condition

$$u'(0) = u'(1) = 0, \tag{6.136}$$

it is easily seen that the eigenvalues are again (6.134) and the corresponding eigenfunctions are

$$u_n = \cos n\pi x. \tag{6.137}$$

In this instance the index n can be zero. Again we have an infinitude of eigenvalues and eigenfunctions.

Example 2. As another example consider

$$\frac{d^2u}{dx^2} = -\lambda u, \quad u(0) = 0, \quad u(1) + u'(1) = 0. \tag{6.138}$$

Since the boundary conditions satisfy (6.132), this is still a regular S–L problem. Arguing as we did in the previous example shows that only positive values of λ are possible. Thus a solution is again trigonometric and the condition at the origin leads to

$$u = \sin \sqrt{\lambda} x.$$

A multiplicative constant can be introduced but plays no role. Why? The second boundary condition then gives

$$\sin \sqrt{\lambda} + \sqrt{\lambda} \cos \sqrt{\lambda} = 0$$

or

$$\sqrt{\lambda} = -\tan \sqrt{\lambda}. \tag{6.139}$$

This is easily solved by graphical means as indicated in Figure 6.11. As Figure 6.11 implies, if λ is large,

$$-\sqrt{\lambda} \approx \frac{(2n+1)}{2}\pi$$

or

$$\lambda_n \approx \frac{(2n+1)^2}{4}\pi^2.$$

(For $n = 0$ this gives $\lambda_0 \approx 2.5$, compared with a numerical calculation of $\lambda_0 \approx 4.1$. For $n > 0$ the estimate for λ_n becomes quite good.) And again the S–L problem generates an infinity of eigenvalues and corresponding eigenfunctions.

These simple examples serve as a guide when it comes to what may be expected in the general case as specified by (6.131), (6.132). In the notation of the examples we pose the general problem as

$$\begin{cases} Lu_n = -\lambda_n r(x)u_n, & a \le x \le b, \\ A_1 u_n(a) + B_1 u_n(a) = 0, \\ A_2 u_n(b) + B_2 u_n(b) = 0. \end{cases} \tag{6.140}$$

Then, if p, q, and r fulfill the three conditions of the definition of the S–L problem as stated earlier, the following theorems are valid.

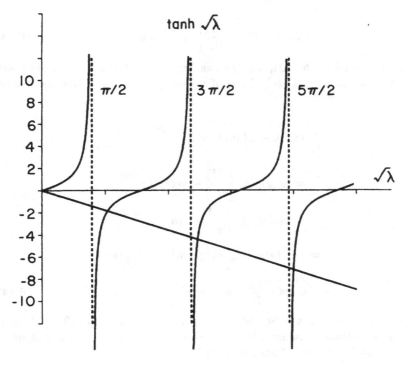

FIGURE 6.11.

Theorem 1. *All eigenvalues λ_n are real.*

Theorem 2. *If $\lambda_n \neq \lambda_m$, then the corresponding eigenfunctions u_n, u_m are orthogonal under the inner product*

$$(u_n, u_m) = \int_a^b u_n(x)u_m(x)r(x)dx = 0. \tag{6.141}$$

Theorem 3. *To each eigenvalue λ_n there corresponds a unique eigenfunction.*

Theorem 4. *The eigenvalues λ_n can be arranged so that n refers to the number of zeros of u_n in the interval and further*

$$\lambda_0 < \lambda_1 < \lambda_2 < \lambda_3 < \ldots$$

with $\lambda_n \uparrow \infty$ for $n \uparrow \infty$.

As preparation for the demonstration of these theorems, consider

$$wLu = w\frac{d}{dx}\left(p\frac{du}{dx}\right) - wqu = \frac{d}{dx}\left(wp\frac{du}{dx} - up\frac{dw}{dx}\right) + u\frac{d}{dx}\left(p\frac{dw}{dx}\right) - uqw$$

or

$$wLu - uLw = \frac{d}{dx}\left[wp\frac{du}{dx} - up\frac{dw}{dx}\right]. \tag{6.142}$$

This last relation, (6.142), is known as *Lagrange's Identity*. Next suppose u and w both satisfy (6.140). Then, if we integrate (6.142) over the interval (a, b), we obtain

$$\int_a^b (wLu - uLw)dx = \left[wp\frac{du}{dx} - up\frac{dw}{dx}\right]_a^b$$

$$= w(b)p(b)\frac{du(b)}{dx} - u(b)p(b)\frac{dw(b)}{dx}$$

$$- w(a)p(a)\frac{du(a)}{dx} + u(a)p(a)\frac{dw(a)}{dx}$$

$$= -w(b)p(b)\frac{A_2}{B_2}u(b) + u(b)p(b)\frac{A_2}{B_2}w(b)$$

$$+ w(a)p(a)\frac{A_1}{B_1}u(a) - u(a)p(a)\frac{A_1}{B_1}w(a) = 0, \tag{6.143}$$

where we have used the boundary conditions stated in (6.140). (Although this demonstration assumes that B_1 and B_2 are not zero, it is not necessary.) Another way of stating (6.143) is

$$(w, Lu)_0 = (Lw, u)_0, \tag{6.144}$$

where the zero subscript signifies an inner product of the form (6.141) but with $r = 1$. In analogy with the case of matrices as mentioned above, when (6.144) holds we say that L is (formally) *self-adjoint* or *symmetric*.

To prove Theorem 1 we take $w = \bar{u}$, the complex conjugate of u. If we conjugate the first form in (6.140) such that

$$L\bar{u} = -\bar{\lambda}\,\bar{u}r(x)$$

and substitute this result into (6.144), then

$$(\lambda - \bar{\lambda})\int_a^b r(x)u(x)\bar{u}(x)dx = (\bar{u}, Lu)_0 = (L\bar{u}, u)_0 = 0. \tag{6.145}$$

Because $ru\bar{u} > 0$, this implies that

$$\lambda = \bar{\lambda},$$

and the theorem is proven. It is left as an exercise to prove that the eigenfunctions are real.

To prove Theorem 2, we take w and u to belong to different eigenvalues λ and μ, $\lambda \neq \mu$, and substitute these in (6.144) to give

$$(\lambda - \mu)(w, u) = (\lambda - \mu) \int_a^b r(x)w(x)u(x)dx = 0.$$

Hence

$$\int_a^b r(x)w(x)u(x)dx = (w, u) = 0, \tag{6.146}$$

which is the required result.

To prove Theorem 3, suppose u_1 and u_2 are both linearly independent and both belong to the same eigenvalue λ. Now from elementary theory any solution to (6.140) can be written in the form

$$U(x) = \alpha u_1 + \beta u_2,$$

where α and β are constants. In particular, we can arbitrarily specify $U(a)$ and $U'(a)$. On the other hand, (6.140) states that $U(a)/U'(a) = -B_1/A_1$, which is a contradiction, and hence Theorem 3 is proven.

The proof of Theorem 4, although not difficult, involves more time than seems warranted. (For details of this and related proofs see Birkhoff & Rota.) However, the kernel of the idea to the proof is quite simple. To see this, we first return to Example 1 of this section. As we observed, the solution of the differential equation (6.133), after imposing the boundary condition at $x = 0$, is

$$u = \sin \sqrt{\lambda} x.$$

From this we notice that as λ is increased past an eigenvalue, λ_n, the zero crossing at $x = 1$ moves into the interval. As λ is further increased, we soon reach another value of λ, viz., λ_{n+1}, for which $x = 1$ is a zero crossing, thereby furnishing us with another eigenvalue and another eigenfunction as illustrated in Figure 6.12.

The behavior just outlined is generic. To motivate this, consider (6.131) under the transformation

$$t(x) = \int_a^x \frac{ds}{p(s)}.$$

Thus

$$p(x)\frac{d}{dx} = \frac{d}{dt}$$

so that the differential equation (6.131) becomes

$$\frac{d^2u}{dt^2} + p(\lambda r - q)u = 0. \tag{6.147}$$

Since p and r are positive, if λ is large enough the coefficient of u is positive throughout the interval. Therefore, in the neighborhood of any point

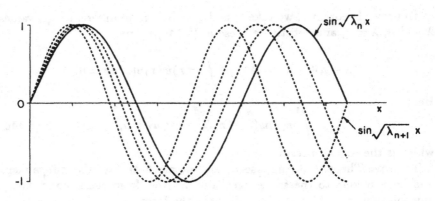

FIGURE 6.12.

of (a, b), (6.147) can be approximated by a trigonometric differential equation. Thus solutions of (6.147) are oscillatory and become more so as λ is increased. On general grounds we can suppose that u depends continuously on λ. Therefore, if we start with a solution of (6.147) which is fixed to be zero at $x = a$ ($t = 0$), then as λ increases beyond λ_n the zero crossing enters the interval (since the oscillations increase)¿ As λ is further increased, we reach a *magical value:* the next eigenvalue, λ_{n+1}, at which u has a zero crossing at $x = b$. We are then given u_{n+1}. This is the germ of the proof and as a bonus we have the following theorem:

Theorem 5. *The zeros of u_{n+1} interlace those of u_n.*

Finally we state without proof that (see Birkhoff & Rota)

Theorem 6. *The eigenfunctions of a regular Sturm–Liouville problem are complete. By this is meant that if $f(x)$, $x \in (a, b)$, is such that*

$$\int_a^b f^2(x)r(x)dx < \infty \tag{6.148}$$

and if

$$a_n = \int_a^b f(x)u_n(x)r(x)/\int_a^b u_n^2(x)r(x)dx$$

are Fourier coefficients, then

$$f_N = \sum_{n=1}^{N} a_n u_n \tag{6.149}$$

converges to f in mean, i.e.,

$$\lim_{N\uparrow\infty} \int_a^b (f - f_N)^2 r(x)dx = 0. \tag{6.150}$$

In particular, if f is piecewise differentiable, then

$$\sum_{n=1}^{\infty} a_n u_n = \frac{1}{2}\{f^+(x) + f^-(x)\} \qquad (6.151)$$

at any point.

Thus we have accomplished what we have set out to do, namely to give a general setting to the various systems of orthogonal functions by which a function can be represented. But if we refer back to our list of equations given at the very beginning of this section, we see that of all the associated S–L problems only the trigonometric cases are regular. Both the Hermite and Laguerre fail to be, because of an infinite interval, while both the Legendre and Tchebycheff cases have a coefficient $p(x)$ which vanishes at the endpoints $(-1, 1)$ (and in a limiting sense this is also true of the Laguerre and Hermite cases). Two immediate questions arise in connection with the singular case: What condition should replace the boundary conditions in (6.140), and what results of a general type hold in the singular case?

An examination of the proof of Theorems 1 and 2 shows that in both cases the proof rests on the fact that

$$p\left(w\frac{du}{dx} - u\frac{dw}{dx}\right)$$

vanishes at each endpoint (see (6.143)). In the cases of Legendre and Tchebycheff differential equations, p vanishes at the endpoints. Therefore the proofs of Theorems 1 and 2 follow in this instance even if the boundary conditions in (6.140) are replaced simply by the conditions that u and u' are bounded at the endpoints.

In the case of Hermite polynomials, if the boundary conditions in (6.140) are replaced by the condition that

$$\lim_{|x|\uparrow\infty} e^{-x^2/2}\left[u\frac{dw}{dx} - w\frac{du}{dx}\right] = 0$$

for any two solutions (not necessarily different) u and w, then these theorems remain valid. Finally, in the case of the Laguerre operator, in order for the two theorems to be valid, the boundary conditions in (6.140) should be replaced by the conditions that $u'(0)$ is bounded and that

$$\lim_{x\uparrow\infty} xe^{-x}\left[u\frac{dw}{dx} - w\frac{du}{dx}\right] = 0$$

for any two solutions u and w. In this case, as in the Hermite case above, we see that the condition at infinity is quite mild and actually allows a solution to grow rapidly. We will see later that there is a physical basis to these seemingly special end conditions.

Theorems 3–6 can also be proven in certain singular cases, but are not generally true. They can be proven for (6.127)–(6.130) under the above revised end conditions. In general, however, with the singular case, a continuous range of eigenvalues can occur. For example, we can consider bounded solutions of

$$\frac{d^2u}{dx^2} = -\lambda u, \quad -\infty < x < \infty. \tag{6.152}$$

These eigenfunctions are given by

$$\left\{ \begin{array}{l} \sin \sqrt{\lambda}x, \\ \cos \sqrt{\lambda}x \end{array} \right. \tag{6.153}$$

for any positive λ. This is in contrast to the regular case when only discrete values of λ occur.

Finally, consider the problem

$$\frac{d^2u}{dx^2} = -\lambda u, \quad u(0) = u(1), \quad u'(0) = u'(1). \tag{6.154}$$

This is a simple illustration of the general situation in which boundary conditions are linked to each other. These are *periodic* boundary conditions and we say that they are *non-separated*. The solution to (6.154) is immediate and is given by

$$\lambda_n = (2n\pi)^2, \quad u_n = (\sin 2\pi nx, \cos 2\pi nx), \quad n = 0, 1, 2, \ldots. \tag{6.155}$$

It is important to note that the eigenvalues are *degenerate*. To each eigenvalue there corresponds two eigenfunctions. As we showed by direct means earlier, these functions are complete. (This is just the case of Fourier series.)

Although there are still many other nooks and crannies to delve into, we now leave this topic and refer the interested reader to the books by Courant & Hilbert and Birkhoff & Rota.

Exercises

1. Show that if

$$Lu = a_2(x)\frac{d^2u}{dx^2} + a_1(x)\frac{du}{dx} + a_0(x)u,$$

then

$$Lu = \frac{a_2}{p}\left\{ \frac{d}{dx}\left(p\frac{du}{dx}\right) + qu \right\},$$

where

$$p = \exp\left(\int^x \frac{a_1(s)}{a_2(s)}ds \right), \quad q = \frac{a_0(x)p(x)}{a_2(x)}.$$

2. Obtain the Laguerre and Legendre differential equations. Use Exercise 1 to put them in S–L form.

3. Prove that an eigenfunction of an S–L problem is real.

4. If for all u and v which satisfy the boundary conditions (6.132) we have that

$$\int_a^b uLv\,dx = \int_a^b vLu\,dx,$$

then L is said to be self-adjoint. Which of the following problems lead to a self-adjoint L?

  (a) $y'' + y' + 2y = 0$, $y(0) = 0 = y(1)$;

  (b) $(1 + x^2)y'' + 2xy' + y = 0$, $y'(0) = 0 = y(1) + 2y'(1)$;

  (c) $(1+x^2)y''+2xy'+y = \lambda(1+x^2)y$, $y(0)-y'(1) = 0$, $y'(0)+2y(1) = 0$.

5. Find eigenvalues and eigenfunctions for each of the following:

  (a) $y'' = -\lambda y$, $y(0) + y'(0) = 0$, $y(1) = 0$;

  (b) $y'' = -\lambda y$, $y(0) = y'(0)$, $y(1) = -y'(1)$.

6. Show that Liouville's transformation

$$u = w/(p(x)r(x))^{1/4}, \quad t = \int^x \sqrt{r(s)/p(s)}\,ds$$

reduces (6.131) to

$$\frac{d^2}{dt^2}w - \hat{q}(t)w = -\lambda w.$$

What is $\hat{q}(t)$? (Note that under this transformation the eigenvalues remain fixed and the weight function becomes unity.)

7. Prove (6.143) if $B_1 = B_2 = 0$ in (6.140).

8. For $\lambda = 0$ find a second solution for the case of (a) Legendre's equation, (b) Laguerre's equation, and (c) Hermite's equation. [Hint: Use the form (6.131).]

9. Show that (a) the derivative of Legendre's polynomial $P_n'(x)$ satisfies a self-adjoint differential equation, with $\lambda = n(n + 1) - 2$; (b)

$$\int_{-1}^1 P_n'(x)P_m'(x)(1 - x^2)dx = 0, \quad m \neq n.$$

10. Find the eigenvalues and eigenfunctions of $x^2y'' = -\lambda y$; $1 \leq x \leq 2$; $y(1) = y(2) = 0$. [Hint: Try $y = x^p$.]

6.6   Orthogonal Expansions in Higher Dimensions

We pointed out in the development of the Gram–Schmidt procedure that the general formalism given there applies virtually without change to function spaces defined on more than one independent variable. We will illustrate this by some simple examples involving functions of two variables—the generalization to higher dimensions is straightforward.

Trigonometric Series. If $f(x, y)$ is defined on the square, $0 < x, y < 1$, then we can formally expand as follows:

$$f(x, y) = \sum_{m,n=-\infty}^{\infty} a_{mn} e^{2\pi i(mx+ny)},$$

$$a_{mn} = \int_0^1 \int_0^1 e^{-2\pi i(mx+ny)} f(x, y) dx\, dy. \qquad (6.156)$$

This can be thought of as a Fourier series in x, followed by a Fourier expansion in y.

Polynomial Expansions. The Tchebycheff expansion of a function in two independent variables follows from Fourier expansions in a manner analogous to the one-dimensional treatment (Section 6.2). Thus, for $f(x, y)$ defined in the square $-1 \leq x, y \leq 1$, we can introduce $x = \cos\theta$, $y = \cos\phi$ for $-\pi \leq \theta, \phi \leq 0$ and formally write

$$f(\cos\theta, \cos\phi) = F(\theta, \phi) = \sum_{m,n=0}^{\infty} A_{mn} \cos m\theta \cos n\phi$$

$$= \sum_{m,n=0}^{\infty} A_{mn} T_m(x) T_n(y), \qquad (6.157)$$

where T_n is the one-dimensional Tchebycheff polynomial. The coefficients A_{mn} are given by

$$A_{00} = \frac{1}{\pi^2} \int_{-1}^1 \int_{-1}^1 f(x, y) \frac{dx}{(1-x^2)^{1/2}} \frac{dy}{(1-y^2)^{1/2}},$$

$$A_{0n} = \frac{2}{\pi^2} \int_{-1}^1 \int_{-1}^1 f(x, y) T_n(y) \frac{dxdy}{(1-x^2)^{1/2}(1-y^2)^{1/2}}, \quad n > 0,$$

$$A_{m0} = \frac{2}{\pi^2} \int_{-1}^1 \int_{-1}^1 f(x, y) T_m(x) \frac{dxdy}{(1-x^2)^{1/2}(1-y^2)^{1/2}}, \quad m > 0,$$

$$A_{mn} = \frac{4}{\pi^2} \int_{-1}^1 \int_{-1}^1 f(x, y) T_n(y) T_m(x) \frac{dxdy}{(1-x^2)^{1/2}(1-y^2)^{1/2}}, \quad m, n > 0.$$

$$(6.158)$$

Another approach to orthogonal expansions in two dimensions is through the Gram–Schmidt procedure (see the exercises). For example, we can seek an orthogonal set for $-1 \leq x, y \leq 1$ from the set $\{x^n y^m\}$. If $0 \leq m, n \leq \infty$, then

$$1, \; x, \; y, \; x^2, \; xy, \; y^2, \; x^3, \; \ldots$$

lead to

$$P_0(x)P_0(y), \; P_0(x)P_1(y), \; P_1(x)P_0(y), \; P_0(x)P_2(y),$$
$$P_1(x)P_1(y), \; P_2(x)P_0(y), \ldots,$$

which clearly form an orthogonal set. More generally,

$$\{P_m(x)P_n(y)\}, \quad n, m = 0, 1, 2, \ldots \tag{6.159}$$

form an orthogonal set on $-1 < x, \; y < 1$, and for a function defined in that region we can formally expand and obtain

$$f(x,y) = \sum_{m,n=0}^{\infty} A_{mn} P_m(x) P_n(y) \tag{6.160}$$

with

$$A_{mn} = \frac{(2n+1)(2m+1)}{4} \int_{-1}^{1} \int_{-1}^{1} f(x,y) P_m(x) P_n(y) dx dy. \tag{6.161}$$

Mixed Expansions. It is clearly the case that a function can be expanded in different orthogonal functions for different variables—as the case may require. For example, if $f(x,y)$ is defined for $-\infty < x < \infty$ and $0 \leq y \leq \infty$, then we formally expand as follows:

$$f(x,y) = \frac{e^{-x^2/4}}{(2\pi)^{1/4}} e^{-y/2} \sum_{n,m=0}^{\infty} B_{mn} He_m(x) L_n(y)$$

with

$$B_{mn} = \frac{1}{m!} \int_{-\infty}^{\infty} \frac{e^{-x^2/4}}{(2\pi)^{1/4}} He_m(x) dx \int_{0}^{\infty} e^{-y/2} L_n(y) f(x,y) dy.$$

Exercises

1. Use the Gram–Schmidt procedure to find the set (6.159).

2. $f(x,y)$ is defined for $-1 < x, y \leq 1$ so that $f = 1$ in the first and third quadrants and $f = -1$ in the second and fourth quadrants. Expand f in a Fourier series.

3. Prove
$$\int_{-\infty}^{\infty} \frac{e^{-t^2}}{\sqrt{\pi}} He_{2n}(xt)\,dt = \frac{(2n)!}{n!}(x^2 - 1).$$

4. Find suitable formal expansions for each of the following cases (give explicit forms for the coefficients):

   (a) $f(x, y, z)$; $-1 \le x, y, z, \le 1$.

   (b) $f(x, y)$; $-\infty < x, y < \infty$; f is one-periodic in x.

   (c) $f(\mathbf{x})$; $-\infty < \mathbf{x} < \infty$; $\mathbf{x} = (x_1, x_2, \ldots, x_N)$.

7

Partial Differential Equations

7.1 Conservation Laws

Many problems drawn from applications can be formulated by appealing to some sort of conservation law. Most frequently it is the conservation of matter which is basic to the theory. This for example is the case for fluid flowing in a blood vessel or on a channel or river; for the description of particles in a suspension; or in an idealized form, even for cars moving down a highway. Additional conservation laws, such as conservation of momentum or energy may also be applicable. In this section we will develop some tools for formulating these principles in mathematical terms and in the process learn a little science.

Diffusion. Imagine a tank of fluid separated into two parts by a vertical diaphragm having, say, tinted fluid on the left and clear fluid on the right (see Figure 7.1). At some instant, which we are free to call the initial moment, the diaphragm is removed. Experience teaches us that the colored and clear fluids *diffuse* into one another and that ultimately the tank becomes uniform or *homogeneous* in color. When this is achieved, the solution is said to be in equilibrium. Our problem is first to formulate the phenomenon in mathematical terms and then to describe or forecast the events leading to equilibrium in mathematical language. To this end let us agree that a dye, which may be small particles or even molecules, gives rise to the observed color and that there is no chemical reaction so that the total number of dye particles in the tank is fixed (since we do not add or remove particles). Further assume that conditions are homogeneous, i.e., uniform across planes parallel to ends of the tank.

Denote by $n(x,t)$ the density of dye particles at position x and time t,

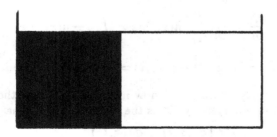

FIGURE 7.1.

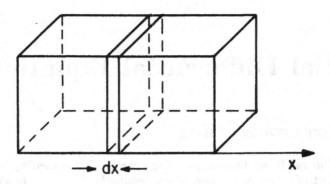

FIGURE 7.2.

defined so that

$$n(x,t)dx$$

is the (average) number of dye particles in a length dx of the tank (see Figure 7.2). Therefore n is the number of particles per unit length. Further, as part of this experiment, we measure the number of particles per unit time crossing a plane at position x at time t. This quantity, denoted by $J(x,t)$, is called the *flux* and has dimensions of number per time. J is reckoned positive if the flux is in the positive x-direction.

In view of the definition of $n(x,t)$, the number of dye particles in a fixed length of tank (x_1, x_2) is given by

$$N(t) = \int_{x_1}^{x_2} n(x,t)dx.$$

As indicated this is a function of time. What causes the change in N with time? In any increment of time Δt, $J(x_1,t)\Delta t$ particles enter at x_1 and $J(x_2,t)\Delta t$ particles leave at x_2; therefore, the net increase of particles due to this flux must equal the change in N so that

$$N(t + \Delta t) - N(t) = J(x_1,t)\Delta t - J(x_2,t)\Delta t.$$

If we divide by Δt and take the limit $\Delta t \to 0$, then the left-hand side is just t1 t1me derivative and we have

$$\frac{\partial N}{\partial t} = \frac{\partial}{\partial t}\int_{x_1}^{x_2} n(x,t)dx = \int_{x_1}^{x_2}\frac{\partial}{\partial t}n(x,t)dx$$

$$= J(x_1,t) - J(x_2,t) = -\int_{x_1}^{x_2}\frac{\partial J(x,t)}{\partial x}dx.$$

The first line simply expresses N now in terms of n, and the second line makes use of a simple identity. Thus the conservation law has the form

$$\int_{x_1}^{x_2}\left[\frac{\partial n(x,t)}{\partial t} + \frac{\partial J(x,t)}{\partial x}\right]dx = 0.$$

If this is divided by $\Delta x = x_2 - x_1$ and the limit $\Delta x \downarrow 0$ is applied, we get

$$\frac{\partial n(x,t)}{\partial t} = -\frac{\partial J(x,t)}{\partial x}. \tag{7.1}$$

This partial differential equation is sometimes called the *continuity equation*.

Very little in the above discussion was specific to suspensions as such. To underline this, let us consider the very different problem of traffic flow. Specifically, consider cars traveling on a stretch of highway which contains neither exits nor entrances. In order to describe the concentration of cars (idealized as points) as a function of position and time, we define $c(x,t)$ so that

$$c(x,t)dx$$

is the average number of cars in an increment length of roadway dx. Next, if we define $f(x,t)$ to be the flux of cars, i.e., the number of cars passing x (at time t) per unit time, then the same deliberations again lead us to a continuity equation, viz.,

$$\frac{\partial c}{\partial t} = -\frac{\partial f}{\partial x}. \tag{7.2}$$

In this instance x need not represent the *x-axis*, but rather is the distance along the highway measured from a reference point.

To return to the case of diffusion, we note that the fluxes must vanish at the ends of the tank. If we designate the ends of the tank by $x = 0$ and $x = L$, then

$$J(0,t) = J(L,t) = 0; \tag{7.3}$$

and if we integrate (7.1) over the length of the tank, then

$$\frac{\partial}{\partial t} \int_0^L n(x,t)dx = -\int_0^L \frac{\partial}{\partial x} J(x,t)dx = J(0,t) - J(L,t) = 0.$$

This implies that the integral on the left is a constant with

$$\int_0^L n(x,t)dx = \int_0^L n(x,0)dx \tag{7.4}$$

or that the total number of dye particles is conserved. Of course this property was built into the derivation of (7.1), and (7.4) serves as an indication that the derivation is correct.

Our goal is to forecast the evolution of dye particles through the use of (7.1), but at this point the description is incomplete, since there is just one equation in the two unknown functions, n and J.

Fick's Law. To overcome this problem we can try to relate J to n. Experience tells us that no flux of particles takes place in a uniform suspension and that inhomogeneities in the dye distribution tend to disappear in time.

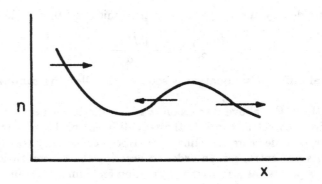

FIGURE 7.3.

That is, the flux appears to act so as to equalize the dye density. Thus there is a migration of particles from high to low density regions. Another observation is that the flux is *local;* i.e., it only depends on the local conditions of the dye density and not on the density at distant locations of the suspension. To illustrate these remarks we indicate a dye density by the curve in Figure 7.3. The flux is indicated by the arrows for this case. Thus, although the overall flux is to the right, there is local flux to the left.

These observations suggest that the flux is proportional to local density differences:

$$J \approx K[n(x + dx) - n(x)],$$

where K is a proportionality constant. On going to the limit $dx \downarrow 0$ we obtain

$$J(x) \approx K\,dx\frac{\partial n}{\partial x} = -\kappa\frac{\partial n}{\partial x}. \tag{7.5}$$

The minus sign is introduced since flux is directed toward low concentrations; i.e., it runs *downhill.* Equation (7.5) is known as *Fick's Law,* and the proportionality coefficient κ as the *diffusivity.*

At this point one must leave the armchair and go to the laboratory first to verify Fick's Law and second to determine the diffusivity κ. Only after this can we feel justified in the use of (7.5). If (7.5) is substituted into (7.1), the *diffusion equation* results:

$$\frac{\partial}{\partial t}n(x, t) = \frac{\partial}{\partial x}\kappa\frac{\partial}{\partial x}n. \tag{7.6}$$

In general the diffusivity, κ, can be dependent on space, time, and n itself. For most practical cases it is a constant to good approximation.

Traffic Flow. Next we turn to the traffic flow equation (7.2). In contrast with the case of diffusing particles, we expect the flux of cars, f, to depend specifically on the concentration of cars, c. If c is constant, the flux f is not

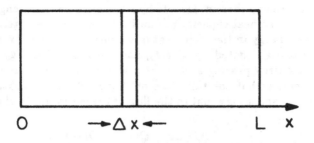

FIGURE 7.4.

expected to be zero—there will still be a steady rate of traffic flow. In two instances f does vanish. As the maximum, bumper-to-bumper density, say $\bar{c}$, is approached, the flux f goes to zero so that we have $f(\bar{c}) = 0$. Obviously the same is true for zero concentration and $f(0) = 0$. From these meager facts, we might model the flux by

$$f = Ac(\bar{c} - c), \tag{7.7}$$

where the positive constant A would have to be determined for given road conditions. If we substitute (7.7) into (7.2), then

$$\frac{\partial c}{\partial t} + A\frac{\partial}{\partial x}[c(\bar{c} - c)] = 0, \tag{7.8}$$

which in form is entirely different than the diffusion equation (7.6). Equations (7.5) and (7.7) in different ways reduce the continuity equation (7.1) to a single equation in one unknown.

Heat Flow. An inquiry into the physical mechanism underlying diffusion shows that it arises from the molecular motion associated with the temperature of the material. In fact for a gas, the temperature (on an absolute scale) measures the average kinetic energy contained in the random internal motion of the molecules. (The diffusivity must therefore depend on temperature, as well as on what solute is in what solvent.) One therefore has the view that diffusion is due to the rapid, frequent collisions undergone naturally by molecules. These random, probabilistic motions produce a gradual migration of the dye molecules. Since this is the case, the conduction of heat should also take place through diffusion. An inhomogeneity in temperature only means that molecules in some region are at a different agitation level; therefore, this too should be passed on by random collisions.

To consider heat flow we consider *conservation of energy* instead of conservation of mass or number. Imagine an infinite solid lying between $x = 0$ and $x = L$ (or, equivalently, an insulated bar of length L), as depicted in Figure 7.4. The energy contained in a slab of thickness Δx is proportional to the mass contained therein. Denote by ρ the mass per unit length (a constant) and by C the heat capacity per unit mass per degree of temperature. Then the heat energy in the increment of *volume*, Δx, is $\rho CT(x,t)\Delta x$. The heat flux vector is denoted by $Q(x,t)$, has the units of energy per time, and gives the energy passing a plane at x at a time t. The conservation of energy then states that the time rate of change of energy contained in a slab between x_1 and x_2 is equal to the flux of energy in minus the flux out:

$$\frac{\partial}{\partial t}\int_{x_1}^{x_2}\rho CT\,dx = Q(x_1) - Q(x_2).\tag{7.9}$$

As before we pass to the limit $(x_2 - x_1 \to 0)$ and obtain

$$\rho C\frac{\partial}{\partial t}T = -\frac{\partial}{\partial x}Q.\tag{7.10}$$

Repeating the argument leading to Fick's Law equation (7.5), we observe that heat flow occurs only for temperature distributions which are not constant, that it flows downhill, and that it is a local phenomenon. It then follows as before that

$$Q = -\lambda\frac{\partial}{\partial x}T.\tag{7.11}$$

This is called the *Fourier heat law* and λ, the *heat conductivity*. If we substitute (7.11) into (7.10), then

$$\frac{\partial T}{\partial t} = \frac{\partial}{\partial x}\frac{\lambda}{\rho C}\frac{\partial T}{\partial x}.\tag{7.12}$$

Note: $\lambda/\rho C$, the *thermal diffusivity*, tells us how well the heat diffuses, but λ, the conductivity, measures the heat flow.

Boundary and Initial Conditions. It should be anticipated that the heat-diffusion equation

$$\frac{\partial}{\partial t}\theta = \frac{\partial}{\partial x}\kappa\frac{\partial}{\partial x}\theta\tag{7.13}$$

has an infinite variety of solutions. To fix on the one which fits a particular problem, we apply conditions on the problem and thereby select an appropriate solution. For the time being we pursue what is plausible and leave a more rigorous discussion for later.

For the problem involving suspensions, it seems evident that the initial density

$$n(x,t)|_{t=0} = n^0(x)$$

should be supplied. Also there should be zero flux condition at the ends, which from Fick's Law, (7.5), says

$$\frac{\partial n}{\partial x}(x,t)|_{x=0} = \frac{\partial n}{\partial x}(x,t)|_{x=L} = 0$$

if the bounding walls are *impermeable*. More generally the zero end conditions can be replaced by specified nonzero fluxes if the walls are *permeable*. For both the diffusion and heat flow problems, it is therefore plausible that we can control the flow at the boundaries.

For the heat flow problem intuition suggests that in addition to the initial temperature distribution

$$T(x,t)|_{t=0} = T^0(x),$$

we can control the temperatures at the ends

$$T(x,t)|_{x=0} = T_1(t), \quad T(x,t)|_{x=L} = T_2(t).$$

All of these seem to be reasonable conditions under which to obtain a solution of (7.13).

Finally, since the mathematics is in the end oblivious to the physical derivation, either problem allows both kinds of boundary conditions and, in general, a mixture of these conditions can apply; i.e., we can seek solution of (7.13) under the boundary conditions

$$\theta(0,t) + \alpha\theta_x(0,t) = \theta_1(t), \quad \theta(L,t) + \beta\theta_x(L,t) = \theta_2(t), \qquad (7.14)$$

with α, β, θ_1, θ_2 all given in addition to the initial condition

$$\theta(x,t)|_{t=0} = \theta^0(x). \qquad (7.15)$$

Heat Flow and Diffusion in Space. Although our considerations have focused on one space dimension, it is not difficult to extend the ideas to an arbitrary number of dimensions. For this purpose it will suffice if we just consider heat flow.

The definition of temperature T clearly does not depend on the number of space dimensions. On the other hand, the density as used above does. We now define the density ρ to be the mass per unit volume, and therefore

$$\rho C T \, d\mathbf{x},$$

with

$$d\mathbf{x} = dx_1 dx_2 \ldots dx_N$$

(N is the number of dimensions and in practice $N = 1$ or 2 or 3), is the amount of energy contained in the element of volume $d\mathbf{x}$. Generalization of the flux is somewhat more complicated. The flux is a vector quantity (as it was before) and in general can flow in any direction.

Define $\mathbf{Q}$ to be the energy flux per unit area so that if dS is an element of area having a normal $\mathbf{n}$, then

$$\mathbf{Q} \cdot (\mathbf{n} \, dS)$$

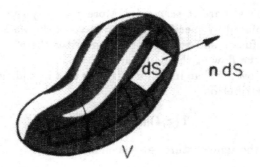

FIGURE 7.5.

is the energy flux across this element of area.

Consider an arbitrary but fixed sample volume V of the material under investigation having a bounding surface denoted by S (see Figure 7.5). The energy contained in V is

$$\int_V \rho C T(x,t)dx.$$

The time rate of change of this quantity is equal to the net flux of energy into the volume. If we denote an element of area of S by $\mathbf{n}\,dS$, where n is the *outward* normal, then the flux across dS into the volume V is

$$-\mathbf{Q}\cdot\mathbf{n}\,dS.$$

Integrating over the entire surface, we obtain

$$\frac{\partial}{\partial t}\int_V \rho C T\,dx = -\int_S \mathbf{Q}\cdot\mathbf{n}\,dS.$$

Gauss's Theorem applied to the right-hand side allows us to write

$$\int_V \left(\rho C \frac{\partial T}{\partial t} + \nabla\cdot\mathbf{Q}\right) dx = 0.$$

Since V is arbitrary, the claim is that the integrand itself is zero. For if this were not true, the integrand would have to carry one signature in some neighborhood. Given that the volume V is arbitrary, we could take it to be this neighborhood. But then the above integral would not equal zero and we would have a contradiction. It therefore follows that the integrand is zero so that

$$\rho C \frac{\partial T}{\partial t} = -\nabla\cdot\mathbf{Q}.$$

The generalization of Fourier's heat law is

$$\mathbf{Q} = -\kappa\,\nabla T. \tag{7.16}$$

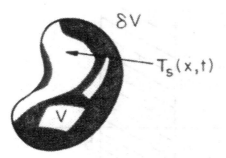

FIGURE 7.6.

This states that heat flow is perpendicular to surfaces of constant temperature. In addition, since $\kappa > 0$, it runs *downhill* and the heat equation is

$$\rho C \frac{\partial T}{\partial t} = \nabla \cdot (\kappa \nabla T). \tag{7.17}$$

Laplace's Equation. The solution of any particular heat conduction problem is determined from a specification of the conditions of the problem. For example, it is reasonable that the evolution of temperature in a finite object is fixed by specifying the temperature initially in the volume such that

$$T(\mathbf{x}, t = 0) = T^0(\mathbf{x}), \quad \mathbf{x} \in V \tag{7.18}$$

($\mathbf{x} = (x, y, z)$) and, in addition, by specifying the temperature on the surface δV of V for all time such that

$$T(\mathbf{x}, t) = T_s(\mathbf{x}, t), \quad \mathbf{x} \in \delta V \tag{7.19}$$

(see Figure 7.6). The properties of the material are determined by ρC and κ in (7.17). If the material is homogeneous, these quantities are to good approximation constants, and we can take $\kappa/\rho C$ constant so that

$$\frac{\partial T}{\partial t} = k \nabla^2 T, \quad k = \kappa/\rho C \tag{7.20}$$

subject to (7.18) and (7.19).

If the prescribed temperature on the surface does not depend on time, i.e.,

$$T(\mathbf{x}, t) = T_s(\mathbf{x}), \quad \mathbf{x} \in \delta V, \tag{7.21}$$

then one can expect the temperature in the solid to reach an equilibrium as time tends to infinity, $t \uparrow \infty$. But this implies that

$$\lim_{t \uparrow \infty} \frac{\partial T}{\partial t} = 0;$$

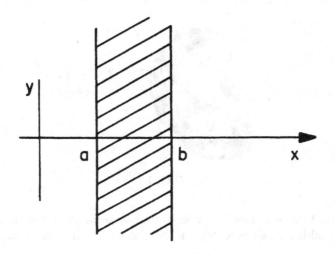

FIGURE 7.7.

i.e., the temperature is only a function of position. If this condition is applied to (7.20), it is seen that the equilibrium solution is governed by

$$0 = \nabla^2 T = \left(\frac{\partial}{\partial x^2} + \frac{\partial^2}{\partial y^2} + \frac{\partial^2}{\partial z^2} \right) T. \tag{7.22}$$

Equation (7.22) is then to be solved subject to the boundary condition (7.21). The limit equation, (7.22), is known as Laplace's equation, and according to our physical arguments is solvable given one boundary datum, for example, (7.21).

As a simple example, consider heat conduction in a slab lying between $x = a$ and $x = b$ (Figure 7.7). The boundary is the pair of points $x = a$ and $x = b$. The specification of the temperature on the boundary is

$$T(x = a) = T_a, \quad T(x = b) = T_b.$$

To achieve an equilibrium, T_a and T_b must be constant. We must also specify an initial temperature distribution,

$$T(x, t = 0) = T^0(x).$$

It is clear that there is no variation in either the y or z directions so that (7.20) reduces to

$$\frac{\partial T}{\partial t} = k \frac{\partial^2 T}{\partial x^2}.$$

This problem will be solved later but for the moment we consider the temperature distribution, in the asymptotic limit, $t \uparrow \infty$. Under this limit,

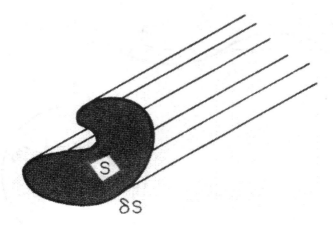

FIGURE 7.8.

$\partial T/\partial t \rightarrow 0$ and we are left with the one-dimensional form of (7.22), namely

$$\frac{\partial^2 T}{\partial x^2} = 0.$$

Under the above boundary conditions, this yields the linear temperature distribution

$$T = T_a + (T_b - T_a)(x - a)/(b - a).$$

If the solid, V, is an infinite cylinder with cross section S, for which the temperature is constant on its generators, then the Laplace equation becomes *two-dimensional* with

$$0 = \nabla^2 T = \frac{\partial^2 T}{\partial x^2} + \frac{\partial^2 T}{\partial y^2}, \quad x, y \in S.$$

The boundary data is then to be specified on the perimeter, δS, of S:

$$T = T_s(x, y), \quad x, y \in \delta S$$

(see Figure 7.8).

Wave Equation. The diffusion and Laplace equations are two of three major equations found in the study and application of partial differential equations. The third is the wave equation. In the remainder of this section we obtain the wave equation as a consequence of modeling in three different situations.

1. Vibrating String. Consider a string of uniform thickness and of uniform properties, stretched between $x = 0$ and $x = 1$. Denote the string's unperturbed position by the x-axis and the distance from this equilibrium

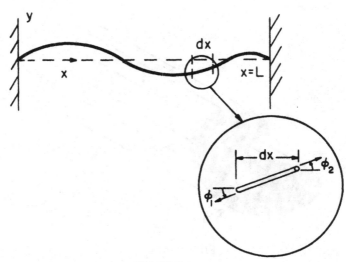

FIGURE 7.9.

position by $y(x,t)$. (It is assumed that the motion takes place in the (x, y)-plane.) Denote the mass per length of the string by ρ so that if the angles ϕ_1, ϕ_2 indicated in Figure 7.9 are not large, $\rho\,dx$ is the mass of the section of string under study. Newton's second law, which equates mass times acceleration to applied forces, states

$$\rho\,dx\frac{\partial^2 y}{\partial t^2} = \text{net force in y-direction.}$$

The tension in the string is denoted by T and is assumed to be constant. Therefore the net force in the y-direction is

$$T\sin\phi_2 - T\sin\phi_1.$$

But

$$\sin\phi_1 = \frac{\partial y/\partial x}{(1 + (\partial y/\partial x)^2)^{1/2}} \approx \frac{\partial y(x,t)}{\partial x},$$

where to get the second expression we have assumed that the *slope* of the string is small. In particular

$$\sin\phi_2 = \frac{\partial y/\partial x|_{x+dx}}{(1 + (\partial y/\partial x)^2|_{x+dx})^{1/2}} \approx \frac{\partial y(x + dx,t)}{\partial x};$$

hence

$$\sin\phi_2 - \sin\phi_1 \approx \frac{\partial y(x + dx)}{\partial x} - \frac{\partial y(x)}{\partial x} \approx \frac{\partial^2 y}{\partial x^2}dx$$

so that finally

$$\frac{\partial^2 y}{\partial t^2} = \frac{T}{\rho}\frac{\partial^2 y}{\partial x^2} = c^2\frac{\partial^2 y}{\partial x^2} \tag{7.23}$$

with

$$c = \sqrt{T/\rho}. \tag{7.24}$$

Equation (7.23) is the *wave equation in one dimension* and c is known as the *speed of propagation*. (A justification of the latter terminology will be given later.) Since the ends of the string do not move, the boundary conditions are

$$y(0) = y(L) = 0,$$

and, as usual, with Newton's equations of motion, we must supply two initial conditions,

$$y(x,0) = y_0(x), \quad 0 \le x \le 1,$$

$$\frac{\partial y(x,0)}{\partial t} = \dot{y}_0(x), \quad 0 \le x \le 1,$$

the initial position and velocity, respectively, of each *particle* of string. We remark that the main condition under which the wave equation was obtained is that $|\partial y/\partial x| << 1$. *This does not require that* $|y|$ *be small but rather that* $|y/L|$ *be small.*

2. Sound Waves. Next we consider sound waves or rather conditions under which we believe these to occur—that is, small oscillations in a gas. One-dimensional gas motions as, for example, would be generated by an infinite plane oscillating *piston* will be considered first. As in our previous treatment of one-dimensional heat conduction, we idealize the problem by imagining that all relevant quantities are constant on planes $x = $ constant. The gas density $\rho(x,t)$ (mass per unit volume) is a function of position, x, and time, t. In speaking of the gas velocity, we also use a field description; i.e., we write $u(x,t)$ for the velocity of a *particle* of fluid passing the point x at time t.

Consider a slab of fluid between two moving planes $x_1(t)$ and $x_2(t)$, each always containing the same fluid particles. The conservation of mass states (compare this with the derivation of (7.1))

$$\frac{d}{dt} \int_{x_1(t)}^{x_2(t)} \rho(x,t)dx = 0. \tag{7.25}$$

(Further imagine that we are considering a cross section of unit area between x_1 and x_2.) If we carry out the differentiation in (7.25), we obtain

$$\rho(x_2,t)\frac{dx_2}{dt} - \rho(x_1,t)\frac{dx_1}{dt} + \int_{x_1(t)}^{x_2(t)} \frac{\partial \rho}{\partial t}dx = 0.$$

But

$$\frac{dx_2}{dt} = u(x_2,t), \quad \frac{dx_1}{dt} = u(x_1,t),$$

and hence the equation can be written as

$$\rho(x_2,t)u(x_2,t) - \rho(x_1,t)u(x_1,t) + \int_{x_1(t)}^{x_2(t)} \frac{\partial \rho}{\partial t} dx = 0$$

since by assumption the endpoints are moving with the fluid. This can also be put in integral form

$$\int_{x_1}^{x_2} \frac{\partial \rho}{\partial t} dx + \int_{x_1}^{x_2} \frac{\partial}{\partial x}(\rho u) dx = 0,$$

which, since $x_1(t)$ and $x_2(t)$ are arbitrary, by an earlier argument yields

$$\frac{\partial \rho}{\partial t} + \frac{\partial \rho u}{\partial x} = 0. \tag{7.26}$$

(In our earlier terminology ρu is the mass flux.) This expresses the conservation of mass (also known as the *continuity equation*) and is one equation in *two unknowns*.

Another principle governing the movement of a fluid is Newton's second law of motion, which in the above notation states (again we consider a cross section of unit area)

$$\frac{d}{dt} \int_{x_1(t)}^{x_2(t)} \rho u \, dx = \text{Force}$$

since

$$\int_{x_1}^{x_2} \rho u \, dx$$

is the momentum contained in the slab. The force per unit area is pressure, which is denoted by p. Hence Force $= p(x_1,t) - p(x_2,t)$. We substitute and carry out the differentiation as in the case of the continuity equation:

$$\frac{d}{dt} \int_{x_1}^{x_2} \rho u \, dx = \rho(x_2)u(x_2)u(x_2) - \rho(x_1)u(x_1)u(x_1) + \int_{x_1}^{x_2} (\rho u)_t dx$$

$$= \int_{x_1}^{x_2} \left(\frac{\partial}{\partial t}\rho u + \frac{\partial}{\partial x}\rho u^2 \right) dx = F = p(x_1) - p(x_2)$$

$$= -\int_{x_1}^{x_2} \frac{\partial p}{\partial x} dx.$$

Since the slab of gas between x_1 and x_2 is arbitrary, it follows as before that

$$\frac{\partial}{\partial t}\rho u + \frac{\partial}{\partial x}(\rho u^2) + \frac{\partial p}{\partial x} = 0, \tag{7.27}$$

which is known as the *momentum equation*. There are still more unknowns, namely three (ρ, u, p), then equations, namely two ((7.26) and (7.27)). To remedy this we adopt a standard approximation that pressure is a function of density; i.e.,

$$p = p(\rho).$$

This we regard as coming either from other theory or experiment. For air we find

$$\frac{p}{p_0} = \left(\frac{\rho}{\rho_0}\right)^{\gamma}, \tag{7.28}$$

where p_0 and ρ_0 are the ambient atmospheric values of pressure and density and $\gamma \approx 1.4$. The system of equations (7.26) and (7.27) along with (7.28) therefore constitutes a consistent system of equations. They are nonlinear and are actually more general than necessary for describing sound waves.

For actual sound waves, relative pressure changes are small; e.g.,

$$\delta p = \frac{p - p_0}{p_0}$$

is about 10^{-2} for loud noises such as a sonic boom at ground level (from an SST). If we write

$$\rho = \rho_0(1 + s)$$

and insert this expression into (7.28), then

$$\frac{p}{p_0} = 1 + \delta p = (1 + s)^{\gamma} \approx 1 + \gamma s$$

or

$$\delta p \approx \gamma s,$$

which implies that s is also small for sound waves. As we will see shortly, u is also proportional to s. If we ignore quadratic and higher order terms (use $p \approx p_0(1 + \gamma s)$), then

$$\frac{\partial s}{\partial t} + \frac{\partial u}{\partial x} = 0, \quad \rho_0 \frac{\partial u}{\partial t} + \gamma p_0 \frac{\partial s}{\partial x} = 0. \tag{7.29}$$

We can eliminate u from the two equations, for example, by differentiating the first equation with respect to time and the second with respect to x, to obtain

$$\frac{\partial^2 s}{\partial t^2} = c^2 \frac{\partial^2 s}{\partial x^2}, \quad c = (\gamma p_0/\rho_0)^{1/2}. \tag{7.30}$$

To solve (7.30), $s(t = 0) = s^0(x)$ and $\partial s/\partial t(t = 0) = -u_x(t = 0)$ are assumed as given initial data.

3. Vascular Pulse Wave. As a last example we consider a liquid-filled elastic vessel, e.g., a water hose filled with water or, closer to home, blood in a blood vessel. In the latter case blood is pumped through the arteries under the action of the heart—which can be regarded as the source of a periodic pressure change. Each such cycle sends a pulsatile wave through

FIGURE 7.10.

the arterial system. These waves are responsible for what is sensed as our *pulse*.

To model this system consider an elastic tube of thickness h and radius r. In general, the thickness varies from point to point, i.e., $h(x)$ and also $r = r(x, t)$ (see Figure 7.10). Denote the equilibrium radius by $r_0(x)$ (no flow state) and the pressure by p $(= p(x, t))$. The equilibrium pressure p_0 is the same as the ambient pressure, and without loss of generality we take $p_0 = 0$. We consider the motion of an incompressible fluid, which has density ρ, through the tube under the action of relatively small pressure changes. In the derivation we will assume that relative changes in the radius are small with

$$\left| \frac{r - r_0}{r_0} \right| \ll 1$$

and also that the fluid velocity $u(x, t)$ is small. Actually the precise meaning of this last statement will not be clear until we have solved our problem.

A word about the velocity: Due to the effects of geometry and viscosity we expect that the velocity will vary across a cross section of tube. This unnecessarily complicates matters and we assume instead that

$$u = u(x, t),$$

i.e., that it is a constant over a cross section. One interpretation of this is that

$$u(x, t) = \frac{1}{A} \int_A u(x, y, z, t) \, dy \, dz,$$

that is, the average value of the true velocity over the cross-sectional area A. A further approximation is that the tube is circular so that

$$A = \pi r^2.$$

To start the derivation of the governing equations consider the conservation of mass. Because $\rho A\,dx$ is the mass contained in an increment dx (and because $\rho\,Au$ is the flux of mass), we obtain

$$\frac{\partial}{\partial t}\int_{x_1(t)}^{x_2(t)}\rho\,A\,dx = 0$$

and, in analogy with (7.26),

$$\frac{\partial}{\partial t}(\rho A) + \frac{\partial}{\partial x}(\rho Au) = 0. \tag{7.31}$$

By the same token, the momentum in dx is $\rho u A\,dx$ and Newton's law states

$$\frac{d}{dt}\int_{x_1(t)}^{x_2(t)}\rho u A\,dx = \text{Force} = (pA)_1 - (pA)_2 = -\int_{x_1}^{x_2}\frac{\partial pA}{\partial x}\,dx.$$

So, by a now standard argument, one obtains

$$\frac{\partial}{\partial t}(\rho u A) + \frac{\partial}{\partial x}(\rho u^2 A) + \frac{\partial}{\partial x}(pA) = 0. \tag{7.32}$$

For convenience it is assumed that pressure is that above equilibrium (i.e., $p_0 = 0$ as mentioned above). If we carry out the linearization mentioned above, then

$$uA \approx A_0 u, \quad u^2 A \approx 0, \quad pA \approx pA_0,$$

and therefore

$$\frac{\partial \tilde{A}}{\partial t} + \frac{\partial}{\partial x}(uA_0) = 0, \quad \tilde{A} = A - A_0, \quad \rho A_0\frac{\partial u}{\partial t} + \frac{\partial pA_0}{\partial x} = 0. \tag{7.33}$$

We must now relate p, the excess pressure, to $\tilde{A}$. To do this, consider a half cross section. It is customary to speak of the tension per unit thickness (per unit length) τ (see Figure 7.11). The downward force as indicated in Figure 7.11 is $2\tau h$ and the upward force is $2rp$. Therefore

$$pr = \tau h$$

or

$$p = \tau h/r.$$

Finally, it is assumed that the first term of the Taylor expansion in τ around r_0 suffices (which is consistent with the linearization) and we can write

$$\tau(r) \approx \left(\frac{\partial \tau}{\partial r}\right)_{r=r_0}(r - r_0) = E\frac{(r - r_0)}{r_0}$$

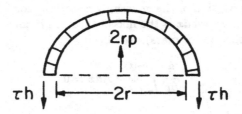

FIGURE 7.11.

(since $\tau(r_0) = 0$), where E is a constant and is referred to as *Young's modulus*. Therefore

$$p = \frac{Eh}{r}\frac{(r - r_0)}{r_0} = \frac{Eh}{r_0}\left(1 - \frac{r_0}{r}\right) = \frac{Eh}{r_0}\left(1 - \frac{(A_0)^{1/2}}{(A_0 + \tilde{A})^{1/2}}\right)$$

$$= \frac{Eh}{r_0}\left(1 - \frac{1}{(1 + \tilde{A}/A_0)^{1/2}}\right) \approx \frac{Eh}{2r_0}\frac{\tilde{A}}{A_0}. \tag{7.34}$$

Hence, if we substitute (7.34) into equations (7.33), then

$$\frac{\partial \tilde{A}}{\partial t} + \frac{\partial}{\partial x}(uA_0) = 0,$$

$$\rho\frac{\partial}{\partial t}(A_0 u) + \frac{\partial}{\partial x}\left(\frac{Eh}{2r_0}\right)\tilde{A} = 0. \tag{7.35}$$

We will suppose $h(x)/r_0(x)$ is constant. Therefore, if we write

$$\begin{cases} (uA_0) = \phi, \\[2mm] c^2 = \dfrac{Eh}{2r_0}, \\[2mm] \dfrac{\partial \tilde{A}}{\partial t} + \dfrac{\partial \phi}{\partial x} = 0, \\[2mm] \rho\dfrac{\partial \phi}{\partial t} + c^2\dfrac{\partial \tilde{A}}{\partial x} = 0 \end{cases} \tag{7.36}$$

and eliminate, say, $\tilde{A}$, it then follows that

$$\frac{\partial^2 \phi}{\partial t^2} = c^2\frac{\partial^2 \phi}{\partial x^2}. \tag{7.37}$$

In this case as opposed to the previous two, the "material" is not homogeneous; i.e., the tube changes properties along its length.

Exercises

1. Obtain a model for traffic flow on a one-dimensional road which has exits *only*.

2. Obtain the heat conduction equation for a finite wire immersed in a reservoir at constant temperature T_0. Each point of the wire loses energy to the surroundings at a rate which is proportional to the temperature difference between it and the reservoir.

3. Furnish the details for the derivation of the system (7.29).

4. Derive the wave equation for a nonuniform string.

5. In the derivation of (7.23) it was assumed that the motion of the string is restricted to the (x, y)-plane. What is the form of equation if the motion of the string takes place in three dimensions?

6. Derive the equation of motion for a circular drumhead.

7. Consider a stretched string immersed in oil. Each portion of the string experiences a drag force proportional to its velocity. Derive the equation of motion.

8. Consider the transformation from

    (a) Cartesian coordinates x, y to cylindrical (polar) coordinates

    $$x = r \cos \theta, \quad y = r \sin \theta$$

    and show

    $$\frac{\partial^2 T}{\partial x^2} T + \frac{\partial^2 T}{\partial y^2} \rightarrow \frac{1}{r} \frac{\partial}{\partial r} \left(r \frac{\partial T}{\partial r} \right) + \frac{1}{r^2} \frac{\partial^2 T}{\partial \theta^2};$$

    (b) Cartesian coordinates x, y, z to spherical coordinates

    $$x = r \sin \phi \cos \theta, \quad y = r \sin \phi \sin \theta, \quad z = r \cos \phi$$

    and show

    $$\nabla^2 T \rightarrow \frac{1}{r^2} \frac{\partial}{\partial r} \left(r^2 \frac{\partial T}{\partial r} \right) + \frac{1}{r^2 \sin \phi} \frac{\partial}{\partial \phi} \left(\sin \phi \frac{\partial T}{\partial \phi} \right)$$

    $$+ \frac{1}{r^2 \sin^2 \phi} \frac{\partial^2 T}{\partial \theta^2}.$$

9. Show that if
$$\kappa \nabla^2 T = \frac{\partial T}{\partial t}, \quad \mathbf{x} \in V,$$

then

$$\kappa \int\int\int_V (\nabla T)^2 d\mathbf{x} = -\frac{1}{2}\frac{\partial}{\partial t}\int\int\int_V T^2 d\mathbf{x} + \kappa \int\int_{\delta V} T\frac{\partial T}{\partial n} dS.$$

[Hint: Multiply the equation by T, integrate over V, and parts integrate.]

10. (a) Consider
$$\kappa \nabla^2 T + U\frac{\partial T}{\partial x} = \frac{\partial T}{\partial t}$$

and write
$$T = e^{\alpha x + \beta t}\phi(x, t)$$

to find constants α and β such that

$$\kappa \nabla^2 \phi = \frac{\partial \phi}{\partial t}.$$

(b) Consider
$$\kappa \nabla^2 T + bT = \frac{\partial T}{\partial t}$$

and write $T = e^{\alpha t}\phi$ to find α such that

$$\kappa \nabla^2 \phi = \frac{\partial \phi}{\partial t}.$$

11. Consider (7.22) under the condition that the volume V is insulated. This means that there is no heat flow at the boundary and therefore that
$$\mathbf{n} \cdot \mathbf{Q} = \mathbf{n} \cdot (\kappa \nabla T) = 0 \quad \text{on } \delta V.$$

In this case show that the final temperature is the average value of the initial temperatures. [Hint: Prove that energy is conserved; viz., $\int_V \rho C T\, d\mathbf{x}$ is a constant.]

12. A fluid is said to be incompressible if its mass density per unit volume is constant. Prove that if a fluid is incompressible, then its velocity is divergence-free, i.e.,
$$\nabla \cdot \mathbf{u} = 0.$$

[Hint: Derive the continuity equation in three dimensions.]

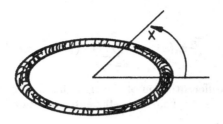

FIGURE 7.12.

7.2 Elementary Problems

In the previous section we derived the three basic partial differential equations: the wave equation, the diffusion equation, and the Laplace equation. The purpose of this section is to become familiar with the kinds of behavior these lead to, through consideration of elementary problems. In doing this we lay the foundation for treating partial differential equations in more general settings.

The Diffusion Equation

We consider a homogeneous wire ring of circumference unity (Figure 7.12). Heat conduction in the ring is governed by (7.12) (with $k = \lambda/\rho c$, a constant), where x denotes circumferential distance. There are no endpoints so it would seem that there are no boundary conditions. (In this model we ignore heat transfer to the *air* in which the ring sits.) This is replaced by the fact that the temperature $T(x, t)$ must be a one-periodic function

$$T(x + n, t) = T(x, t)$$

for all integers n. In particular, the initial data $T^0(x)$ is one-periodic and therefore has the Fourier series expansion

$$T^0 = \sum_{n=-\infty}^{\infty} \tau_n^0 e_n(x), \quad \tau_n^0 = (e_n, \tau_0) = \int_0^1 T^0(x) e^{-2\pi i n x} dx, \qquad (7.38)$$

as does the time-dependent temperature where

$$T(x, t) = \sum_{n=-\infty}^{\infty} \tau_n(t) e_n(x), \quad \tau_n = \int_0^1 (e_n, T). \qquad (7.39)$$

If we formally substitute (7.39) into the diffusion equation

$$\frac{\partial}{\partial t} T = k \frac{\partial^2}{\partial x^2} T,$$

we obtain,

$$\sum_{n=-\infty}^{\infty} \frac{\partial \tau_n}{\partial t} e_n(x) = k \sum_{n=-\infty}^{\infty} (-(2\pi n)^2) \tau_n e_n(x).$$

The term-by-term differentiation of the right-hand side still has to be justified. Next we equate the Fourier coefficients to obtain

$$\frac{\partial \tau_n}{\partial t} = -(2\pi n)^2 k \tau_n, \quad n = 0, \pm 1, \ldots, \tag{7.40}$$

which is an infinite system of *decoupled* equations. Each of these is immediately integrable and we get

$$\tau_n = \tau_n^0 e^{-k(2\pi n)^2 t} \tag{7.41}$$

in terms of the initial data τ_n^0, (7.38). (Note that each τ_n vanishes exponentially in n^2 which justifies the above term-by-term differentiation.) Thus, with the use of Fourier series, the solution resolves itself, in a natural way, into a series of harmonics or *modes*. Each mode has its own trigonometric spatial dependence, $e_n(x) = \exp(2\pi i n x)$, and decays at a fixed exponential rate, $\exp(-k(2\pi n)^2 t)$. The more highly oscillatory modes ($|n| \uparrow \infty$) vanish most rapidly, and ultimately as $t \uparrow \infty$ we are left only with the zeroth mode,

$$\lim_{t \uparrow \infty} T(x,t) = \tau_0^0 = \int_0^1 T^0(x) dx. \tag{7.42}$$

Thus the constant final temperature is the average of the initial temperature distribution. (This is a variant of the conservation of energy.)

If we substitute (7.41) and (7.38) into (7.39), we obtain

$$T = \sum_{n=-\infty}^{\infty} e^{-kt(2\pi n)^2} \int_0^1 e^{2\pi i n(x-y)} T^0(y) dy$$

$$= \int_0^1 \left(\sum_{n=-\infty}^{\infty} e^{-kt(2\pi n)^2} e_n(x-y) \right) T^0(y) dy. \tag{7.43}$$

We recall the definition of convolution for periodic functions (5.21) and write (7.43) as

$$T = G \star T^0 \tag{7.44}$$

with

$$G(x,t) = \sum_{n=-\infty}^{\infty} e^{-kt(2\pi n)^2} e_n(x). \tag{7.45}$$

In keeping with an earlier definition, $G(x,t)$ may be regarded as the solution operator or fundamental solution for the problem of heat conduction on a ring. For any initial temperature $T^0(x)$, (7.44) furnishes the solution for all time. Since $G(x,0) \star T^0 = T^0$, $G(x,t)$ must have the delta function property as $t \to 0$. This is seen to be the case if we return to our discussion of the Dirichlet kernel in Section 5.3; i.e.,

$$G_N = \sum_{n=-N}^{N} e^{2\pi i n x - (2\pi n)^2 kt}$$

for $t = 0$ is just the Dirichlet kernel, $D_N(x)$, as given by (5.17).

An Eigenfunction Approach

We digress briefly to view the above method of solution in a different light, one which will deeply change the way in which we view such problems.

The question to be asked is why was the above approach so effective in solving the problem. The central reason is that for the ring geometry just discussed, $\{e_n(x)\} = \{e^{2\pi i n x}\}$ are the eigenfunctions of the operator d^2/dx^2 (see (6.154)), which is the operator on the right-hand side of the diffusion equation, (7.39). In particular,

$$\frac{d^2}{dx^2} e_n = -\lambda_n e_n$$

under the periodic boundary conditions

$$e_n(0) = e_n(1), \quad e_n'(0) = e_n'(1)$$

has the sinusoids $\{\exp(2\pi i n x)\}$ for eigenfunctions and $\lambda_n = (2\pi n)^2$ for eigenvalues. It is this property which lies at the heart of why our method of solution worked so well.

As will be seen, the eigenfunction approach is both general and powerful. To underline this generality and power, we now outline the method somewhat abstractly but keep in mind that the ring problem is a special case. Consider

$$\frac{\partial u}{\partial t} = Lu, \quad u(t = 0) = f, \tag{7.46}$$

where L is a self-adjoint linear operator and u satisfies homogeneous boundary conditions, such as (6.132). If the eigenfunctions of L are denoted by u_n, then under the same boundary conditions of the problem,

$$Lu_n = -\lambda_n u_n.$$

$\{u_n\}$ is a complete orthonormal set of functions, and we can formally express the solution of (7.46) as

$$u = \sum_{n=0}^{\infty} a_n(t) u_n, \tag{7.47}$$

where
$$a_n(t) = (u_n, u). \tag{7.48}$$

In particular the initial data can also be represented in this way such that
$$a_n(0) = (u_n, f). \tag{7.49}$$

The inner product is the one which arises from the operator L. If we take the inner product of (7.46) with u_n, then
$$\left(u_n, \frac{\partial u}{\partial t}\right) = \frac{\partial a_n}{\partial t} = (u_n, Lu) = (Lu_n, u) = -\lambda_n a_n. \tag{7.50}$$

The solution to this simple first order ordinary differential equation under the initial data (7.49) is
$$a_n = e^{-\lambda_n t}(u_n, f). \tag{7.51}$$

Thus the solution to the problem (7.46) is
$$u = \sum_{n=0}^{\infty} e^{-\lambda_n t}(u_n, f)u_n. \tag{7.52}$$

The derivation of the temperature evolution in a ring is seen to be a case in point of this abstract procedure (the only variation being that periodic boundary conditions applied in that case).

It is important and essential to observe that the eigenfunction approach allows us to replace a differential operator like $\partial^2/\partial x^2$ by algebraic multiplication by the factor $-\lambda$.

Heat Condition on an Interval

We now present another illustration of the eigenfunction method. Consider the heat conduction problem
$$\begin{cases} \dfrac{\partial T}{\partial t} = k\dfrac{\partial^2 T}{\partial x^2}, \\[2mm] T(x,0) = T^0(x), \\[1mm] T(0,t) = T(1,t) = 0. \end{cases} \tag{7.53}$$

The corresponding eigenvalue problem based on the boundary conditions and the spatial part of the diffusion equation is
$$\frac{d^2 u}{dx^2} = -\lambda u, \quad u(0) = u(1) = 0.$$

This is a regular Sturm-Liouville problem and has eigenfunctions
$$u_n = \sin n\pi x$$

and eigenvalues

$$\lambda_n = n^2\pi^2.$$

The initial temperature is expanded in the eigenfunctions with

$$T^0 = \sum_{n=1}^{\infty} \tau_n^0 \sin n\pi x, \quad \tau_n^0 = 2 \int_0^1 T^0(x) \sin n\pi x \, dx,$$

as is the time-dependent temperature with

$$T = \sum_{n=1}^{\infty} \tau_n(t) \sin n\pi x, \quad \tau_n(t) = 2 \int_0^1 T(x,t) \sin n\pi x \, dx.$$

The diffusion equation in (7.53) is therefore reduced to

$$\frac{\partial \tau_n}{\partial t} = -n^2 \pi^2 \tau_n,$$

which has the solution

$$\tau_n = \tau_n^0 e^{-n^2\pi^2 t}.$$

Thus the solution to the problem is

$$T = \sum_{n=1}^{\infty} e^{-n^2\pi^2 kt} \tau_n^0 \sin n\pi x. \tag{7.54}$$

Earth's Temperature Profile

An example of a different sort involving the diffusion equation is the following. Consider the distribution of temperature below the earth's surface due to annual temperature variations. We idealize the problem by considering a flat earth, the temperature of which is governed by the equation

$$\frac{\partial T}{\partial t} = \kappa \frac{\partial^2 T}{\partial x^2}, \tag{7.55}$$

and x measures the distance down from the earth's surface. The thermal diffusivity, in cgs units, is estimated to be

$$\kappa \approx 2 \times 10^{-3} \, \text{cm}^2/\text{sec}.$$

The temperature at the surface of the earth $T_0(t)$ is in some average sense a periodic function of period one year. It therefore can be expanded in a Fourier series:

$$T_0 = \sum_{n=-\infty}^{\infty} a_n^0 e^{2\pi i n t/\tau}, \tag{7.56}$$

where one year τ, expressed in seconds, is

$$\tau \approx 3.15 \times 10^7 \text{sec}. \tag{7.57}$$

It is plausible that the temperature below the surface is also τ-periodic so that the solution to (7.55) has the form

$$T(x,t) = \sum_{n=-\infty}^{\infty} a_n(x)e^{2\pi int/\tau}. \tag{7.58}$$

It is important to note that at this point we abandon the notion of an initial value problem. Rather, we take it that the *initial data*, which was presented in the infinite past, is now irrelevant and that a *steady state oscillation* controls the time dependence.

If we substitute (7.58) into the temperature equation (7.55), then each harmonic component $a_n(x)$ is governed by

$$\frac{2\pi in}{\tau} a_n = \kappa \frac{d^2 a_n}{dx^2}.$$

The solution to this is given by

$$a_n = A_n e^{\lambda_n x} + B_n e^{-\lambda_n x}$$

with

$$\lambda_n = (\pi|n|/\kappa\tau)^{1/2}(1 \pm i) = \mu_n(1 \pm i),$$

where the plus sign goes with index $n > 0$ and the minus sign with index $n < 0$ (The $n = 0$ mode is excluded by assuming that temperature is measured relative to the yearly average.) Since the temperature is bounded as $x \uparrow \infty$, the first term of a_n cannot appear, i.e., $A_n = 0$. Finally, if we apply the surface condition (7.56), we obtain

$$T(x,t) = \sum_{n=-\infty}^{\infty} a_n^0 e^{(2\pi int/\tau)-(1+i\,sgn\,n)\mu_n x}.$$

If we consider just the first harmonic (where most of the signal resides), then

$$T \approx a_1^0 e^{(2\pi it/\tau)-(1+i)\mu_1 x} + \overline{a}_1^0 e^{(-2\pi it/\tau)-(1-i)\mu_1 x}.$$

(We have $a_{-1}^0 = \overline{a}_1^0$ in order to insure that the temperature is a real quantity.) It is convenient to write

$$a_1^0 = Ae^{i\gamma}, \quad A = |a_1^0|$$

and then

$$T \approx 2Ae^{-(\pi/\kappa\tau)^{1/2}x} \cos\left(\frac{2\pi t}{\tau} - (\pi/\kappa\tau)^{1/2}x + \gamma\right).$$

This has a very interesting conclusion, for when $x = \pi/(\pi/\kappa\tau)^{1/2}$, this states that the temperature is 180° behind that at the surface. If, for example, the surface is at *midsummer*, this point is at *midwinter*, and vice versa. The location of this point according to the specified constants is given by

$$x \approx 4.4\,\text{m}.$$

This explains the use of deep cellars in former times for food storage.

Exercises

1. Reconsider the problem of the earth's temperature distribution for the case of daily temperature variations (ignore yearly variations). At what location is the daily temperature twelve hours out of phase with the temperature at the surface.

2. Solve

$$\frac{\partial T}{\partial t} = \kappa \frac{\partial^2 T}{\partial x^2}, \quad 0 \le x \le 1,$$

$$T(x = 0) = 0, \quad T(x = 1) = T^0, \quad \text{a constant},$$

$$T(x,0) = 0.$$

[Hint: Use the steady solution to reduce the problem to the case of homogeneous boundary conditions.]

3. Consider the temperature distribution on a ring of circumference 2π and for which initially

$$T = 1, \quad 0 < x < \pi/4,$$

$$T = 0, \quad \pi/4 < x < 2\pi.$$

Find the temperature distribution for all time.

4. An infinite slab, $-1 < x < 1$, is at a constant temperature T^0. If at $t = 0$ the faces of the slab, $x = \pm 1$, are brought to zero temperature and held there, then find the temperature evolution in the slab.

5. Consider a rod $0 < x < 1$ with

$$T(0,t) = 0, \quad T(x,0) = 1,$$

and

$$T(1,t) = \kappa \frac{\partial T}{\partial x}(1,t).$$

Solve this for all time.

6. Use the eigenfunction method to solve

$$\frac{\partial u}{\partial t} + u = \frac{\partial^2 u}{\partial x^2}; \quad u(0) = u(1) = 0$$

with

$$u(t = 0) = \sin \pi x + \frac{1}{2} \sin 2\pi x + \frac{1}{3} \sin 3\pi x.$$

7. Use the eigenfunction method to solve

$$\frac{\partial T}{\partial t} = \kappa \frac{\partial^2 T}{\partial x^2} + F(x,t)$$

with

$$T(0,t) = T(1,t) = 0, \quad T(x,0) = 0.$$

[Hint: Expand $F(x,t)$ in the eigenfunctions.]

8. A rod is heated at $x = 1$ with a constant heat flux

$$\frac{\partial T(1,t)}{\partial x} = c.$$

The end at $x = 0$ is maintained at $T = 0$. If the temperature is initially T_0, what is the solution for all time?

The Wave Equation

In this section we consider the wave equation

$$\phi_{tt} = c^2 \phi_{xx}. \tag{7.59}$$

Perhaps the simplest example is obtained if we consider wave propagation on a ring of unit circumference. As before, the *boundary condition* is the requirement of periodicity. Since the right-hand side of (7.59) is $\partial^2/\partial x^2$, the appropriate eigenfunctions are again $\{e_n(x)\} = \{e^{2\pi i n x}\}$. Thus we express ϕ in terms of this complete set of functions such that

$$\phi(x,t) = \sum_{n=-\infty}^{\infty} \phi_n(t) e_n(x), \tag{7.60}$$

so that we again have a Fourier series. Also we expand the initial data:

$$\phi(x,0) = \phi^0(x) = \sum_{n=-\infty}^{\infty} \phi_n^0 e_n(x),$$

$$\frac{\partial}{\partial t}\phi(x,0) = \dot{\phi}^0(x) = \sum_{n=-\infty}^{\infty} \dot{\phi}_n^0 e_n(x) \tag{7.61}$$

with the implied relations

$$\begin{cases} \phi_n(t=0) = \phi_n^0, \\[2mm] \dfrac{\partial}{\partial t}\phi_n(t=0) = \dot{\phi}_n^0. \end{cases} \tag{7.62}$$

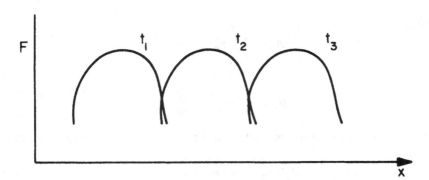

FIGURE 7.13.

If we take the inner product of the wave equation, (7.59), with $e_n(x)$, we see that each Fourier coefficient, ϕ_n, satisfies the equation

$$\frac{d^2\phi_n}{dt^2} = -c^2(2\pi n)^2\phi_n. \tag{7.63}$$

Alternatively, each term of the series (7.60) satisfies (7.59). In any case, (7.63) is easily integrated in terms of complex exponentials:

$$\phi_n = A_n e^{-ct2\pi in} + B_n e^{ct2\pi in}. \tag{7.64}$$

If we substitute these back into the series (7.60), we obtain

$$\phi = \sum_{n=-\infty}^{\infty} A_n e^{2\pi in(x-ct)} + \sum_{n=-\infty}^{\infty} B_n e^{2\pi in(x+ct)}, \tag{7.65}$$

which states that ϕ can be expressed as a function of $(x-ct)$ plus a function of $(x+ct)$. Formally (7.65) states that we can write

$$\phi = F(x - ct) + G(x + ct), \tag{7.66}$$

provided F and G are each twice differentiable. The individual terms of (7.66) satisfy the wave equation (7.59).

Consider, for example, $F(x - ct)$. We see that F is constant for $x - ct$ constant. Thus the same shape translates to the right with a speed c (see Figure 7.13). This justifies the name of wave equation and also the reference to c as the wave speed.

To return to the problem at hand we impose the conditions (7.62). This states that

$$A_n + B_n = \phi_n^0$$

and

$$B_n - A_n = \frac{\dot\phi_n^0}{2\pi icn}.$$

Thus

$$A_n = \frac{1}{2}\left\{\phi_n^0 - \frac{\dot{\phi}_n^0}{2\pi i c n}\right\}$$

and

$$B_n = \frac{1}{2}\left\{\phi_n^0 + \frac{\dot{\phi}_n^0}{2\pi i c n}\right\}.$$

When these are substituted back into the series solution (7.65), we obtain

$$\phi(x,t) = \frac{1}{2}\left(\sum_{n=-\infty}^{\infty} \phi_n^0 e^{2\pi i n(x-ct)} + \sum_{n=-\infty}^{\infty} \phi_n^0 e^{2\pi i n(x+ct)}\right)$$

$$-\frac{1}{2}\left(\sum_{n=-\infty}^{\infty} \frac{\dot{\phi}_n^0 e^{2\pi i n(x-ct)}}{2\pi i c n} - \sum_{n=-\infty}^{\infty} \frac{\dot{\phi}_n^0 e^{2\pi i n(x+ct)}}{2\pi i c n}\right). \qquad (7.67)$$

The first term, from comparison with (7.61), is just $\frac{1}{2}[\phi^0(x-ct)+\phi^0(x+ct)]$.
To obtain a comparable form for the second term, observe that

$$-\int_{-t}^{t} e^{2\pi i n c s}\,ds = \frac{1}{2\pi i n c}[e^{-2\pi i n c t} - e^{2\pi i n c t}].$$

Therefore, instead of (7.67), we can write

$$\phi(x,t) = \frac{1}{2}[\phi^0(x-ct) + \phi^0(x+ct)] + \frac{1}{2}\int_{-t}^{t} \dot{\phi}^0(x+cs)\,ds$$

$$= \frac{1}{2}[\phi^0(x-ct) + \phi^0(x+ct)] + \frac{1}{2c}\int_{x-ct}^{x+ct} \dot{\phi}^0(\eta)\,d\eta. \qquad (7.68)$$

The second form results from a simple variable change. Equation (7.68)
is known as *D'Alembert's Formula*. It can be verified directly that (7.68)
satisfies the wave equation (7.59) and the initial condition (7.61).

Although D'Alembert's form of the solution resulted from the periodic
initial value problem, we see that periodicity does not enter into the form
of (7.68). Therefore D'Alembert's solution holds without restriction for the
pure initial value problem, i.e., in the absence of boundaries.

Wave Propagation on an Interval

As a second example of the solution of the wave equation, consider (7.59)
with the boundary conditions

$$\phi(0) = 0 = \phi(L) \qquad (7.69)$$

and the initial conditions (7.61). This we recall corresponds to the physical
problem of a plucked string—see Figure 7.9. To solve this problem, we again

seek an approach based on eigenfunctions. Consider the eigenfunctions u_n and the eigenvalues λ_n such that

$$\frac{d^2 u_n}{dx^2} = -\lambda_n u_n, \quad u_n(0) = u_n(L) = 0.$$

The solution is immediate and is given by

$$u_n = \sin \frac{n\pi x}{L}, \quad \lambda_n = \frac{n^2 \pi^2}{L^2}. \tag{7.70}$$

The solution to the problem in terms of the u_n is

$$\phi = \sum_{n=1}^{\infty} b_n(t) \sin \frac{n\pi x}{L}. \tag{7.71}$$

The next step is to take the inner product of the wave equation, (7.59), with u_n to obtain

$$\frac{d^2 b_n}{dt^2} = -c^2 \lambda_n b_n = -\left(\frac{cn\pi}{L}\right)^2 b_n. \tag{7.72}$$

Alternatively, this expresses the idea that each term of (7.71) must satisfy the wave equation, (7.59). Appropriate initial conditions for solution of this equation are obtained by expanding the initial data (7.61) in the eigenfunctions u_n:

$$\phi^0(x) = \sum_{n=1}^{\infty} \phi_n^0 \sin \frac{n\pi x}{L}, \quad \phi_n^0 = \frac{2}{L} \int_0^L \phi^0(x) \sin \frac{n\pi x}{L} dx;$$

$$\dot{\phi}^0(x) = \sum_{n=1}^{\infty} \dot{\phi}_n^0 \sin \frac{n\pi x}{L}, \quad \dot{\phi}_n^0 = \frac{2}{L} \int_0^L \dot{\phi}^0(x) \sin \frac{n\pi x}{L} dx. \tag{7.73}$$

In terms of the coefficients ϕ_n^0 and $\dot{\phi}_n^0$, the solution to (7.72) is given by

$$b_n = \phi_n^0 \cos \frac{n\pi ct}{L} + \dot{\phi}_n^0 \frac{L}{n\pi c} \sin \frac{n\pi ct}{L};$$

and if we substitute this expression into the series (7.71), then

$$\phi = \sum_{n=1}^{\infty} \phi_n^0 \cos \frac{n\pi ct}{L} \sin \frac{n\pi x}{L} + \sum_{n=1}^{\infty} \frac{\dot{\phi}_n^0 L}{n\pi c} \sin \frac{n\pi ct}{L} \sin \frac{n\pi x}{L}. \tag{7.74}$$

To discuss this solution, first consider the case $\dot{\phi}^0 = 0$, i.e., consider a string which is initially displaced but is not initially in motion. If the trigonometric identity

$$\cos \frac{n\pi ct}{L} \sin \frac{n\pi x}{L} = \frac{1}{2} \left[\sin \frac{n\pi}{L}(x + ct) + \sin \frac{n\pi}{L}(x - ct) \right] \tag{7.75}$$

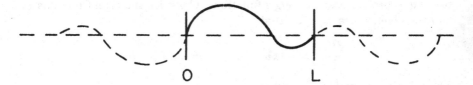

O   L

FIGURE 7.14.

is substituted into the solution (7.74), we obtain

$$\phi(x,t) = \frac{1}{2}\left(\sum_{n=1}^{\infty} \phi_n^0 \sin \frac{n\pi}{L}(x + ct) + \sum_{n=1}^{\infty} \phi_n^0 \sin \frac{n\pi}{L}(x - ct)\right). \quad (7.76)$$

This affords us an interesting interpretation of our solution. If we take the initial displacement $\phi^0(x)$, which is only defined for $0 \leq x \leq L$, and redefine it for the entire real line as a periodic function, odd in the origin and in the point L, then the solution to our problem is given by D'Alembert's Formula, (7.68),

$$\phi(x,t) = \frac{1}{2}(\phi^0(x + ct) + \phi^0(x - ct)), \quad (7.77)$$

which has the Fourier decomposition (7.76). After a little thought, we see that the reason for this is that the extension of the initial data to the entire real line, in the way indicated in Figure 7.14, *forces* $x = 0$ and $x = L$ to remain always at rest. (Such points are called *nodal points*.) Therefore the pure initial value problem and the boundary value problem have the same solution on the interval $0 \leq x \leq L$.

The above extension procedure is referred to as the *Method of Images*. It can also be extended to the full problem if $\dot{\phi}^0$ is also extended as a periodic function odd in the points $x = 0$ and $x = L$. Specifically D'Alembert's Formula, (7.68), is valid for the finite interval problem if both $\phi^0(x)$ and $\dot{\phi}^0(x)$ are redefined for $-\infty < x < \infty$ as $2L$-periodic and odd.

There is an interesting and useful interpretation to (7.75) which we mention before leaving the topic. For the sake of simplicity, consider solutions of (7.59) and (7.69) subject to the initial data

$$\phi^0 = \sin \frac{\pi x}{L}, \quad \dot{\phi}^0 = 0.$$

Then the corresponding solution from (7.74) is

$$y(x,t) = \cos \frac{\pi ct}{L} \sin \frac{\pi x}{L},$$

and from (7.75) this can also be written as

$$y(x,t) = \frac{1}{2}\left(\sin \frac{\pi}{L}(x - ct) + \sin \frac{\pi}{L}(x + ct)\right).$$

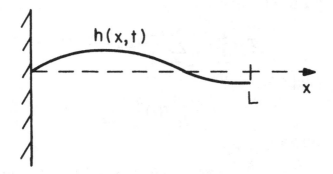

FIGURE 7.15.

In the first form we see that the string always has the form $\sin \pi x/L$, with an amplitude that varies sinusoidally in time between $+1$ and -1 with a period $2L/c$. This is referred to as a *standing wave*. On the other hand, the second form says this can also be interpreted as two *traveling waves* of amplitude $1/2$ passing to the right and to the left with a speed c.

Exercises

1. Verify that D'Alembert's Formula (7.68) satisfies the wave equation (7.59) and the initial data (7.62).

2. The deflection $h(x,t)$ of a *leaf spring* embedded in a wall can be shown to be described by the wave equation $h_{tt} = c^2 h_{xx}$, with the boundary conditions $h(0) = 0$, $(\partial h/\partial x)(L) = 0$ (Figure 7.15). Find the solution for

$$h(x,0) = f(x), \qquad \frac{\partial h}{\partial t}(x,0) = 0.$$

   Interpret your solution as being given by the Method of Images.

3. Solve the problem

$$\phi_{tt} = c^2 \phi_{xx}, \quad 0 \leq x < \infty,$$

$$\phi(0,t) = 0, \quad \phi(x,0) = f(x), \quad \phi_t(x,0) = g(x).$$

   [Hint: Use the Method of Images and D'Alembert's Formula.]

4. (a) Show that

$$\frac{\partial T}{\partial t} = \kappa \frac{\partial^2 T}{\partial x^2}, \quad T(0,t) = 0, \quad T(\ell,t) = 1, \quad T(x,0) = 0$$

is solved by

$$T = \frac{x}{\ell} + \frac{2}{\pi} \sum_{n=1}^{\infty} \frac{(-)^n}{n} \sin \frac{n\pi x}{\ell} e^{-\kappa n^2 \pi^2 t/\ell^2}.$$

(b) Prove

$$T = \int_0^t f(\tau) \frac{\partial T(x, t-\tau)}{\partial t} d\tau$$

satisfies

$$\frac{\partial T}{\partial t} = \kappa \frac{\partial^2 T}{\partial x^2}, \quad T(0,t) = 0, \quad T(\ell,t) = f(t), \quad T(x,0) = 0,$$

where $f(0) = 0$ but $f'(0)$ is not necessarily zero.

5. Demonstrate that each of the following standing waves can be represented as a superposition of traveling waves:

$$\cos \frac{\omega x}{c} \cos \omega t, \quad \sin \frac{\omega x}{c} \sin \omega t, \quad \sin \frac{\omega x}{c} \cos \omega t.$$

6. The motion of a vibrating string immersed in a viscous fluid is governed by

$$c^2 \phi_{xx} = \phi_{tt} + k\phi_t, \quad k > 0.$$

Suppose that

$$\phi(0,t) = \phi(1,t) = 0, \quad \phi(x,0) = f(x), \quad \dot\phi(x,0) = 0.$$

Use the eigenfunction approach to solve this problem.

7. Show directly that (7.74) can be put into the form of D'Alembert's solution, (7.68), if $\phi^0(x)$ and $\dot\phi^0(x)$ are extended by the Method of Images.

Laplace's Equation

As a last case, we consider the Laplace equation in two dimensions:

$$\nabla^2 T = \left(\frac{\partial^2}{\partial x^2} + \frac{\partial^2}{\partial y^2} \right) T = 0. \tag{7.78}$$

I   ...ly, this can be regarded as a description of a two-dimensional region in the sense described in Figure 7.8. As a specific example, we attempt to describe the temperature T in the interior of a circle of radius R, given the temperature on the circle (see Figure 7.16). In view of the circular

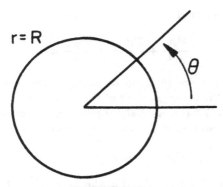

r = R

θ

FIGURE 7.16.

symmetry, it is advisable to use Laplace's equation in polar coordinates (see Exercise 8 at the end of Section 7.1):

$$\frac{\partial^2 T(r,\theta)}{\partial r^2} + \frac{1}{r}\frac{\partial T(r,\theta)}{\partial r} + \frac{1}{r^2}\frac{\partial^2 T(r,\theta)}{\partial \theta^2} = 0 \qquad (7.79)$$

with

$$T(R,\theta) = f(\theta). \qquad (7.80)$$

The eigenfunction approach for this problem is not entirely obvious. Recall that by this we mean the eventual replacement of a differential operator by a multiplication operator. Equation (7.79) contains three terms and the question is *"Which term(s) should we pick on?"* In view of the nature of the problem, $f(\theta)$ and hence $T(r,\theta)$ must be 2π-periodic functions on θ. It therefore seems appropriate that the presence of $\partial^2/\partial\theta^2$ along with this periodicity condition recommends use of the eigenfunctions $\{e^{in\theta}\}$. Thus we expand

$$T(r,\theta) = \sum_{n=-\infty}^{\infty} \rho_n(r)e^{in\theta}. \qquad (7.81)$$

We also expand (7.80) as

$$f(\theta) = \sum_{n=-\infty}^{\infty} f_n e^{in\theta}, \quad f_n = \frac{1}{2\pi}\int_0^{2\pi} f(\theta)e^{-in\theta}d\theta, \qquad (7.82)$$

so that

$$\rho_n(R) = f_n. \qquad (7.83)$$

To solve the problem we substitute the expansion (7.81) into (7.79). Upon interchange of summation and differentiation, this yields

$$\sum_{n=-\infty}^{\infty} \left(\frac{d^2\rho_n}{dr^2} + \frac{1}{r}\frac{d\rho_n}{dr} - \frac{n^2\rho_n}{r^2} \right) e^{in\theta} = 0,$$

which in turn implies that

$$\frac{d^2\rho_n}{dr^2} + \frac{1}{r}\frac{d\rho_n}{dr} - \frac{n^2\rho_n}{r^2} = 0. \tag{7.84}$$

As usual, these steps are justified on the grounds that the same result appears if we take the inner product of (7.79) with a typical *eigenfunction*, $\exp(in\theta)$. Equation (7.84) is an *Euler differential equation* and we therefore know that it has solutions of the form r^α. Thus we write

$$\rho_n = r^\alpha$$

and substitute this in (7.84) to obtain the *indicial relation*

$$\alpha(\alpha - 1) + \alpha - n^2 = 0 \rightarrow \alpha = \pm n.$$

The negative exponent is unacceptable since the temperature must be bounded at the origin. (If $n = 0$, the solution is $\ln r$, which must be excluded on the same grounds.) If we impose the boundary condition (7.83), we have

$$\rho_n = f_n \frac{r^{|n|}}{R^{|n|}}$$

and the solution for the temperature in the disc is given by

$$T = \sum_{n=-\infty}^{\infty} f_n \frac{r^{|n|}}{R^{|n|}} e^{in\theta}. \tag{7.85}$$

For example, if $f(\theta) = a \cos\theta$, then

$$T = a\frac{r}{R}\cos\theta$$

is the solution.

This solution can be given a more compact form by expressing f_n in terms of the integral in (7.82). Then, formally interchanging orders of summation and integration, we obtain

$$T = \frac{1}{2\pi}\int_0^{2\pi} f(\phi) \sum_{n=-\infty}^{\infty} \frac{r^{|n|}}{R^{|n|}} e^{i(\theta-\phi)n} d\phi. \tag{7.86}$$

This we see is in convolution form. To explore this further consider

$$\sum_{n=-\infty}^{\infty} \left(\frac{r}{R}\right)^{|n|} e^{in\theta} = \sum_{n=0}^{\infty} \left(\frac{r}{R}\right)^n e^{in\theta} + \sum_{n=0}^{\infty} \left(\frac{r}{R}\right)^n e^{-in\theta} - 1$$

$$= \frac{1}{1 - (r/R)e^{i\theta}} + \frac{1}{1 - (r/R)e^{-i\theta}} - 1 = \frac{1 - (r/R)^2}{1 - (2r/R)\cos\theta + (r/R)^2}. \tag{7.87}$$

Therefore

$$T(r,\theta) = \frac{1}{2\pi} \int_0^{2\pi} \frac{f(\phi)(1 - ((r/R)^2)d\phi}{1 - (2r/R)\cos(\theta - \phi) + (r/R)^2}. \tag{7.88}$$

This is known as *Poisson's Integral*. It represents the temperature at any interior point of the disc $r < R$ in terms of the temperature on the circumference of the disc.

Exercises

1. Solve for the steady temperature distribution in a square of side one if the temperature is zero on three sides and unity on the fourth.

2. Consider the (x, y)-plane with a circle of radius R punched out of it. The temperature is specified on $r = R$ and $T \to 0$ as $r \uparrow \infty$. Find the Poisson Integral representation of the steady temperature distribution in the plane.

3. Consider the annulus $a < r < A$. The temperature on $r = a$ is $\sin 4\theta$ and the temperature on $r = A$ is $\frac{1}{2}\cos 3\theta$. Find the steady temperature distribution in the annulus.

4. Prove

$$\frac{1}{2\pi} \int_0^{2\pi} \frac{R^2 - r^2}{R^2 - 2Rr\cos(\theta - \phi) + r^2} d\phi = 1.$$

$$\left[\text{Hint:} \quad \int_0^{2\pi} \frac{d\theta}{1 - A\cos\theta} = \frac{2\pi}{1 - A^2}. \right]$$

From this show

$$\min T(R, \theta) \le T(r, \theta) \le \max T(R, \theta).$$

5. Find the Poisson Integral representation for the annulus $a < r < A$.

6. Solve $u_{xx} + u_{yy} = 0$ in $x^2 + y^2 < 1$ and

   (a) $u = y^2x$ on $x^2 + y^2 = 1$,
   (b) $u = 1 + x + y$ on $x^2 + y^2 = 1$.

7. Solve $u_{xx} + u_{yy} = 0$ for the problem shown in Figure 7.17. What is the answer if $u = F_0$, a constant?

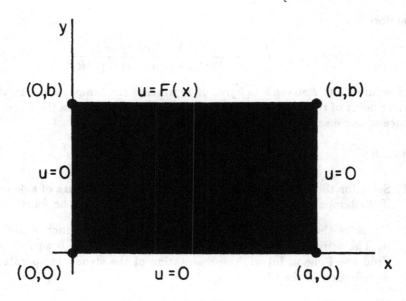

FIGURE 7.17.

8

The Fourier and Laplace Transforms

We pause in our discussion of partial differential equations to develop two techniques for treating problems in the infinite domain. One of these, the Laplace transform, is the continuous analog of the Z-transform, which we recall was developed for treating difference equations. The other technique, the (infinite) Fourier transform, is the extension to the infinite domain of Fourier series. The last we recall was itself the continuous version of the discrete Fourier transform. Since the Z- and the discrete Fourier transforms are themselves related, all these methods are connected to one another. These connections will be further discussed in this chapter.

8.1   Fourier Integral

In our discussion thus far Fourier series have been used to represent functions which are either periodic or finitely defined and then periodically extended to the entire real line. We next attempt to represent an arbitrarily defined function on the entire real line in terms of sinusoids.

To accomplish this we start with a function, $f(t)$, which is absolutely integrable over the real line, i.e.,

$$\int_{-\infty}^{\infty} |f(t)|dt < \infty. \tag{8.1}$$

As a first step consider the restriction of $f(t)$ to the interval $(-T/2, T/2)$,

$$f_T = f(t), \quad -\frac{T}{2} < t < \frac{T}{2}.$$

Next f_T is extended to be a T-periodic function by expanding in a Fourier series:

$$f_T = \sum_{n=-\infty}^{\infty} a_n e^{2\pi i n t/T}, \tag{8.2}$$

$$a_n = \frac{1}{T} \int_{-T/2}^{T/2} f_T(t)e^{-2\pi i n t/T} dt. \tag{8.3}$$

The strategy is to take the limit $T \uparrow \infty$. However, an immediate problem is that a_n vanishes under this limit. To repair this difficulty, we consider instead the product $a_n T$, which is seen to exist under this limit.

Next set

$$\Delta\omega = 1/T$$

and define the function $F(\omega)$ through

$$a_n T = F(n\Delta\omega).$$

Equation (8.3) can now be rewritten as

$$F(n\Delta\omega) = \int_{-T/2}^{T/2} f_T(t)e^{-2\pi i t(n\Delta\omega)}dt \qquad (8.4)$$

and (8.2) as

$$f_T = \Delta\omega \sum_{n=-\infty}^{\infty} F(n\Delta\omega)e^{2\pi i t(n\Delta\omega)}. \qquad (8.5)$$

If we now regard $F(n\Delta\omega)$ as the sampled form of a function $F(\omega)$ and then proceed to the limit $T \uparrow \infty$ in (8.4), we obtain

$$F(\omega) = \int_{-\infty}^{\infty} e^{-2\pi i \omega t}f(t)dt. \qquad (8.6)$$

The right-hand side of (8.5) under the limit $T \uparrow \infty$ or $\Delta\omega \to 0$ is clearly the discrete approximation, i.e., the Riemann sum, of

$$f(t) = \int_{-\infty}^{\infty} e^{2\pi i \omega t}F(\omega)d\omega. \qquad (8.7)$$

Equations (8.6) and (8.7) are referred to as a *Fourier pair*. Equation (8.7) is the sought-after representation of a function in terms of sinusoids, while (8.6), the *Fourier transform* of $f(t)$, can be viewed as its inversion, thus giving the amplitudes of the sinusoids for the synthesis (8.7). We continue to proceed formally and substitute (8.6) into (8.7) to find that

$$f(t) = \int_{-\infty}^{\infty} \int_{-\infty}^{\infty} e^{2\pi i \omega(t-s)}f(s)ds d\omega. \qquad (8.8)$$

This is the content of Fourier's Theorem in the present situation. We next sketch a mathematically more respectable proof of this result.

As a first step observe that by virtue of (8.1), the Fourier transform (8.6) exists. For (8.1) insures that $f(t)$ vanishes as $|t| \uparrow \infty$ and the sinusoidal coefficient in (8.6) causes the integrand to oscillate about zero, which can only enhance convergence. The demonstration given below requires that

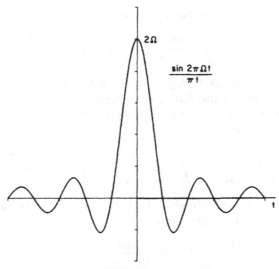

FIGURE 8.1.

$f(t)$ be *Hölder continuous*. This simply means that at each point t, there exists a constant K such that

$$|f(t + \delta) - f(t)| < K|\delta|^{\alpha}, \quad \alpha > 0.$$

(Actually f can be discontinuous but must be Hölder continuous to the right and left of a discontinuity.) Next consider

$$f_\Omega(t) = \int_{-\Omega}^{\Omega} d\omega \int_{-\infty}^{\infty} f(s)e^{2\pi i\omega(t-s)}ds. \tag{8.9}$$

Then under the hypothesis on f we wish to show

$$\lim_{\Omega \uparrow \infty} f_\Omega(t) = f(t). \tag{8.10}$$

Since the integrand of (8.9) is now absolutely integrable, the orders of integration can be interchanged:

$$f_\Omega(t) = \int_{-\infty}^{\infty} f(s) \int_{-\Omega}^{\Omega} e^{2\pi i\omega(t-s)}d\omega ds$$

$$= \int_{-\infty}^{\infty} f(s)\frac{\sin 2\pi\Omega(t - s)}{\pi(t - s)}ds, \tag{8.11}$$

where a simple integration has been performed in the last.

The even function $(\sin 2\pi\Omega t)/\pi t$ is sketched in Figure 8.1. The peak height is 2Ω and the frequency of oscillation is Ω. As we have seen earlier in our discussion of Gibbs' phenomenon,

$$\int_0^\infty \frac{\sin 2\pi\Omega t}{\pi t}\,dt = \int_{-\infty}^0 \frac{\sin 2\pi\Omega t}{\pi t}\,dt = \frac{1}{\pi}\int_0^\infty \frac{\sin s\,ds}{s} = \frac{1}{2}. \qquad (8.12)$$

(A simple contour integration gives this result.) Thus it can be concluded that $(\sin 2\pi\Omega t)/\pi t$ has the *delta function property*, and the proof of (8.10) now seems clear. To fill in the details, we choose a small $\epsilon > 0$ and decompose (8.11) as follows:

$$f_\Omega = \left(\int_{|s-t|>\epsilon} ds + \int_t^{t+\epsilon} ds + \int_{t-\epsilon}^t ds\right) f(s)\frac{\sin 2\pi\Omega(t-s)}{\pi(t-s)}.$$

Since $f(s)/(t-s)$ is absolutely integrable in the first integral, it follows from the Riemann–Lebesgue Lemma that

$$\int_{|s-t|>\epsilon} f(s)\frac{\sin 2\pi\Omega(t-s)}{\pi(t-s)}\,ds = o(1)$$

as $\Omega \uparrow \infty$. Next consider the second integral and rewrite it as

$$\int_t^{t+\epsilon} f(s)\frac{\sin 2\pi\Omega(t-s)}{\pi(t-s)}\,ds = f^+(t)\int_t^{t+\epsilon} \frac{\sin 2\pi\Omega(t-s)}{\pi(t-s)}\,ds$$

$$+ \int_t^{t+\epsilon} (f(s) - f^+(t))\frac{\sin 2\pi\Omega(t-s)}{\pi(t-s)}\,ds,$$

where we have adopted the notation that

$$f^\pm(t) = \lim_{\substack{\delta\to 0 \\ \delta \gtrless 0}} f(t+\delta).$$

In the first of these integrals set $-2\pi\Omega(t-s) = x$; then

$$\int_t^{t+\epsilon} \frac{\sin 2\pi\Omega(t-s)}{\pi(t-s)}\,ds = \frac{1}{\pi}\int_0^{2\pi\epsilon\Omega} \frac{\sin x}{x}\,dx \xrightarrow{\Omega\uparrow\infty} \frac{1}{2}.$$

In the second integral Hölder continuity is used to estimate it as follows:

$$\left|\int_t^{t+\epsilon} (f(s) - f^+(t))\frac{\sin 2\pi\Omega(t-s)}{\pi(t-s)}\,ds\right|$$

$$\le \int_t^{t+\epsilon} |f(s) - f^+(t)|\frac{|\sin 2\pi\Omega(t-s)|}{\pi|t-s|}\,ds \le \frac{1}{\pi}\int_t^{t+\epsilon} \frac{K|t-s|^\alpha}{|t-s|}\,ds = \frac{K}{\pi\alpha}\epsilon^\alpha,$$

which, since $\alpha > 0$, is small if ϵ is small. A similar discussion holds for the third integral, and putting this together we arrive at

$$f(t) = \frac{1}{2}\{f^+(t) + f^-(t)\} + o(1).$$

Thus, if we redefine the function $f(t)$ to be its arithmetic mean at a discontinuity, then the Fourier Theorem is proven. With the use of a more powerful mathematical base, the Fourier Theorem can be proven under weaker hypotheses on the function $f(t)$.

A number of properties now follow, more or less in analogy with the results obtained for Fourier series and for discrete Fourier series. The Fourier *cosine* and *sine transforms* are defined by

$$a(\omega) = \int_{-\infty}^{\infty} f(t) \cos 2\pi\omega t \, dt = a(-\omega),$$

$$b(\omega) = \int_{-\infty}^{\infty} f(t) \sin 2\pi\omega t \, dt = -b(-\omega) \tag{8.13}$$

so that

$$F(\omega) = a(\omega) - ib(\omega) \quad (= \overline{F}(-\omega), \text{ if } f \text{ is real}). \tag{8.14}$$

For square integrable f and g, we can define the inner product

$$(f, g) = \int_{-\infty}^{\infty} \overline{f}(t)g(t)dt \tag{8.15}$$

(the existence of which follows from Schwarz's Inequality). It is then left as an exercise to demonstrate that

$$(f, g) = \int_{-\infty}^{\infty} \overline{F}(\omega)G(\omega)d\omega = (F, G) \tag{8.16}$$

and, in particular,

$$(f, f) = \|f\|^2 = (F, F) = \|F\|^2. \tag{8.17}$$

Thus the *distance* of a function and its transform from the *origin* is the same.

Of special interest is the convolution integral, which is now defined as

$$f \star g = \int_{-\infty}^{\infty} f(t - s)g(s)ds \xrightarrow{t-s=u} \int_{-\infty}^{\infty} f(u)g(t - u)du = g \star f. \tag{8.18}$$

As was the case for Fourier series and discrete Fourier series, the convolution product becomes ordinary multiplication under Fourier transformation with

$$\int_{-\infty}^{\infty} e^{-2\pi i \omega t}(f \star g)dt = F(\omega)G(\omega). \tag{8.19}$$

The inversion of this is

$$f \star g = \int_{-\infty}^{\infty} e^{2\pi i \omega t} F(\omega)G(\omega)d\omega. \tag{8.20}$$

Proofs of these relations are left to the exercises.

Linear Systems. In Section 4.1 it was pointed out that a *black box* in certain instances can be regarded as a system which is linear and translationally invariant in time. We now reopen this discussion. Denote inputs to such a system by f and g and their respective outputs by $\hat{f}$ and $\hat{g}$. The system (or black box) will be represented by $\mathcal{L}$. It follows from linearity that

$$\mathcal{L}(af + bg) = a\hat{f} + b\hat{g}, \quad a, b \text{ constant},$$

and from translational invariance in time that

$$\mathcal{L}(f(t + \tau)) = \hat{f}(t + \tau).$$

Under these conditions it was shown in Section 4.1 that $\mathcal{L}$ can be written as a convolution product,

$$\mathcal{L}f = \int_{-\infty}^{\infty} \ell(t - s)f(s)ds = \ell \star f. \tag{8.21}$$

With the use of (8.20), (8.21) can be written as

$$\mathcal{L}f = \ell \star f = \int_{-\infty}^{\infty} L(\omega)F(\omega)e^{2\pi i \omega t}d\omega. \tag{8.22}$$

In this form the operator $\mathcal{L}$ is seen to have an important interpretation. The Fourier representation, (8.7), can be regarded as a resolution of the signal, or input, f into an infinite sum of sinusoids, densely spaced on the ω or frequency axis. The convolution, (8.22), which gives the output, multiplies each amplitude $F(\omega)$ of this sum by a factor $L(\omega)$. This allows us to regard $\mathcal{L}$ as a *filter*. It *amplifies* (or deamplifies) each individual frequency component, and since $L(\omega)$ is complex, it can be written as

$$L(\omega) = \lambda(\omega)e^{i\theta(\omega)}, \quad \lambda = |L(\omega)|,$$

so that a *phase* change $\theta(\omega)$, as well as multiplication by λ, is introduced by the filtering process. Thus the black box is entirely characterized by $L(\omega)$.

As an example, consider the simple linear, time-invariant operator

$$\tilde{g}(t) = \frac{1}{T} \int_{t-T/2}^{t+T/2} g(\tau)d\tau. \tag{8.23}$$

This operator *smooths* a function by taking a running average over a *window* of size T. The claim is that (8.23) can be regarded as a filtering operation.

For this purpose we first define

$$h_T(t) = \begin{cases} 1/T, & |t| \leq T/2, \\ 0, & \text{otherwise}, \end{cases}$$

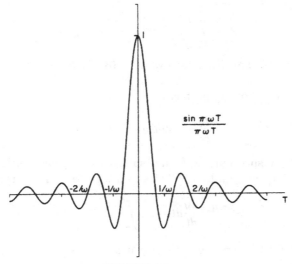

FIGURE 8.2.

from which it may be seen that (8.23) can also be written as $h_T \star g = \tilde{g}$ or

$$\tilde{g} = \int_{-\infty}^{\infty} h_T(t-s)g(s)ds = \frac{1}{T} \int_{t-T/2}^{t+T/2} g(s)ds.$$

From (8.22) $\tilde{g}$ can also be written as

$$\tilde{g} = \int_{-\infty}^{\infty} e^{2\pi i\omega t} H_T(\omega)G(\omega)d\omega$$

with

$$H_T(\omega) = \int_{-\infty}^{\infty} e^{-2\pi i\omega t} h_T(t)dt = \int_{-T/2}^{T/2} \frac{e^{-2\pi i\omega t}}{T} dt = \frac{\sin \pi\omega T}{\pi\omega T},$$

which is essentially the same function encountered in (8.11). A sketch of the *filter* $H_T(\omega)$ is given in Figure 8.2. (Note that $H_T(\omega)$ is real so that no phase change appears.) In a widely used terminology one says that the running average is a *narrowband lowpass filter* if T is large and a *wideband* (or *all-pass*) *filter* if T is small (i.e., as $T \downarrow 0$ we have a delta function). The term *pass* refers to the *frequencies*, ω, that are not significantly removed.

The Diffusion Equation. Each time we carry out a Fourier transform we obtain a Fourier transform pair and in this way a table is compiled. For example, $h_T(t)$ and $H_T(\omega)$ give us one such pair, and after the exercises we will be able to add several additional pairs to our table. Books listing such pairs are available. A particularly important transform is that of $y =$

$\exp[-at^2]$, i.e.,

$$Y = \int_{-\infty}^{\infty} e^{-2\pi i\omega t} e^{-at^2} dt, \quad a > 0. \tag{8.24}$$

For purposes of exposition we evaluate (8.24) by a slightly unorthodox method.

Consider the following problem:

$$\frac{dy}{dt} = -2aty, \quad y(0) = 1.$$

As seen by inspection, $\exp[-at^2]$ satisfies this problem. If we multiply the differential equation by $e^{-i2\pi\omega t}$ and integrate over the real axis, then

$$\int_{-\infty}^{\infty} e^{-2\pi i\omega t} \frac{dy}{dt} dt + 2a \int_{-\infty}^{\infty} e^{-2\pi i\omega t} ty \, dt = 0.$$

Since $y \to 0$ for $|t| \uparrow \infty$, a parts integration yields

$$2\pi i\omega \int_{-\infty}^{\infty} e^{-2\pi i\omega t} y \, dt - \frac{a}{\pi i} \frac{d}{d\omega} \int_{-\infty}^{\infty} e^{-2\pi i\omega t} y \, dt$$

$$= 2\pi i\omega Y - \frac{a}{\pi i} \frac{d}{d\omega} Y = 0.$$

Note that differentiation with respect to ω brings down a factor of t in the second integral. Therefore the transformed differential equation can be written as

$$\frac{d}{d\omega} Y = -\frac{2\omega\pi^2}{a} Y.$$

This is easily integrated and gives

$$Y = Ae^{-\omega^2\pi^2/a},$$

where the multiplicative constant, A, remains to be evaluated. From (8.24) we observe

$$Y(0) = \int_{-\infty}^{\infty} e^{-at^2} dt = (\pi/a)^{1/2}$$

and hence

$$Y = (\pi/a)^{1/2} e^{-\omega^2\pi^2/a}. \tag{8.25}$$

Therefore $Y(\omega)$, as given by (8.25), and $y = \exp[-at^2]$ constitute a transform pair.

Example. An application of this result is the problem of diffusion in an infinite domain,

$$\partial_t T = \kappa \partial_x^2 T, \quad T(t = 0) = T^0(x), \quad -\infty \leq x \leq \infty. \tag{8.26}$$

As we have mentioned on a number of previous occasions, ∂_x^2 is a self-adjoint linear operator, and sinusoids are formal eigenfunctions of it. In particular,

$$\partial_x^2 e^{2\pi inx} = -(2n\pi)^2 e^{2\pi inx}$$

for all n (i.e., n is not restricted to integers and can take on any real value). Therefore, in analogy with the periodic case (7.39), one should expand $T(x, t)$ in terms of the *eigenfunctions* of ∂_x^2, specifically,

$$T(x, t) = \int_{-\infty}^{\infty} e^{2\pi ikx} T(k, t) dk, \tag{8.27}$$

which is just another way of introducing the Fourier transform. We also write

$$T^0(x) = \int_{-\infty}^{\infty} e^{2\pi ikx} T^0(k) dk, \tag{8.28}$$

where

$$T^0(k) = \int_{-\infty}^{\infty} e^{-2\pi ikx} T^0(x) dx. \tag{8.29}$$

If (8.27) is substituted into the equation (8.26), we obtain

$$\int_{-\infty}^{\infty} e^{2\pi ikx} \frac{\partial T(k, t)}{\partial t} dk = \kappa \int_{-\infty}^{\infty} e^{2\pi ikx} (-(2\pi k)^2) T(k, t)) dk.$$

where the right hand follows from repeated parts integration. (See Exercise 7.) Then, since zero is the only function which transforms as zero, we arrive at the transformed equation

$$\frac{\partial T}{\partial t} = -\kappa (2\pi k)^2 T. \tag{8.30}$$

The solution to this equation, subject to the data (8.29), is given by

$$T = e^{-\kappa k^2 4\pi^2 t} T^0(k),$$

while the solution to the problem originally posed, (8.26), is given by

$$T(x, t) = \int_{-\infty}^{\infty} e^{2\pi ikx} e^{-\kappa k^2 4\pi^2 t} T^0(k) dk. \tag{8.31}$$

From (8.20), (8.31) can be rewritten in convolution form

$$T(x) = U \star T^0(x), \tag{8.32}$$

where

$$U(x, t) = \int_{-\infty}^{\infty} e^{2\pi ikx} e^{-\kappa k^2 4\pi^2 t} dk.$$

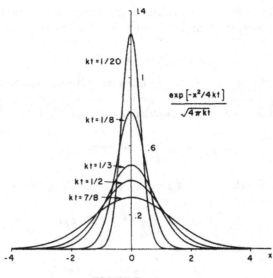

FIGURE 8.3.

But U can be evaluated by the transform pair (8.24), (8.25) and is given by

$$U(x,t) = \frac{e^{-(x^2/4kt)}}{\sqrt{4\pi kt}};\qquad(8.33)$$

hence,

$$T(x,t) = \int_{-\infty}^{\infty} \frac{e^{-(x-y)^2/4kt}}{\sqrt{4\pi\kappa t}} T^0(y)dy.\qquad(8.34)$$

$U(x,t)$ is the *solution operator* or *fundamental solution* for the diffusion equation. Since $T(x,t)$ in (8.34) must tend to $T^0(x)$ as $t \downarrow 0$, U must have the delta function property as $t \downarrow 0$. This can be seen directly. First observe

$$\int_{-\infty}^{\infty} \frac{e^{-x^2/4kt}}{\sqrt{4\pi kt}} dx \xrightarrow{\ s=x/\sqrt{2kt}\ } \frac{1}{\sqrt{2\pi}} \int_{-\infty}^{\infty} e^{-s^2/2} ds = 1.\qquad(8.35)$$

A plot of U for various values of kt is shown in Figure 8.3.

$U(x,t)$ itself is a solution of the diffusion equation (Exercise 8) and gives the temperature evolution of an initial point source of heat (and thus an infinite temperature) at the origin. We notice from (8.34) that even the most rayy itial temperature becomes infinitely smooth (i.e., is infinitely differentiable) after the initial instant. This was also observed in the periodic case (7.43).

A more exhaustive discussion of this solution will follow in the next chapter. In the meantime, some additional features are discussed in the exercises.

Exercises

1. Find the Fourier transform of each of the following and discuss their filter properties:

   (a)
   $$\begin{cases} e^{-at}, & t > 0, \ a > 0, \\ 0, & t < 0; \end{cases}$$

   (b)
   $$e^{-|t|}, \quad -\infty < t < \infty;$$

   (c)
   $$\begin{cases} \cos 2\pi\omega_0 t, & |t| < T, \\ 0, & |t| > T; \end{cases}$$

   (d)
   $$\begin{cases} 1, & 0 < t < T, \\ 0, & t < 0, \ t > T; \end{cases}$$

   (e)
   $$f(x) = \begin{cases} A(a - |x|), & |x| < a, \\ 0, & |x| > a. \end{cases}$$

2. Find $f(t)$ for each of the following:

   (a) $F(\omega) = \int_{-\infty}^{\infty} e^{i\omega t} f(t) dt$;

   (b) $\mathcal{F}(\omega) = \frac{1}{2\pi} \int_{-\infty}^{\infty} e^{-i\omega t} f(t) dt$;

   (c) $F_c(\omega) = \frac{1}{2\pi} \int_{0}^{\infty} \cos\omega t \, f(t) dt$ \quad (cosine transform);

   (d) $F_s(\omega) = \frac{1}{2\pi} \int_{0}^{\infty} \sin\omega t \, f(t) dt$ \quad (sine transform).

3. Evaluate the Fourier transform of $\exp(-ax^2)$, i.e.,

   $$\int_{-\infty}^{\infty} e^{-2\pi i k x - a x^2} dx,$$

   by completing the square in the exponents. [Hint: Make use of $\int_{-\infty}^{\infty} \exp(-x^2/2) dx = \sqrt{2\pi}$, but be careful in regard to the path of integration.]

4. Solve
   $$\frac{\partial u}{\partial t} + V_0 \frac{\partial u}{\partial x} + ku = \frac{\partial^2 u}{\partial x^2}, \quad -\infty < x < \infty,$$
   $$u(x,0) = u^0(x), \quad V_0 \text{ and } k \text{ are constants} > 0,$$

   by means of Fourier transforms. Express the solution in the form

   $$U \star u^0,$$

   give U explicitly, and show that it has the delta function property as $t \downarrow 0$.

5. Solve

$$\int_{-\infty}^{\infty} e^{-|x-y|} \phi(y)dy + 2\phi(x) = h_T(x)$$

for ϕ by Fourier transforms (h_T is defined in the above subsection entitled "Linear Systems"). You should find an explicit form of the solution ϕ.

6. Formally demonstrate (8.17), (8.19), and (8.20).

7. If $f(x)$ and its derivatives are absolutely integrable, demonstrate that

$$\int_{-\infty}^{\infty} e^{-ikx} \frac{d^n f}{dx^n} dx = (ik)^n \int_{-\infty}^{\infty} e^{-ikx} f(x)dx.$$

8. Show that $U(x,t)$ defined by (8.33) satisfies the diffusion equation, (8.26).

9. Show that $F(-k) = \overline{F}(k)$ is a necessary and sufficient condition for $f(x)$ to be real and that $F(-k) = -\overline{F}(k)$ is a necessary and sufficient condition that $f(x)$ be imaginary.

10. What is $f(x)$ if

   (a) $F(k) = k/(k^2 + a^2)$,

   (b) $F(k) = 1/(k^2 + a^2)$? [Hint: Use a contour integral to evaluate the inversion integral.]

11. Solve

$$\frac{\partial^2 \phi}{\partial t^2} = c^2 \frac{\partial^2 \phi}{\partial x^2}, \quad -\infty < x < \infty,$$

$$\phi(0) = f(x), \quad \frac{\partial \phi}{\partial t}(0) = g(x)$$

by means of Fourier transforms and obtain D'Alembert's Formula, (7.68).

12. Show Parseval's Formula, (8.16),

$$(f,g) = \int_{-\infty}^{\infty} \overline{f}(x)g(x)dx = \int_{-\infty}^{\infty} \overline{F}(k)G(k)dk = (F,G).$$

8.2   Laplace Transform

The Fourier transform can fail to converge if the function $f(t)$ does not tend to zero as $t \uparrow \infty$. For this and other reasons we consider the *Laplace transform*. Suppose we have a function $f(t)$ defined for $0 \le t \le \infty$. Its *Laplace transform* $F(p)$ is defined to be

$$F(p) = \mathcal{L}(f) = \int_0^\infty e^{-pt} f(t) dt. \tag{8.36}$$

p is a complex variable, $p = r + is$, and it is assumed that the real part, r, is sufficiently large so that the integral in (8.36) converges.

The Laplace transform is the continuous analogue of the Z-transform discussed earlier (see Section 4.5). To see this we approximate the integral (8.36) by its Riemann sum:

$$F(p) \approx \Delta t \sum_{n=0}^\infty e^{-pn\Delta t} f(n\Delta t) \approx \Delta t \sum_{n=0}^\infty z^{-n} f_n = \Delta t G(z), \tag{8.37}$$

where we have set

$$z = e^{p\Delta t} \tag{8.38}$$

and

$$f_n = f(n\Delta t). \tag{8.39}$$

From (8.37) it is seen that $G(z)$ is the Z-transform of the sequence $\{f_n\}$, i.e., $G(z) = Z[f_n]$.

This heuristic analogy also indicates the proper way in which to invert a Laplace transform, i.e., to find $f(t)$ from $F(p)$. From the theory of the Z-transform we know

$$f_n = \frac{1}{2\pi i} \oint_{|z|=R} z^{n-1} G(z) dz, \tag{8.40}$$

where $|z| = R$ is an appropriate circle about the origin. If we now use (8.38) to transform back to the Laplace variable, p, and (8.39) to reintroduce $f(t)$, then

$$f(n\Delta t) = \frac{\Delta t}{2\pi i} \int_{\text{path}} e^{p(n-1)\Delta t} G(z) e^{p\Delta t} dp = \frac{1}{2\pi i} \int_{\text{path}} e^{pn\Delta t} F(p) dp, \tag{8.41}$$

where (8.37) has been substituted to obtain the last integral and the *path* is still to be determined. Then if we set $n\Delta t = t$,

$$f(t) = \frac{1}{2\pi i} \int_B e^{pt} F(p) dp, \tag{8.42}$$

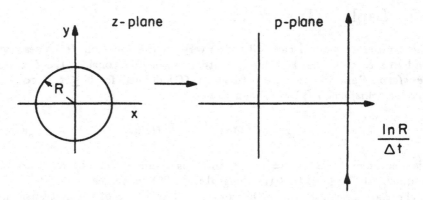

FIGURE 8.4.

where the still to be determined path of integration is denoted by B. In order to specify B, observe that from (8.38)

$$p = \frac{\ln z}{\Delta t};$$

and if we write $z = Re^{i\theta}$, $-\pi < \theta < \pi$, the path of integration is

$$p = \frac{\ln R}{\Delta t} + \frac{i\theta}{\Delta t}, \quad -\pi < \theta < \pi.$$

We recall that the inversion (8.40) requires R to be large enough to contain the singularities of $G(z)$; hence, R can be taken to be greater than one. Therefore $\ln R/\Delta t$ as $\Delta t \to 0$ is a large positive quantity, while $i\theta/\Delta t$ varies between $-i\infty$ and $i\infty$ in the limit $\Delta t \to 0$. As indicated in Figure 8.4, the circle $|z| = R$ maps into a vertical line which in the limit, $\Delta t \to 0$, goes between $\pm i\infty$.

The path B is referred to as a *Bromwich path* and is usually represented by a vertical arrow. Thus, instead of (8.42), we write

$$f(t) = \frac{1}{2\pi i} \int_{\uparrow} e^{pt} F(p)dp = \mathcal{L}^{-1}F. \tag{8.43}$$

In writing (8.43) there is the fine print that the Bromwich path lies sufficiently far to the right in the p-plane. This follows from the fact that R in the above heuristic treatment must be large enough to contain all singularities of $G(z)$. Equations (8.36) and (8.43) constitute a Laplace transform pair so that all singularities of F lie to the left of the Bromwich path. Our demonstration was heuristic and we now give a rigorous though limited proof that f and $F(p)$ are inverses of one another.

Consider the function $f(t)$ defined to be piecewise analytic on the positive real axis. By this we mean that f is complex analytic except possibly at isolated

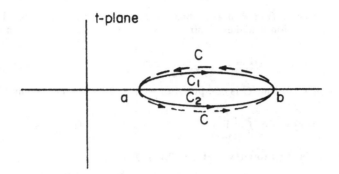

FIGURE 8.5.

points. Therefore f can be regarded as a sum of functions, each analytic in some interval and zero elsewhere. Thus, without too much loss of generality, we fix attention on a function $f(t)$ which is analytic for $t \in (a, b)$ and zero elsewhere on the real axis. As a function of z, $f(z)$ is analytic in the neighborhood of this segment, and $f \equiv 0$ elsewhere on the real line. The transform of f is

$$F(p) = \int_0^\infty f(t)e^{-pt}dt = \int_a^b f(t)e^{-pt}dt.$$

Consider

$$\tilde{f}(x) = \frac{1}{2\pi i}\int_{\alpha-i\beta}^{\alpha+i\beta} e^{px}\left(\int_a^b e^{-pt}f(t)dt\right)dp.$$

Since $F(p)$ itself is clearly an analytic function of p (it has a derivative with respect to the complex variable p), the contour between the endpoints $\alpha \pm i\beta$ can be distorted and to fix matters we take it to be the vertical line $Re\, p = \alpha$. Next write

$$\tilde{f}(x) = \frac{1}{2\pi i}\int_{\alpha}^{\alpha+i\beta} e^{px}F(p)dp + \frac{1}{2\pi i}\int_{\alpha-i\beta}^{\alpha} e^{px}F(p)dp. \qquad (8.44)$$

Since $f(x)$ is analytic, we can distort the path of integration in the definition of $F(p)$. Considering Figure 8.5, we employ C_2 in the first integral of (8.44) and C_1 in the second integral. If we introduce this and interchange orders of integration (since we have finite integrals, which are absolutely integrable, this presents no problem), then

$$\tilde{f}(x) = \frac{1}{2\pi i}\int_{C_2} dt\, f(t)\int_{\alpha}^{\alpha+i\beta} e^{p(x-t)}dp$$

$$+ \frac{1}{2\pi i}\int_{C_1} dt\, f(t)\int_{\alpha-i\beta}^{\alpha} e^{p(x-t)}dp.$$

The integration is carried out to obtain

$$\tilde{f}(x) = \frac{1}{2\pi i}\int_{C_2} \frac{f(t)}{x-t}\{e^{(\alpha+i\beta)(x-t)} - e^{\alpha(x-t)}\}dt$$

$$+ \frac{1}{2\pi i}\int_{C_1} \frac{f(t)}{x-t}\{e^{\alpha(x-t)} - e^{(\alpha-i\beta)(x-t)}\}dt.$$

In the first integral $Im\, t < 0$ and thus $-i\beta t < 0$. Hence, as $\beta \uparrow \infty$ the corresponding contribution vanishes. A similar remark follows for the second integral. Therefore

$$\lim_{\beta \uparrow \infty} \tilde{f} = \frac{1}{2\pi i} \int_C \frac{f(t)e^{\alpha(x-t)}}{t-x} dt,$$

where the closed contour C is composed of $-C_1$ and C_2. From Cauchy's Integral Formula

$$f(x) = \lim_{\beta \uparrow \infty} \tilde{f} = \lim_{\beta \uparrow \infty} \frac{1}{2\pi i} \int_{\alpha-i\beta}^{\alpha+i\beta} dp\, e^{px} \int_{-\infty}^{\infty} f(t)e^{-pt} dt \qquad (8.45)$$

(included here is the property that $f = 0$ if $x \notin (a, b)$).

This concludes the proof in the special case. No general proof will be given here. We simply state the following:

Inversion Theorem. *If* $e^{-pt}f(t)$ *is absolutely integrable for some* $p = p_0$, *say, then* (8.43) *is valid for any Bromwich path on which the real part is greater than* $Re\, p_0$.

Properties of the Laplace Transform

1. *Convergence Abscissa.* Suppose $f(t)$ is such that $f(t)e^{-pt}$ is absolutely integrable for $Re\, p = r_0$. Then, since

$$|f(t)e^{-pt}| = |f|e^{-rt} = |f|e^{-r_0 t}e^{-(r-r_0)t}, \quad Re\, p = r,$$

$f(t)e^{-pt}$ is absolutely integrable for all $Re\, p > r_0$. Clearly there is a minimum R_0 (possibly $-\infty$) such that $f(t)e^{-pt}$ is absolutely integrable for

$$Re\, p > R_0.$$

R_0 is known as the *abscissa of (absolute) convergence* and is analogous to the radius of convergence for power series.

2. *The Class of Laplace Transformable Functions is Linear.* For each function f_i there is an abscissa of (absolute) convergence R_i. Therefore, for arbitrary constants a_1 and a_2,

$$f = a_1 f_1 + a_2 f_2$$

is transformable for

$$Re\, p = r > \max(R_1, R_2).$$

To see this, observe that

$$|a_1 f_1 + a_2 f_2| \le |a_1|\, |f_1| + |a_2|\, |f_2|$$

and

$$\left| \int_0^{\infty} f e^{-pt} dt \right| = \left| \int_0^{\infty} (a_1 f_1 + a_2 f_2)e^{-pt} dt \right|$$

$$\leq \int_0^\infty |e^{-pt}|\, |a_1 f_1 + a_2 f_2|\, dt \leq \int_0^\infty e^{-rt}(|a_1|\,|f_1| + |a_2|\,|f_2|)\, dt,$$

which converges for $r > \max(R_1, R_2)$.

3. *Laplace Transforms are Analytic (for Re $p > R$).* Consider f with abscissa of convergence R such that

$$F(p) = \int_0^\infty e^{-pt} f(t)\, dt.$$

The formal derivative is

$$\int_0^\infty (-t)e^{-pt} f(t)\, dt, \tag{8.46}$$

and by definition the derivative of F is

$$\lim_{\Delta p \to 0} \frac{F(p + \Delta p) - F(p)}{\Delta p} = \lim_{\Delta p \to 0} \int_0^\infty f(t) \left(\frac{e^{-(p+\Delta p)t} - e^{-pt}}{\Delta p} \right) dt.$$

We therefore consider the difference between this and the formal derivative (8.46):

$$I = \int_0^\infty e^{-pt} f(t) \left\{ \frac{e^{-\Delta pt} - 1}{\Delta p} + t \right\} dt. \tag{8.47}$$

Using simple estimates, we have that

$$\left| \frac{e^{-\Delta pt} - 1}{\Delta p} + t \right| = \left| \Delta p \frac{t^2}{2!} - \frac{(\Delta p)^2}{3!} t^3 \pm \cdots \right| = |\Delta p| \frac{t^2}{2} \left| 1 - \Delta p \frac{t}{3} \pm \cdots \right|$$

$$\leq |\Delta p| \frac{t^2}{2} \left| 1 + |\Delta p|t + |\Delta p|^2 \frac{t^2}{2!} + \cdots \right| = |\Delta p| \frac{t^2}{2} e^{|\Delta p|t}.$$

Substituting this into (8.47) then gives us

$$|I| \leq \frac{|\Delta p|}{2} \int_0^\infty |f(t)|\, |e^{-pt}|t^2|e^{|\Delta p|t}|\, dt.$$

The integral clearly converges and I tends to zero as $|\Delta p| \to 0$. This proves that F has a derivative with respect to the complex variable p and

$$F'(p) = \int_0^\infty (-t)e^{-pt} f(t)\, dt, \tag{8.48}$$

which is independent of Δp so that F is analytic for Re $p > R$.

4. $F(p) \to 0$ for $|p| \to \infty$ and Re $p > R$. This corresponds to the property of Z-transforms that they vanish at infinity. It plays an equally important role in revealing when a function is not a Laplace transform. To show this

property, note that if R is the abscissa of convergence of $f(t)$, then for all $r > R$

$$f(t)e^{-rt}$$

is absolutely integrable. Therefore, by the Riemann–Lebesgue Lemma,

$$\int_0^\infty e^{-pt} f(t)dt = \int_0^\infty (e^{-rt} f(t))e^{-ist} dt, \quad p = r + is,$$

tends to zero for $|s| \uparrow \infty$. (If $|p| \to \infty$ but s is fixed, $F(p) \to 0$. Why?)

5. $\int_0^\infty e^{-pt} f(t)dt = 0 \Rightarrow f(t) \equiv 0$ and

$$\frac{1}{2\pi i} \int_\uparrow e^{pt} F(p)dp = 0 \Rightarrow F(p) = 0.$$

This follows immediately from the inversion theorem.

6. *Transform of Derivatives and Integrals.*

$$\mathcal{L}(f') = \int_0^\infty e^{-pt} f'(t)dt = \int_0^\infty \frac{d}{dt}(e^{-pt} f)dt + p \int_0^\infty f(t)e^{-pt} dt.$$

Therefore

$$\mathcal{L}(f') = p\mathcal{L}(f) - f(0) = pF(p) - f(0)$$

and

$$\mathcal{L}(f'') = p\mathcal{L}(f') - f'(0) = p^2\mathcal{L}(f) - pf(0) - f'(0)$$
$$= p^2 F(p) - pf(0) - f'(0).$$

In general

$$\mathcal{L}(f^{(n)}) = p^n F(p) - p^{n-1}f(0) - p^{n-2}f'(0) - \cdots - f^{n-1}(0), \qquad (8.49)$$

where we have used the notation that

$$f^{(n)} = \frac{d^n f}{dt^n}.$$

This notation can be extended to *negative derivatives* which are defined by

$$f^{(-1)} = \int_0^t f(s)ds, \ldots, f^{(-n)} = \int_0^t f^{(-n+1)}(s)ds.$$

To see this, consider

$$\mathcal{L}(f^{(-1)}) = \int_0^\infty e^{-pt} \int_0^t f(s)ds\, dt = \int_0^\infty \left(\frac{d}{dt}\frac{e^{-pt}}{-p}\right) \int_0^t f(s)ds\, dt$$

$$= -\frac{1}{p} \int_0^\infty \frac{d}{dt}\left(e^{-pt} \int_0^t f(s)ds\right) dt + \frac{1}{p} \int_0^\infty e^{-pt} f(t)dt = \frac{1}{p}F(p).$$

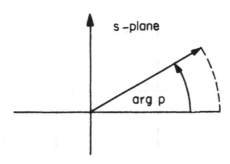

FIGURE 8.6.

In general

$$\mathcal{L}(f^{(-n)}) = p^{-n}\mathcal{L}(f), \quad n > 0. \tag{8.50}$$

Note that from the definition of a negative derivative, $f^{(-n)}(0) = 0$, $n > 0$.

Examples.

1. We start with the simple case of $f = t^n$, $t > 0$, for n an integer. Then

$$\mathcal{L}(t^n) = \int_0^\infty e^{-pt} t^n \, dt.$$

Set

$$pt = s.$$

Then since p is complex with $Re\, p > 0$, the limits of integration become

$$t = 0 \rightarrow s = 0,$$

$$t = \infty \rightarrow s = \infty \exp(i \arg p).$$

Therefore

$$\mathcal{L}(t^n) = \frac{1}{p^{n+1}} \int_0^{\infty \exp(i\, arg\, p)} e^{-s} s^n \, ds,$$

where the path of integration is a ray in the complex s-plane at an angle $\arg p$ (see Figure 8.6). However, we can bring this down to the real axis since the contribution on the dotted path in Figure 8.6 becomes vanishingly small as R, in the figure, tends toward ∞. Thus

$$\mathcal{L}(t^n) = \frac{1}{p^{n+1}} \int_0^\infty e^{-t} t^n \, dt = \frac{n!}{p^{n+1}}, \tag{8.51}$$

where the function denoted by $n!$ is called the *gamma function* (see Exercise 3). Let us also verify the inversion formula.

Consider the integral

$$\frac{1}{2\pi i} \int_{\uparrow} e^{pt} \frac{n!}{p^{n+1}} \, dp.$$

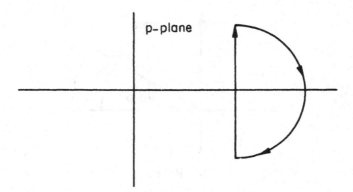

FIGURE 8.7.

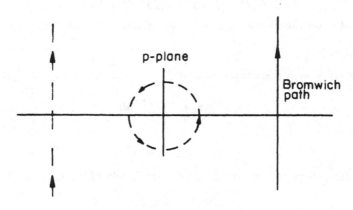

FIGURE 8.8.

From the general inversion formula, we know that the Bromwich path lies to the right of all singularities of the integrand. (Recall that we have proven analyticity for $Re\, p > R$. $R = 0$ for this case.)

In the above, for $t < 0$, we can close the path in the right half-plane and this gives zero as it should (see Figure 8.7). For $t > 0$, we can *push* the path to the left. As indicated in Figure 8.8 (dotted paths), this results in

$$\frac{n!}{2\pi i} \oint_{|p|=\epsilon} \frac{e^{pt}}{p^{n+1}} dp = n!\, Res\left[\frac{e^{pt}}{p^{n+1}}; 0\right] = t^n.$$

2. As a second example we consider $\sqrt{t}$, in which case

$$\mathcal{L}(t^{1/2}) = \int_0^\infty e^{-pt} \sqrt{t}\, dt.$$

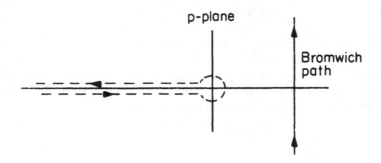

FIGURE 8.9.

The reduction to the gamma function proceeds as above. We first set $pt = s$ and obtain

$$\mathcal{L}(t^{1/2}) = \frac{(1/2)!}{p\sqrt{p}}, \quad \left(\frac{1}{2}\right)! = \int_0^\infty e^{-s} s^{1/2} ds = \sqrt{\pi}/2. \tag{8.52}$$

(This last relation is left as an exercise.) Since the square root of a complex number appears, some thought should be given to the branch. This is easy since for p real, $\mathcal{L}(t^{1/2})$ is real and positive. This places us on the principal branch.

Consider next the inversion

$$\mathcal{F} = \frac{1}{2\pi i} \int_\uparrow e^{+pt} \frac{(1/2)!}{p^{3/2}} dp. \tag{8.53}$$

The path of integration is again the Bromwich path indicated in Figure 8.9. For $t < 0$ it is clear that we can *push* to the right and we get zero for the value for (8.53). For $t > 0$ we *push* to the left and obtain the dashed contour of Figure 8.9. Since a root of p occurs in the integrand of (8.53), a branch cut is required and it is natural to take it along the negative real axis as shown.

It is convenient first to parts integrate (8.53):

$$\mathcal{F} = \frac{(1/2)!}{2\pi i} \int_\uparrow e^{+pt} \left(-2 \frac{d}{dp} \frac{1}{\sqrt{p}} \right) dp$$

$$= \frac{(1/2)!}{2\pi i} \int_\uparrow \left(-2 \frac{d}{dp} \frac{e^{+pt}}{\sqrt{p}} \right) dp + \frac{(1/2)!}{2\pi i} \int_\uparrow t\, 2 \frac{e^{+pt}}{\sqrt{p}} dp.$$

With either path in Figure 8.9 the first integral vanishes since the integrand vanishes at the endpoints. Therefore

$$\mathcal{F} = t \frac{(1/2)!}{\pi i} \int_\uparrow e^{pt} \frac{dp}{\sqrt{p}}.$$

If we go to the *dotted* path in Figure 8.9, then

$$\mathcal{F} = t\frac{(1/2)!}{\pi i}\left\{\int_0^{-\infty} e^{pt}\frac{1}{(\sqrt{p})^+}\,dp + \int_{-\infty}^0 e^{pt}\frac{1}{(\sqrt{p})^-}\,dp\right\},$$

where the $\pm$ signifies that $\arg p = \pm\pi$. (It is left to the exercises, specifically Exercise 5, to show that the integral around the origin is zero.) In each integral set $p = -s$ and observe that in the first integral

$$(\sqrt{p})^+ = (\sqrt{-s})^+ = \sqrt{s}\,i$$

and in the second

$$(\sqrt{p})^- = (\sqrt{-s})^- = -\sqrt{s}\,i.$$

Therefore

$$\mathcal{F} = \frac{t(1/2)!}{\pi i}\left\{-\int_0^{\infty} e^{-st}\frac{ds}{i\sqrt{s}} - \int_{\infty}^0 e^{-st}\frac{1}{-i\sqrt{s}}\,ds\right\}$$

$$= \frac{2t(1/2)!}{\pi}\int_0^{\infty} e^{-st}s^{-1/2}\,ds = \sqrt{t}\frac{2(1/2)!(-1/2)!}{\pi} = \sqrt{t}$$

since (see Exercise 3)

$$\left(\frac{1}{2}\right)! = \frac{\sqrt{\pi}}{2} \quad \& \quad \left(\frac{1}{2}\right)! = \frac{1}{2}\left(-\frac{1}{2}\right)!.$$

3. As another example, consider $\sin\omega t$. The Laplace transform is given by

$$F(p) = \mathcal{L}(\sin\omega t) = \int_0^{\infty} e^{-pt}\sin\omega t\,dt = \int_0^{\infty} e^{-pt}\frac{e^{i\omega t} - e^{-i\omega t}}{2i}\,dt$$

$$= \frac{1}{2i}\left\{\frac{1}{p - i\omega} - \frac{1}{p + i\omega}\right\} = \frac{\omega}{p^2 + \omega^2}. \tag{8.54}$$

To invert (8.54) we have

$$\frac{1}{2\pi i}\int_{\uparrow} e^{pt}\frac{\omega}{p^2 + \omega^2}\,dp = \frac{\omega}{2\pi i}\int_{\uparrow} e^{pt}\frac{1}{(p - i\omega)(p + i\omega)}\,dp.$$

For $t < 0$ this vanishes, and for $t > 0$ we push the path to the left, as indicated in Figure 8.10. We are then left with a residue calculation,

$$\omega\left\{Res\left[\frac{e^{pt}}{p^2 + \omega^2}; i\omega\right] + Res\left[\frac{e^{pt}}{p^2 + \omega^2}; -i\omega\right]\right\}$$

$$= \omega\left\{\frac{e^{i\omega t}}{2i\omega} + \frac{e^{-i\omega t}}{-2i\omega}\right\} = \sin\omega t.$$

The remaining integration, along the vertical dotted path, is seen to vanish as it is pushed to the left since $\exp(pt) \downarrow 0$.

It should be noted that with each new transform calculation we obtain a transform pair and therefore a table of transforms. Thus far we have

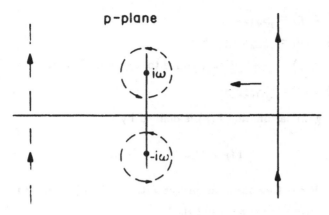

FIGURE 8.10.

$f(t)$	$F(p)$
t^n	$\dfrac{n!}{p^{n+1}}$
$\sqrt{t}$	$\dfrac{\sqrt{\pi}}{2p^{3/2}}$
$\sin \omega t$	$\dfrac{\omega}{p^2 + \omega^2}$

There are books of transform tables. For each of the above examples it was unnecessary to calculate the inverse transforms. This was done only as practice.

Exercises

1. Find the Laplace transform of each of the following functions:

   (a) $H(t) = 1$ for $t > 0$ and $H(t) = 0$ otherwise. This is known as the *Heaviside function*.
   (b) $H(t - b)$, $b > 0$.
   (c) $H(t - b)e^{-at}$, $b > 0$.
   (d) $\cos \omega t$.
   (e) $H(t - b) \sin \omega t$, $b > 0$.

2. Find the inverse Laplace transform of the following. In each case use the inversion integral (8.43) and a Bromwich path to the right of all singularities of $F(p)$.

(a) $F(p) = \frac{p}{p^2+k^2}$.

(b) $F(p) = \frac{1}{(p-a)(p-b)(p-c)}$.

(c) $F(p) = \frac{1}{p^{4/3}}$. [Hint: Reduce to a gamma function.]

(d) $F(p) = \frac{1}{\sqrt{p}(p^2+\omega^2)}$.

3. The gamma function $\Gamma(p)$ is defined by

$$\Gamma(p+1) = \int_0^\infty e^{-t} t^p \, dt \quad (= p!).$$

(a) Prove that for p an integer with $p = n$, $\Gamma(1+n) = n!$.

(b) Prove $\Gamma(1+p) = p\Gamma(p)$.

(c) Show $\Gamma(1/2) = \sqrt{\pi}$. [Hint: Set $t = s^2$ in $\Gamma(1/2)$.]

4. Prove

$$\mathcal{L}^{-1}\left(\frac{1}{p}\frac{1-e^{-p\pi}}{1+e^{+p\pi}}\right) = \frac{4}{\pi}\sum_{n=0}^\infty \frac{1}{2n+1}\sin(2n+1)t,$$

which is the square wave of period 2π. Show by direct Laplace transform of the square wave for $t > 0$ that

$$\frac{1}{p}\frac{1-e^{-p\pi}}{1+e^{-p\pi}}$$

results.

5. Show that

$$\lim_{\epsilon\downarrow 0}\int_{|p|=\epsilon} e^{pt}\frac{dp}{\sqrt{p}} = 0.$$

6. Prove $\lim_{p\to\infty} pF(p) = \lim_{t\downarrow 0} f(t)$.

7. Find the inverse transform of

(a) $1/[(s+a)(s+b)]$,

(b) $s/[(s+a)(s+b)]$,

(c) $s/[(s^2+a^2)(s^2+b^2)]$

for $a \neq b$.

8. Find the Laplace transform of

(a) $\sinh ax$,

(b) $\exp(ax)\sin(bx)$,

(c) $x^n \exp(ax)$.

8.3 Convolution Products

We continue to restrict our attention to functions defined on the half line $t > 0$ and now consider the convolution product which is given by

$$f \star g = \int_0^t f(t - \tau)g(\tau)d\tau. \tag{8.55}$$

This is the same as the infinite convolution product, for if $f = 0, g = 0$ for $t < 0$, it then follows that

$$f \star g = \int_{-\infty}^{\infty} f(t - \tau)g(\tau)d\tau = \int_0^t f(t - \tau)g(\tau)d\tau.$$

If both fe^{-pt} and ge^{-pt} are absolutely integrable on $(0, \infty)$ for $Re\, p > R$, then

$$\mathcal{L}(f \star g) = \int_0^{\infty} e^{-pt}dt \int_{-\infty}^{\infty} g(\tau)f(t - \tau)d\tau.$$

If we interchange orders of integration,

$$\mathcal{L}(f \star g) = \int_{-\infty}^{\infty} d\tau\, g(\tau) \int_0^{\infty} e^{-pt}f(t - \tau)dt,$$

and since $f(t - \tau) = 0$ for $t - \tau < 0$,

$$\mathcal{L}(f \star g) = \int_{-\infty}^{\infty} g(\tau)d\tau \int_{\tau}^{\infty} e^{-pt}f(t - \tau)dt.$$

Finally, if we set $t - \tau = s$, then

$$\mathcal{L}(f \star g) = \int_0^{\infty} e^{-p\tau}g(\tau)d\tau \int_0^{\infty} e^{-ps}f(s)ds = G(p)F(p). \tag{8.56}$$

Example. As an application of (8.56), consider

$$\int_0^t \frac{g(\tau)d\tau}{(t - \tau)^{\mu}} = f(t), \quad 0 < \mu < 1. \tag{8.57}$$

This is known as *Abel's integral equation*. $f(t)$ is regarded as a known function and $g(t)$ is to be determined. If we Laplace transform (8.57) and use (8.56), it follows that

$$\mathcal{L}(g)\mathcal{L}(1/t^{\mu}) = \mathcal{L}(f). \tag{8.58}$$

The Laplace transform of $1/t^{\mu}$ is (according to the same reasoning that went into finding $\mathcal{L}(t^{1/2})$ given by (8.52))

$$\mathcal{L}(1/t^{\mu}) = \int_0^{\infty} e^{-pt}\frac{1}{t^{\mu}}dt = p^{-1+\mu}(-\mu)!.$$

Therefore (8.58) becomes

$$G(p)p^{-1+\mu}(-\mu)! = F(p).$$

This can be solved for the unknown transform $G(p)$ and we have

$$G(p) = \frac{F(p)}{(-\mu)!p^{-1}p^{\mu}} = \frac{pF(p)}{(-\mu)!p^{\mu}}.$$

Finally we formally invert this expression to find the solution to the problem (8.57)

$$g(t) = \frac{1}{2\pi i}\int_{\uparrow} e^{+pt}\frac{pF(p)}{(-\mu)!p^{\mu}}dp. \qquad (8.59)$$

We must be a little careful at this point. It is tempting to view the term

$$\frac{pF(p)}{p^{\mu}}$$

as a product and then write the inverse transform as a convolution product—*and this is the correct strategy*. Two groupings suggest themselves: $F(p) \cdot (p^{1-\mu})$ and $pF(p) \cdot (p^{-\mu})$. Both are wrong in that they lead to embarrassment as we now see. In the first instance, the second factor has the inverse transform

$$\int_{\uparrow} e^{pt}p^{1-\mu}dp = \infty,$$

which is divergent as indicated. This could have been anticipated since $p^{1-\mu}$ *is not* a Laplace transform. (Recall that Property 4 requires that $F(p) \to 0$ as $|p| \uparrow \infty$.) In the second instance the first factor has the inverse transform

$$\int_{\uparrow} e^{+pt}pF(p)dp.$$

This too does not exist unless $pF(p) \to 0$ for $|p| \uparrow \infty$ on the Bromwich path. Again $pF(p)$ is not necessarily a Laplace transform. This last point is brought out—and the problem resolved—by recalling that (see (8.49))

$$\mathcal{L}(f') = pF(p) - f(0) = \int_0^{\infty} e^{-pt}f'(t)dt.$$

Now the integral vanishes for $|p| \uparrow \infty$ on the Bromwich path (why?), which therefore says that

$$pF(p) \to f(0).$$

The way out is now clear. We write (8.59) as

$$g(t) = \frac{1}{2\pi i}\int_{\uparrow} e^{pt}\frac{pF(p) - f(0)}{(-\mu)!p^{\mu}}dp + \frac{f(0)}{2\pi i}\int_{\uparrow} \frac{e^{pt}dp}{(-\mu)!p^{\mu}}.$$

It is clear from the Laplace transform of t^α ($\mathcal{L}(t^\alpha) = \alpha! p^{-1-\alpha}$) that

$$\frac{1}{2\pi i}\int_\uparrow \frac{e^{pt}\,dp}{p^\mu} = \frac{t^{\mu-1}}{(\mu-1)!}.$$

Finally it follows that the solution of the Abel integral equation is

$$g(t) = \frac{1}{(-\mu)!(\mu-1)!}\int_0^t \frac{f'(\tau)}{(t-\tau)^{1-\mu}}d\tau + \frac{f(0)}{(-\mu)!(\mu-1)!t^{1-\mu}}. \tag{8.60}$$

This is further simplified by the relation

$$(-\mu)!(\mu-1)! = \frac{\pi}{\sin\mu\pi}, \tag{8.61}$$

which appears in the exercises.

A main application of the Laplace transform is to differential equations which we consider next.

Exercises

1. Show
$$t^a \star t^b = t^{a+b+1}\int_0^1 s^a(1-s)^b\,ds.$$

2. Find the Laplace transform of
$$g(t) = \int_0^t f(s)\,ds$$
in terms of the transform of $f(t)$.

3. Prove
$$\int_0^t \int_0^{t_n}\cdots\int_0^{t_2} f(t_1)\,dt_1 dt_2 \ldots dt_n = \int_0^t \frac{(t-\tau)^{n-1}f(\tau)}{(n-1)!}d\tau.$$
[Hint: Take the Laplace transform and use (8.50).]

4. The Beta function is defined to be
$$B(p+1, q+1) = \int_0^1 t^p(1-t)^q\,dt.$$

Prove $B(1+p, 1+q) = p!q!/(1+p+q)!$. [Hint: Define $b(x) = \int_0^x (x-t)^p t^q\,dt$ and note that $b(1) = B(1+p, 1+q)$. Evaluate $b(x)$ by Laplace transform.]

5. Show $B(1+p, 1-p) = p!(-p)! = p\pi/\sin p\pi$ (which after minor changes gives (8.61)). [Hint:
$$p!(-p)! = \int_0^1 \left(\frac{t}{1-t}\right)^p dt \xrightarrow{x=t/(1-t)} \int_0^\infty \frac{x^p}{(1+x^2)}dx,$$
which can be evaluated by contour integration methods.]

8.4   Differential Equations with Constant Coefficients

Consider the problem posed by the following first order equation and initial condition:

$$\frac{dy}{dt} + ay = f(t), \quad y(t = 0) = y^0.$$

To solve, we formally Laplace transform the differential equation to get

$$pY(p) - y^0 + aY = F(p),$$

where $\mathcal{L}(f) = F(p)$ and (8.49) have been used to obtain the Laplace transform of the first derivative. On solving for Y we obtain

$$Y = \frac{1}{p+a}y^0 + \frac{1}{p+a}F(p).$$

Since

$$\mathcal{L}^{-1}\left(\frac{1}{p+a}\right) = e^{-at},$$

we have

$$y(t) = y^0 e^{-at} + \int_0^t e^{-a(t-\tau)}f(\tau)d\tau.$$

This can be immediately generalized to nth order equations with constant coefficients. Thus we consider

$$\begin{cases} a_n y^n + a_{n-1}y^{(n-1)} + \cdots + a_0 y = \sum_{j=0}^n a_j y^{(j)} = f(t), \\ y^{(k)}(0) = y_0^{(k)}, \quad k = 0, \ldots, n-1. \end{cases} \tag{8.62}$$

We recall (8.49), which states that

$$\mathcal{L}(y^{(j)}) = p^j Y - \sum_{m=0}^{j-1} y_0^{(m)} p^{(j-1-m)}.$$

Therefore the Laplace transform of the problem (8.62) is

$$\sum_{j=0}^n a_j \left(p^j Y - \sum_{m=0}^{j-1} y_0^{(m)} p^{j-1-m}\right) = F(p).$$

This can be easily solved for Y:

$$Y = T(p)\{F(p) + U(p)\}, \tag{8.63}$$

where

$$T(p) = \frac{1}{\sum_{j=0}^n a_j p^j} \tag{8.64}$$

and

$$U(p) = \sum_{j=0}^{n} \sum_{m=0}^{j-1} a_j y_0^{(m)} p^{j-1-m}. \tag{8.65}$$

If (8.63) is inverted, we find

$$y(t) = \mathcal{L}^{-1}(T(p)F(p)) + \mathcal{L}^{-1}(T(p)U(p)). \tag{8.66}$$

We have therefore solved the problem (8.62) in the sense that the problem is reduced to the evaluation of integrals, and in fact we know that the evaluation of these integrals is especially simple by virtue of the theory of residues. In particular, suppose the polynomial in (8.64) is factorable into simple roots, i.e.,

$$T = \frac{1}{(p - p_1)(p - p_2) \cdots (p - p_n)} = \frac{1}{P(p)} \tag{8.67}$$

(without loss of generality we have taken $a_n = 1$). Then T has the partial fraction decomposition

$$T = \frac{\alpha_1}{p - p_1} + \frac{\alpha_2}{p - p_2} + \cdots + \frac{\alpha_n}{p - p_n},$$

where

$$\alpha_1 = \frac{1}{(p_1 - p_2) \cdots (p_1 - p_n)}, \quad \alpha_2 = \frac{1}{(p_2 - p_1)(p_2 - p_3) \cdots (p_2 - p_n)},$$

$$\alpha_k = \frac{1}{\prod_{j \neq k}(p_k - p_j)}.$$

If we take the inverse Laplace transform of T, we obtain the *solution operator*

$$S(t) = \mathcal{L}^{-1}(T) = \sum_{j=1}^{n} \alpha_j e^{p_j t}. \tag{8.68}$$

This terminology is used since we can write the solution (8.66) as

$$y = S \star f + U\left(\frac{d}{dt}\right) S. \tag{8.69}$$

The last term in (8.69) needs some explanation.
  Consider

$$\mathcal{L}^{-1}(TU) = \frac{1}{2\pi i} \int_\uparrow \frac{U(p)}{P(p)} e^{pt} \, dp.$$

The denominator is a polynomial of degree n (see (8.67)). The numerator is also a polynomial, which by inspection of (8.65) is of degree $n - 1$. This being the case, there is no difficulty with respect to convergence, and any monomial p^k in the numerator can be taken outside as $(d/dt)^k$. Thus

$$\mathcal{L}^{-1}(TU) = \sum_{j=0}^{n} \sum_{m=0}^{j-1} a_j y_0^{(m)} \left(\frac{d}{dt}\right)^{j-1-m} \mathcal{L}^{-1}(T),$$

and hence the shorthand given in (8.69) is clarified.

The case of multiple roots contains a slight additional complexity of a sort already encountered in residue theory. A detailed account of second order equations reveals all. Consider

$$\frac{d^2 y}{dt^2} + 2a\frac{dy}{dt} + a^2 y = f(t).$$

Then

$$S = \mathcal{L}^{-1}(T) = \frac{1}{2\pi i}\int_{\uparrow} \frac{e^{pt}}{(p+a)^2}dp = Res\left[\frac{e^{pt}}{(p+a)^2}; -a\right] = te^{-at},$$

which must be used in (8.69) instead of (8.68).

Frequently in applications a system of ordinary differential equations is obtained. For example,

$$\begin{cases} \dfrac{d}{dt}\mathbf{x} = \mathbf{A}\mathbf{x}, \\ \mathbf{x}(t=0) = \mathbf{x}^0, \end{cases} \tag{8.70}$$

where $\mathbf{x}$ is an n-vector and $\mathbf{A}$ is a constant $n \times n$ matrix. If the Laplace transform is applied to (8.70), then

$$p\mathbf{X}(p) = \mathbf{A}\mathbf{X} + \mathbf{x}^0.$$

This can be rewritten as

$$(p\mathbf{1} - \mathbf{A})\mathbf{X} = \mathbf{x}^0, \tag{8.71}$$

where $\mathbf{1}$ represents the unit matrix. The solution therefore (not unexpectedly) depends on the eigenvalues of $\mathbf{A}$, which arise from the formal inversion of (8.71) with

$$\mathbf{X} = (p\mathbf{1} - \mathbf{A})^{-1}\mathbf{x}^0. \tag{8.72}$$

From this the solution to (8.70) can be represented by

$$\mathbf{x} = \frac{1}{2\pi i}\int_{\uparrow} e^{pt}(p\mathbf{1} - \mathbf{A})^{-1}dp\,\mathbf{x}^0 = \mathbf{S}(t)\mathbf{x}^0,$$

where

$$\mathbf{S}(t) = \frac{1}{2\pi i}\int_{\uparrow} e^{pt}(p\mathbf{1} - \mathbf{A})^{-1}dp \tag{8.73}$$

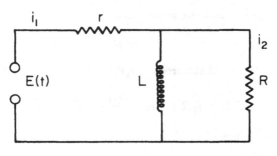

FIGURE 8.11.

is referred to as the *solution operator* or *matrix*. Cramer's Rule can be used to evaluate the elements of the matrix, $(p\mathbf{1} - \mathbf{A})^{-1}$ and hence the integral. Specific calculations will be discussed in the exercises.

It should be noted that any system of ordinary differential equations can be reduced to first order form. For example,

$$\frac{d^2y}{dt^2} + 2a\frac{dy}{dt} + by = f$$

can be written as

$$\frac{du}{dt} + 2au + by = f, \quad \frac{dy}{dt} = u.$$

The second of these equations defines u. In general, any system of (not necessarily first order) equations can be written as

$$\mathbf{A}\frac{d\mathbf{x}}{dt} = \mathbf{B}\mathbf{x}, \quad \mathbf{x}(0) = \mathbf{x}^0,$$

which, in turn, can be written as

$$\frac{d\mathbf{x}}{dt} = \mathbf{A}^{-1}\mathbf{B}\mathbf{x}$$

if $\mathbf{A}^{-1}$ exists.

If

$$\det |\mathbf{A}| = 0,$$

then $\mathbf{A}^{-1}$ does not exist, which implies degeneracy and requires special attention. Rather than look at the general case, we consider a special case—which is simple and transparent. Consider the circuit shown in Figure 8.11. This contains the inductance L, two resistance elements r and R, and an electromotive force (emf) $\mathcal{E}(t)$. A few simple relations give rise to the governing equations. Denote by i_1 the current passing through r and by i_2 that passing through R. The current passing through the inductance is therefore $i_1 - i_2$ and hence the voltage across the inductance is

$$E = L\frac{d}{dt}(i_1 - i_2).$$

The voltage across R has the same value

$$E = i_2 R.$$

Also, this voltage plus that across r, $i_1 r$, is equal to the emf

$$\mathcal{E}(t) = i_2 R + i_1 r, \quad \frac{d(i_1 - i_2)}{dt} = \frac{i_2 R}{L}.$$

In matrix form we have

$$\begin{pmatrix} 0 & 0 \\ 1 & -1 \end{pmatrix} \frac{d}{dt} \begin{pmatrix} i_1 \\ i_2 \end{pmatrix} = \begin{pmatrix} r & R \\ 0 & R/L \end{pmatrix} \begin{pmatrix} i_1 \\ i_2 \end{pmatrix} + \begin{pmatrix} -\mathcal{E}(t) \\ 0 \end{pmatrix}.$$

In the symbolism of the above general discussion, $\det |A| = 0$.

Actually, what is being said by the condition $\det |A| = 0$ is that the system is in fact equivalent to a smaller system. For the case under discussion, we may substitute for i_1 to obtain

$$-\frac{d}{dt}i_2 + \frac{d}{dt}\left\{\frac{\mathcal{E}}{r} - i_2\frac{R}{r}\right\} = \frac{i_2 R}{L}$$

or

$$\frac{d}{dt}i_2 + \left(1 + \frac{R}{r}\right)^{-1}\frac{R}{L}i_2 = \left(1 + \frac{R}{r}\right)^{-1}\frac{\mathcal{E}'}{r}.$$

We observe that only one initial condition can be independently given, since $\mathcal{E}(0) = i_2(0)R + i_1(0)r$ must hold initially.

While the Laplace transform method gives a systematic procedure for solving differential equations with constant coefficients, it yields nothing new. The solutions appear in terms of sums of exponentials, a fact which is well known from the elementary theory of ordinary differential equations. In fact it is usually simpler and more direct to solve such systems directly in terms of sums of exponentials. The Laplace transform method reduces a typical problem to algebraic manipulations and as a result is helpful when confronted by new situations or *treacherous waters*.

Exercises

1. Solve the following problems by means of Laplace transforms:

(a) $\ddot{x} + 4\dot{x} + 3x = 0$, $x(0) = 1$, $\dot{x}(0) = 2$.

(b) $\ddot{x} + 4\dot{x} + 3x = e^t(t^3 + t^2 - 1)$, $x(0) = \dot{x}(0) = 0$.

(c)
$$\ddot{x} + 4\dot{x} + 3x = \begin{cases} t, & 0 \leq t \leq 2, \\ 0, & t > 2, \end{cases}$$

$x(0) = \dot{x}(0) = 0$.

(d) $\ddot{x} + 4\dot{x} + 3x = \sin t$, $x(0) = \dot{x}(0) = 0$.

2. Consider

$$\frac{d}{dt}\begin{bmatrix} x_1 \\ x_2 \end{bmatrix} = \frac{dx}{dt} = \begin{bmatrix} a_{11} & a_{12} \\ a_{21} & a_{22} \end{bmatrix}\begin{bmatrix} x_1 \\ x_2 \end{bmatrix} = ax$$

with $x(0) = x^0$. Then, according to (8.71), the solution is $x = S(t)x^0$, where

$$S(t) = \frac{1}{2\pi i}\int_{\uparrow} \frac{e^{pt}}{(p1 - a)}\,dp.$$

Evaluate $S(t)$. [Hint: Find the explicit form of $(p1-a)^{-1}$ from Cramer's formula and perform the integration over each component of this matrix.]

3. Solve the following systems by Laplace transforms:

(a) $2\dot{x} - x + \dot{y} + 13y = 16e^{2t}$, $2\dot{x} - 3x - \dot{y} + 7y = 0$, $x(0) = y(0) = 0$.

(b) $\ddot{x} - x + y = e^t$, $x + \ddot{y} - y = 0$, $x(0) = y(0) = \dot{x}(0) = \dot{y}(0) = 0$.

(c) $2\dot{x} + x - \int_0^t x(s)ds + \dot{y} + 5y - 6\int_0^t y(s)ds = 0$, $\dot{x} - \int_0^t x(s)ds + \dot{y} + y - 2\int_0^t y(s)ds = 0$, $x(0) = 1$, $y(0) = 2$.

4. Solve

$$\ddot{x} + 2n\dot{x} + n^2 x = 0,$$
$$\ddot{y} + 2n\dot{y} + n^2 y = \mu\dot{x}$$

with $x(0) = y(0) = \dot{y}(0) = 0$, $\dot{x}(0) = 1$.

5. Solve

$$\alpha x + \beta\int_0^t x(s)ds = f(t)$$

(α and β constant) by Laplace transformation.

9

Partial Differential Equations (Continued)

We again take up the subject of partial differential equations. The approach in this chapter will be more formal than before.

9.1 Canonical Forms for Second Order Equations

The three equations considered in Chapter 7,

$$\frac{1}{C^2}\frac{\partial^2\phi}{\partial t^2} = \frac{\partial^2\phi}{\partial x^2},\tag{9.1}$$

$$\frac{\partial\phi}{\partial t} = k\frac{\partial^2\phi}{\partial x^2},\tag{9.2}$$

$$\frac{\partial^2\phi}{\partial x^2} + \frac{\partial^2\phi}{\partial y^2} = 0,\tag{9.3}$$

are equations of second order. This terminology is determined by the highest derivative which in each case is of second order. The most general second order partial differential equation in two independent variables can be written as

$$F(x,y,\phi,\phi_x,\phi_y,\phi_{xy},\phi_{xx},\phi_{yy}) = 0.$$

Any serious remarks about such a general equation is beyond our scope. Instead we consider the linear case

$$a\phi_{xx} + 2b\phi_{xy} + c\phi_{yy} + 2d\phi_x + 2e\phi_y + f\phi = 0,\tag{9.4}$$

and in addition we take the coefficients $a,\ldots,f$ to be constants.

Equation (9.4) includes the cases of (9.1)–(9.3). For example, to obtain the wave equation, (9.1), set $a = 1$, $b = 0$, $c = -1/C^2$, $d = e = f = 0$; to obtain the diffusion equation, (9.2), set $a = k, e = -1/2, d = c = b = f = 0$; and to obtain Laplace's equation, (9.3), set $a = c = 1$, $b = d = e = f = 0$. In the first two instances we have set $y = t$.

From D'Alembert's Formula (7.68), a solution to the wave equation is

$$\phi = F(x - Ct) + G(x + Ct).$$

This suggests that we might consider a change of independent variable to

$$\eta = (x - Ct), \quad \zeta = (x + Ct) \tag{9.5}$$

so that

$$\phi = F(\eta) + G(\zeta). \tag{9.6}$$

If we perform the same change of variable, (9.5), on the wave equation (9.1), we obtain

$$\frac{\partial^2 \phi}{\partial \zeta \partial \eta} = 0. \tag{9.7}$$

If (9.6) is substituted into (9.7), it clearly satisfies the equation.

The reduction of (9.1) to (9.7) suggests that we seek more general linear transformations to reduce (9.4) to a simpler form. With this goal in mind we consider the *principal part* of (9.4), L, defined by

$$L\phi = a\frac{\partial^2 \phi}{\partial x_1^2} + 2b\frac{\partial^2 \phi}{\partial x_1 \partial x_2} + c\frac{\partial^2 \phi}{\partial x_2^2}$$

$$= \left(\frac{\partial}{\partial x_1}, \frac{\partial}{\partial x_2} \right) \left(\begin{array}{cc} a & b \\ b & c \end{array} \right) \left(\begin{array}{c} \partial/\partial x_1 \\ \partial/\partial x_2 \end{array} \right) \phi. \tag{9.8}$$

The second matrix form will prove to be useful in a moment. For convenience we have adopted the trivial transformation $(x, y) \to (x_1, x_2)$.

To explore the possibility of simplifying the form of the principal part, L, we consider the transformation of variables

$$\left(\begin{array}{c} y_1 \\ y_2 \end{array} \right) = \left(\begin{array}{cc} T_{11} & T_{12} \\ T_{21} & T_{22} \end{array} \right) \left(\begin{array}{c} x_1 \\ x_2 \end{array} \right)$$

or, in vector notation,

$$\mathbf{y} = \mathbf{T} \; \mathbf{x}. \tag{9.9}$$

This implies that

$$\left(\begin{array}{c} \partial/\partial x_1 \\ \partial/\partial x_2 \end{array} \right) = \left(\begin{array}{cc} T_{11} & T_{21} \\ T_{12} & T_{22} \end{array} \right) \left(\begin{array}{c} \partial/\partial y_1 \\ \partial/\partial y_2 \end{array} \right) = \mathbf{T}^\dagger \frac{\partial}{\partial \mathbf{y}}, \tag{9.10}$$

where as indicated $\mathbf{T}^\dagger$ is the transpose of $\mathbf{T}$. Thus (9.8) under this transformation becomes

$$L\phi = \left(\frac{\partial}{\partial y_1}, \frac{\partial}{\partial y_2} \right) \mathbf{TMT}^\dagger \left(\begin{array}{c} \partial/\partial y_1 \\ \partial/\partial y_2 \end{array} \right) \phi, \tag{9.11}$$

where

$$\mathbf{M} = \left(\begin{array}{cc} a & b \\ b & c \end{array} \right).$$

The matrix M is symmetric and therefore has real eigenvalues, and its eigenvectors can be taken to be *orthonormal*. We denote the eigenvalues and eigenvectors by λ and v, respectively, i.e.,

$$Mv_j = \lambda_j v_j, \quad j = 1, 2.$$

We also define

$$V = (v_1, v_2)$$

to be the matrix whose columns are the orthonormal eigenvectors. It then follows that

$$M = V \begin{pmatrix} \lambda_1 & 0 \\ 0 & \lambda_2 \end{pmatrix} V^\dagger \tag{9.12}$$

since

$$V^{-1} = V^\dagger.$$

Thus if in (9.11) we substitute (9.12) and take

$$T = V^\dagger$$

(this makes (9.9) an orthogonal transformation), we then have

$$L\phi = \lambda_1 \frac{\partial^2 \phi}{\partial y_1^2} + \lambda_2 \frac{\partial^2 \phi}{\partial y_2^2}, \tag{9.13}$$

which is the sought-after simplification.

The eigenvalues λ_1, λ_2 are determined by

$$\det \begin{pmatrix} a - \lambda & b \\ b & c - \lambda \end{pmatrix} = \lambda^2 - (a + c)\lambda + (ac - b^2) = 0. \tag{9.14}$$

Therefore

$$\lambda = \frac{1}{2}(a + c \pm \sqrt{(a+c)^2 + 4(b^2 - ac)}).$$

Since $(a + c)^2 + 4(b^2 - ac) = (a - c)^2 + 4b^2$, this confirms the fact that both eigenvalues are real. There are three cases to consider:

$$b^2 - ac > 0, \quad \lambda_1 \text{ and } \lambda_2 \text{ are of different sign;}$$

$$b^2 - ac < 0, \quad \lambda_1 \text{ and } \lambda_2 \text{ are of the same sign;}$$

$$b^2 - ac = 0, \quad \text{one eigenvalue is zero.}$$

Hyperbolic Case: $b^2 - ac > 0$

When this condition is satisfied, the general second order partial differential equation (9.4) is said to be *hyperbolic*. If we denote the eigenvalues by μ^2 and $-\nu^2$, then the transformed equation takes the form

$$\mu^2 \frac{\partial^2 \phi}{\partial y_1^2} - \nu^2 \frac{\partial^2 \phi}{\partial y_2^2} + \mu D \frac{\partial \phi}{\partial y_1} + \nu E \frac{\partial \phi}{\partial y_2} + F\phi = 0.$$

For discussion purposes, the precise form of the new constants D, E, F is unimportant. Under the further transformation

$$\alpha = \frac{y_1}{\mu}, \quad \beta = \frac{y_2}{\nu} \tag{9.15}$$

we obtain

$$\left(\frac{\partial^2}{\partial \alpha^2} - \frac{\partial^2}{\partial \beta^2}\right)\phi + D\frac{\partial \phi}{\partial \alpha} + E\frac{\partial \phi}{\partial \beta} + F\phi = 0. \tag{9.16}$$

A further reduction is obtained by setting

$$\phi = e^{-(D\alpha - E\beta)/2}\Phi. \tag{9.17}$$

This leaves us with

$$\left(\frac{\partial^2}{\partial \alpha^2} - \frac{\partial^2}{\partial \beta^2}\right)\Phi + k\Phi = 0, \tag{9.18}$$

where the precise form of k is left as an exercise. The principal part of this operator is the same as that of the wave equation (9.1), with *wave speed* equal to unity. We might expect the solution to (9.18) to have properties similar to those of the wave equation (9.1).

An alternative *canonical* form is suggested by (9.5); i.e., if we set

$$\eta = \frac{\alpha - \beta}{2}, \quad \zeta = \frac{\alpha + \beta}{2}, \tag{9.19}$$

we obtain

$$\frac{\partial^2 \Phi}{\partial \eta \partial \zeta} + k\Phi = 0. \tag{9.20}$$

Elliptic Case: $b^2 < ac$

When this condition is satisfied, (9.4) is said to be *elliptic*. In this case both eigenvalues are of the same sign, and under a transformation analogous to (9.15), we get

$$\frac{\partial^2 \phi}{\partial \eta^2} + \frac{\partial^2 \phi}{\partial \rho^2} + D\frac{\partial \phi}{\partial \eta} + E\frac{\partial \phi}{\partial \rho} + F\phi = 0. \tag{9.21}$$

In analogy with (9.17), we write

$$\phi = e^{(D\eta + E\rho)/2}\Phi. \tag{9.22}$$

This leads to

$$\left(\frac{\partial^2}{\partial \eta^2} + \frac{\partial^2}{\partial \rho^2}\right)\Phi + k\Phi = 0, \tag{9.23}$$

which is the *canonical* form of an elliptic equation. The principal part of the operator is the same as that of Laplace's equation (9.3), and we might expect the behavior of solutions to (9.23) to resemble those of the Laplace equation.

Parabolic Case: $b^2 = ac$.

Since one eigenvalue vanishes in this case, the transformed equation is easily placed in the general form

$$\frac{\partial^2 \phi}{\partial \rho^2} - D \frac{\partial \phi}{\partial \eta} + E \frac{\partial \phi}{\partial \rho} + F\phi = 0, \tag{9.24}$$

which under further transformation by

$$\phi = \Phi \exp\left[-\frac{E\rho}{2} - \frac{\eta}{D}\left(\frac{E^2}{4} - F\right)\right] \tag{9.25}$$

becomes

$$\frac{\partial^2}{\partial \rho^2}\Phi = D\frac{\partial \Phi}{\partial \eta}, \tag{9.26}$$

the diffusion equation. This is called the *parabolic* case.

Thus the general second order partial differential equation (9.4) takes on essentially one of the three canonical forms (9.18) (or (9.20)), (9.23), and (9.26). These correspond to the wave equation, the Laplace equation, and the diffusion equation, which were all considered in the last chapter and which we now reconsider in more detail.

When a second order partial differential equation is put into the form of (9.16) or (9.21) or (9.24)—whichever is appropriate—it is said to be in canonical form. As a practical matter, it is not necessary to go through the detailed steps which lead to (9.13) in order to arrive at a canonical form.

Example. Consider

$$\phi_{xx} + 4\phi_{xy} - 2\phi_{yy} = 0.$$

Therefore $a = 1$, $b = 2$, $c = -2$ so that $b^2 - 4ac = 6 > 0$, and the case is hyperbolic. To put the equation in hyperbolic form, we introduce

$$X = x + \gamma y, \quad Y = -\gamma x + y,$$

which form an orthogonal system of coordinates. Then since

$$\frac{\partial}{\partial x} = \frac{\partial X}{\partial x}\frac{\partial}{\partial X} + \frac{\partial Y}{\partial x}\frac{\partial}{\partial Y},$$

we obtain

$$\partial_x = \partial_X - \gamma\partial_Y.$$

Similarly

$$\partial_y = \gamma\partial_X + \partial_Y.$$

By taking appropriate products, the second derivatives are

$$\partial_x^2 = \partial_X^2 - 2\gamma\partial_{XY} + \gamma^2\partial_Y^2,$$

$$\partial_{xy} = \gamma\partial_X^2 + (1 - \gamma^2)\partial_{XY} - \gamma\partial_Y^2,$$

$$\partial_y^2 = \gamma^2\partial_X^2 + 2\gamma\partial_{XY} + \partial_Y^2.$$

Thus the equation then becomes

$$(1 + 4\gamma - 2\gamma^2)\phi_{XX} + (-2\gamma + 4(1 - \gamma^2) - 4\gamma)\phi_{XY} + (\gamma^2 - 4\gamma - 2)\phi_{YY} = 0.$$

Therefore, to put this into canonical form, we first take the coefficient of ϕ_{XY} to be zero:

$$-2\gamma + 4(1 - \gamma^2) - 4\gamma = 4 - 6\gamma - 4\gamma^2 = 0$$

or

$$\gamma = \frac{1}{2}, -2.$$

Thus, if we take $\gamma = -2$ (the other choice, $\gamma = 1/2$, just interchanges X and Y),

$$-15\phi_{XX} + 10\phi_{YY} = 0;$$

and, for example, if we take

$$\alpha = \frac{X}{\sqrt{15}}, \quad \beta = \frac{Y}{\sqrt{10}},$$

it follows that the equation becomes

$$\phi_{\alpha\alpha} = \phi_{\beta\beta}.$$

Exercises

1. Furnish the details for the demonstration of (9.10) and (9.11).

2. Use (9.17) to reduce (9.16) to (9.18). What is k given by?

3. Carry out the reduction of (9.18) to (9.20).

4. Obtain the reduction of the following: (a) (9.21) to (9.23); (b) (9.24) to (9.26).

5. Suppose

$$L_1 u = au_x + bu_y + cu,$$

$$L_2 u = fu_x + gu_y + fu,$$

with $a, b, c, \ldots, f$ all constant. Solve $L_1 L_2 u = 0$ in terms of the solutions

$$L_1 u_1 = 0, \quad L_2 u_2 = 0.$$

6. Show that the condition that (9.4) be factorable into the form

$$L_1 L_2 = 0$$

(where L_1 and L_2 are first order as above) is that

$$\det \begin{pmatrix} a & b & d \\ b & c & e \\ d & e & f \end{pmatrix} = 0.$$

7. Classify each of the following equations, and put them into canonical form:

   (a) $\phi_{xx} - \phi_{xy} + 2\phi_{yy} + \phi_x + \phi_y = 0$,

   (b) $\phi_{xx} - 2\phi_{xy} + \phi_{yy} + \phi_x + 2\phi_y + \phi = 0$,

   (c) $\phi_{xx} - \phi_{yy} - \phi_x + \phi_y = 0$,

   (d) $\phi_{xx} + 4\phi_{xy} - \phi_{yy} + \phi = 0$.

8. Reduce the following to canonical form:

   (a) $\phi_{xx} + 2\phi_{xy} + \phi_{yy} + \phi_x + \phi_y = 0$,

   (b) $\phi_{xx} + 2\phi_{xy} + 5\phi_{yy} + 3\phi_x + \phi = 0$,

   (c) $3\phi_{xx} + 10\phi_{xy} + 3\phi_{yy} = 0$.

9. Consider the equation

$$\phi_{xx} + 2a\phi_{xy} + b\phi_{yy} = 0.$$

For what values of a and b is it: (a) hyperbolic; (b) elliptic; (c) parabolic?

9.2   Hyperbolic Case—The Wave Equation

We refer back to D'Alembert's Formula (7.68) to express the solution to the wave equation (9.1) as

$$\phi(x,t) = \frac{1}{2}\{f(x - Ct) + f(x + Ct)\} + \frac{1}{2C} \int_{x-Ct}^{x+Ct} g(s)ds.$$

It is easily verified that if the limit $t \downarrow 0$ is applied to this formula, we find that $\phi(x,0) = f(x)$ and by differentiating that $\phi_t(x,0) = g(x)$. We now apply some simple geometrical considerations to deepen our understanding of this case.

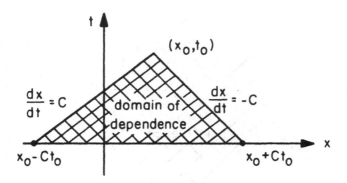

FIGURE 9.1.

Consider any point (x_0, t_0) in the (x,t)-plane and construct the *lines of propagation*, or *characteristic lines* as they are called,

$$\frac{x - x_0}{t - t_0} = \pm C.$$

These are indicated in Figure 9.1. If we consider the first term of D'Alembert's Formula,

$$\frac{1}{2} f(x - Ct),$$

we see that it has the same value for all points as it has at (x_0, t_0), i.e.,

$$x - Ct = x_0 - Ct_0.$$

Therefore it can be evaluated at $t = 0$ where x is just $x_0 - Ct_0$, the intercept of the *positive* characteristic line. Similarly the second term can be evaluated by the intercept, $x_0 + Ct_0$, of the *negative* characteristic line. The integral term depends on g in the interval between these points. For this reason this interval is said to be in the *domain of dependence* on the initial line of the point (x_0, t_0), and the triangular area below (x_0, t_0) is called the domain of dependence of this point.

A related notion concerns the region that a particular point of the initial line can influence. This is the collection of all points which contain the given point in the domain of dependence. A sketch of this is shown in Figure 9.2.

If we imagine the case of sound waves, then the *domain of influence* relates to the fact that a wave generated at some time at a point cannot send a signal to another point in less time than it takes to traverse the intervening distance. Alternatively, the *domain of dependence* originates in the fact that we can only be signaled by soundwaves generated in the past and for any given time lapse these signals cannot have originated beyond the distance that the sound can travel in that time.

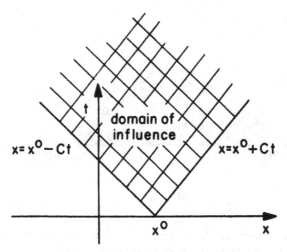

FIGURE 9.2.

Inhomogeneous Problem

These intuitively simple ideas now play a role in the construction of the solution to the inhomogeneous wave equation

$$\frac{\partial^2 \phi}{\partial t^2} = \frac{\partial^2 \phi}{\partial x^2} + f(x,t). \tag{9.27}$$

We also generalize the initial conditions so that ϕ and the normal derivative ϕ_n are specified on a curve Γ^0 in the (x,t)-plane, not necessarily the usual initial line $t = 0$. (For simplicity we have taken $C = 1$ in (9.27)). The characteristic lines are now given by $\Gamma^\pm : x - x_0 = \pm(t - t_0)$. To carry out the construction of the solution at an arbitrary point, which we specify by (x_0, t_0), we form the region R bounded by the initial line Γ^0, and the characteristic lines Γ^+ and Γ^- as indicated in Figure 9.3. As a first step in the solution of (9.27) we integrate this equation in the domain R shown in the figure

$$\int_R \left(\frac{\partial \phi_t}{\partial t} - \frac{\partial \phi_x}{\partial x} \right) dx dt = \int_R f(x,t) dx dt.$$

(We write ϕ_{tt} as $\partial \phi_t / \partial t$ and ϕ_{xx} as $\partial \phi_x / \partial x$ for later purposes.)

The left-hand side of this equation may be viewed as the integral of the divergence operator $(\partial/\partial t, \partial/\partial x)$ acting on the vector $(\phi_t, -\phi_x)$. Therefore the Divergence Theorem can then be applied. Thus

$$\int_R \left[\frac{\partial}{\partial t} \phi_t + \frac{\partial}{\partial x}(-\phi_x) \right] dx dt = \oint_\Gamma (-\phi_x, \phi_t) \cdot \mathbf{n} \, d\ell,$$

where $\Gamma = \Gamma^0 + \Gamma^+ + \Gamma^-$ in the sense shown in Figure 9.2, $\mathbf{n}$ is the outward

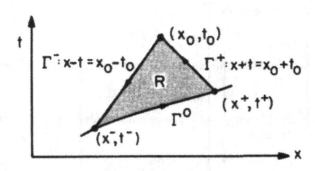

FIGURE 9.3.

normal, and $d\ell$ is the element of arc length. As a little thought shows,

$$\mathbf{n}\, d\ell = (dt, -dx),$$

and we are therefore led to

$$-\oint_\Gamma \phi_t dx - \oint_\Gamma \phi_x dt = \int_R f(x,t)dxdt.$$

Since $dx = dt$ on Γ^-, we can write

$$-\int_{\Gamma^-} (\phi_t dx + \phi_x dt) = -\int_{\Gamma^-} (\phi_x dx + \phi_t dt) = -\int_{\Gamma^-} d\phi \equiv \phi^0 - \phi^-,$$

where $\phi^0 = \phi(x_0, t_0)$ and $\phi^- = \phi(x^-, t^-)$. Similarly,

$$-\int_{\Gamma^+} (\phi_t dx + \phi_x dt) = \phi^0 - \phi^+,$$

and therefore on solving for ϕ^0, we obtain

$$\phi(x_0, t_0) = \frac{1}{2}\{\phi^+ + \phi^-\} + \frac{1}{2}\int_{\Gamma^0} (\phi_x dt + \phi_t dx)$$

$$+ \frac{1}{2}\int_R f(x,t)dxdt. \tag{9.28}$$

This generalizes D'Alembert's Formula to the case of an arbitrary initial curve and an inhomogeneous wave equation. To see this note that if Γ^0 is $t = 0$, then $dt = 0$ and that if $f = 0$, then (9.28) is just (7.68).

It should be observed that (9.28) requires that ϕ_x and ϕ_t be given on the *initial line* Γ^0. This is equivalent to specifying ϕ and its normal derivative on Γ^0. To see this let us suppose that the initial line is specified parametrically as

$$\Gamma^0 : x = x^0(s), \quad t = t^0(s),$$

where s is the arc length along Γ^0 from (x^-, t^-) to (x^+, t^+) of Figure 9.3. The unit vector tangent to Γ^0 is

$$\mathbf{t} = (\dot{x}^0, \dot{t}^0),$$

where the dot denotes differentiation with respect to s. Then the unit normal is

$$\mathbf{n} = (-\dot{t}^0, \dot{x}^0).$$

Thus, if ϕ^0 and ϕ_n^0 denote the given initial values of ϕ and its normal derivative on Γ^0, then

$$\frac{\partial \phi^0}{\partial s} = \dot{x}^0 \frac{\partial \phi}{\partial x} + \dot{t}^0 \frac{\partial \phi}{\partial t},$$

$$\frac{\partial \phi^0}{\partial n} = -\dot{t}^0 \frac{\partial \phi}{\partial x} + \dot{x}^0 \frac{\partial \phi}{\partial t}.$$

These can now be solved to give ϕ_t and ϕ_x on the initial curve Γ^0.

It is now important to note that the above construction requires that the *initial curve* have a slope such that $|dx/dt| > C$. Otherwise, the construction breaks down; viz., the characteristic lines only intersect Γ^0 in one point. When all points of a curve are such that $|dx/dt| < C$, it is said to be *time-like* (since for the time axis $dx/dt = 0$); and when $|dx/dt| > C$ for all points on a curve, it is said to be *space-like* (since for the space axis $dx/dt = \infty$). If we refer back to our formulation of the boundary value problem for the wave equation, Section 7.1, the walls anchoring the string are time-like. As we saw there, just one datum is specified on the time-like curves—whereas the two pieces of data are prescribed on a space-like curve. This specification is the general rule for solving the wave equation.

Energy Integral

It has been shown that the hyperbolic case (Section 9.1) can be reduced to the canonical form

$$\frac{\partial^2 \phi}{\partial t^2} = c^2 \frac{\partial^2 \phi}{\partial x^2} - k\phi. \tag{9.29}$$

If (9.29) is multiplied by ϕ_t, then

$$\frac{\partial \phi}{\partial t} \frac{\partial^2 \phi}{\partial t^2} = \frac{\partial}{\partial t} \frac{\phi_t^2}{2} = c^2 \frac{\partial \phi}{\partial t} \frac{\partial^2 \phi}{\partial x^2} - k \frac{\partial}{\partial t} \frac{\phi^2}{2}$$

$$= c^2 \frac{\partial}{\partial x}(\phi_t \phi_x) - c^2 \frac{\partial}{\partial t} \frac{\phi_x^2}{2} - k \frac{\partial}{\partial t} \frac{\phi^2}{2}.$$

After a little rearrangement of terms, this can be rewritten as

$$\frac{\partial}{\partial t} e = \frac{\partial}{\partial t}\left(\frac{\phi_t^2}{2} + c^2 \frac{\phi_x^2}{2} + k \frac{\phi^2}{2}\right) = c^2 \frac{\partial}{\partial x}(\phi_t \phi_x), \tag{9.30}$$

where the *energy* e is defined by the second expression.

In the context of the stretched string, the speed of propagation is defined by $c^2 = T/\rho$. Also, the term $-k\phi$ can be interpreted as an additional spring force perpendicular to the string. (On physical grounds k must be positive.) The energy e given by

$$e = \frac{1}{2}\{\phi_t^2 + c^2\phi_x^2 + k\phi^2\}$$

can be interpreted as the energy per unit length of string. $(1/2)\phi_t^2$ represents the kinetic energy per unit mass, and the remaining two terms represent the potential energy of position per unit mass.

If we integrate (9.30) over the interval of definition, which we can take to be unity, then

$$\frac{dE}{dt} = \frac{d}{dt}\int_0^1 \frac{1}{2}(\phi_t^2 + c^2\phi_x^2 + k\phi^2)dx = c^2\int_0^1 \frac{\partial}{\partial x}(\phi_t\phi_x)dx$$

$$= c^2(\phi_t(1)\phi_x(1) - \phi_t(0)\phi_x(0)),$$

where E represents the total energy, lying in the interval $(0, 1)$. In the case of the stretched string, $\phi(0) = \phi(1) = 0$ and hence $\phi_t(0) = \phi_t(1) = 0$. For this as well as other circumstances we have

$$\frac{dE}{dt} = 0.$$

This states that the energy of the system is conserved; i.e.,

$$E(\phi) = \frac{1}{2}\int_0^1 (\phi_t^2 + c^2\phi_x^2 + k\phi^2)dx = E_0, \tag{9.31}$$

where the constant E_0 is the energy at the initial instant.

Uniqueness

The energy equation, (9.31), can be immediately applied to show that the problem, as we have specified it, has a unique solution. Toward this end consider the problem

$$\phi_{tt} = c^2\phi_{xx} - k\phi, \quad 0 < x < 1,$$

$$\phi(x, 0) = f(x), \quad \phi_t(x, 0) = g(x),$$

with the boundary conditions (in general, functions of time)

$$\phi \text{ (or } \phi_x) \text{ given at } x = 0, 1.$$

Suppose the solution to this problem is not unique; i.e., there exist, say, solutions ϕ_1 and ϕ_2. Then $\Phi = \phi_1 - \phi_2$ is such that

$$\Phi_{tt} = c^2\Phi_{xx} - k\Phi, \quad 0 < x < 1,$$

$$\Phi(x,0) = 0 = \Phi_t(x,0),$$

$$\Phi \ (\text{or } \Phi_x) = 0 \quad \text{at} \quad x = 0,1.$$

The energy conservation law (9.31) can be applied to the problem specifying Φ. This yields

$$E(\Phi) = E_0 = 0.$$

(Note $\Phi(x,0) = 0$ implies $\Phi_x(x,0) = 0$.) Then, since the integrand of (9.31) is positive, this implies that

$$\Phi = 0.$$

Hence uniqueness follows. With a little extra work the same proof of uniqueness can be applied to the case of a general, space-like initial curve and time-like boundaries.

A Geometrical Construction

We next present a useful and pretty method of solution which is entirely geometrical. For simplicity we consider the wave equation with speed unity:

$$\phi_{tt} = \phi_{xx}. \tag{9.32}$$

Any solution of this equation has the form

$$\phi = F(x - t) + G(x + t).$$

With the use of Figure 9.4, we have

$$F(A) + F(C) = F(x - k - t + h) + F(x + k - t - h) = F(D) + F(B)$$

and in a similar fashion

$$G(A) + G(C) = G(D) + G(B).$$

We have shown therefore that

$$\phi(A) + \phi(C) = \phi(B) + \phi(D). \tag{9.33}$$

Equation (9.33) can be used to solve (9.32) for the problem depicted in Figure 9.5. ϕ and its normal derivative ϕ_n are given on a space-like curve Γ^0 and ϕ is given on the two time-like curves Γ^- and Γ^+. In region I we solve as a pure initial value problem. (We use (9.28) with $f = 0$.) Next ϕ is determined at an arbitrary point C of region II by use of (9.33). In a similar fashion ϕ is also determined everywhere in region III. From these we can next determine ϕ in IV as indicated by the rectangle $\overline{A}\ \overline{B}\ \overline{C}\ \overline{D}$. And so forth.

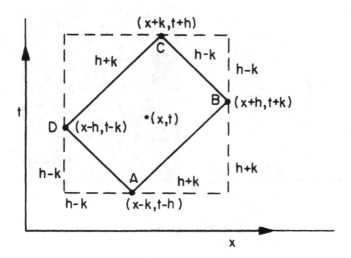

FIGURE 9.4.

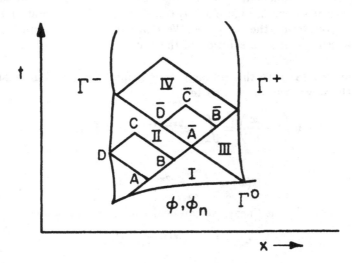

FIGURE 9.5.

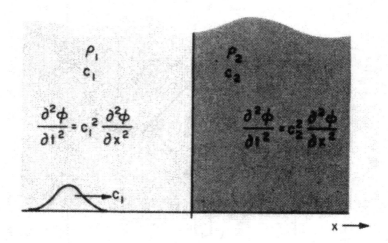

$$\frac{\partial^2 \phi}{\partial t^2} = c_1^2 \frac{\partial^2 \phi}{\partial x^2}$$

$$\frac{\partial^2 \phi}{\partial t^2} = c_2^2 \frac{\partial^2 \phi}{\partial x^2}$$

FIGURE 9.6.

Reflection and Transmission of Waves

We now consider the important problem of a wave passing from one material to another across a sharp boundary. This could be the case, for example, for sound waves passing through an air-water interface.

As indicated in Figure 9.6, a different wave equation applies in each material. At the moment at which we begin to view the problem, we suppose a waveform, as shown in Figure 9.6, traveling to the right in material 1 and not yet encountering the interface. We represent the waveform by $u(t - x/c_1)$—which is clearly a solution of the wave equation in material 1 and is moving to the right.

In order to deal with this problem, it is helpful first to Fourier analyze in time the waveform; i.e., we write

$$u\left(t - \frac{x}{c_1}\right) = \int_{-\infty}^{\infty} e^{2\pi i \omega(t - (x/c_1))} U(\omega) d\omega, \qquad (9.34)$$

where

$$U(\omega) = \int_{-\infty}^{\infty} e^{-2\pi i \omega s} u(s) ds.$$

Since the problem is linear, we can deal with each harmonic

$$u_\omega = U(\omega) e^{2\pi i \omega(t - (x/c_1))} \qquad (9.35)$$

separately and at the end of our deliberations sum the components, i.e., perform the required integration. It is worth noting in passing that although the pulse (9.34) is of finite duration, it can be viewed as a summation of waves of infinite duration, (9.35).

The component, (9.35), is a wave of period $1/\omega$ (frequency ω) and wavelength c_1/ω (wave number of spatial frequency ω/c_1). As a result it is referred to as *a monochromatic wave.*

After the wave strikes the interface we expect, in general, to find a wave moving to the right in region 2 and a wave moving to the left in region 1—briefly said, a transmitted and a reflected wave. At the interface, $x = 0$, (9.35) is

$$U(\omega)e^{2\pi i\omega t}.$$

Therefore, in region 2, we simply see an oscillating source of frequency ω. Consequently the frequency (and not necessarily the spatial frequency or wavelength) is preserved across the interface. This suggests that the transmitted wave has the form

$$T(\omega)e^{2\pi i\omega(t-(x/c_2))} \tag{9.36}$$

and the reflected wave, the form

$$R(\omega)e^{2\pi i\omega(t+(x/c_1))}. \tag{9.37}$$

Note that (9.36) satisfies the wave equation in region 2 and represents a wave moving to the right, while (9.37) satisfies the wave equation in region 1 and represents a wave moving to the left.

The *complex amplitudes* $T(\omega)$ and $R(\omega)$ are still to be determined. In order to calculate these, we return to one of the physical models. To be specific we consider the equations for acoustic disturbances, presented in Section 7.1:

$$\frac{\partial s}{\partial t} + \frac{\partial u}{\partial x} = 0, \tag{9.38}$$

$$\frac{\partial u}{\partial t} + c^2 \frac{\partial s}{\partial x} = 0, \quad c^2 = \frac{\gamma p_0}{\rho_0}, \tag{9.39}$$

$$p - p_0 = \gamma p_0 s. \tag{9.40}$$

We recall that u represents the velocity perturbation, s the density perturbation, p the pressure, p_0 its value at equilibrium ($p - p_0$ is the pressure perturbation), and ρ_0 the equilibrium gas density.

By elimination of s between (9.38) and (9.39), we see that the velocity u satisfies the wave equation. We regard (9.35) as specifying u. Note that from (9.38)

$$s = \frac{1}{c}u$$

for a wave traveling to the right, while

$$s = -\frac{1}{c}u$$

for a wave moving to the left. Also from (9.40) the pressure perturbation is given by

$$p - p_0 = \pm \frac{\gamma p_0}{c} u = \pm \rho_0 c u$$

for a wave moving to the right or left, respectively. (We have used $p_0 = \rho_0 c^2 / \gamma$, from which it additionally follows that $|(p - p_0)/p_0| \ll 1 \rightarrow |u/c| \ll 1$, as was stated earlier.)

To return to the problem at hand, it follows from physical considerations that the velocity of the gas is continuous at $x = 0$. Therefore, from (9.35), (9.36), and (9.37),

$$U + R = T. \tag{9.41}$$

Furthermore, the pressure must be continuous; otherwise infinite accelerations will occur. Hence

$$c_1 \rho_1 U - c_1 \rho_1 R = c_2 \rho_2 T. \tag{9.42}$$

Equations (9.41) and (9.42) represent the two equations in two unknowns which are needed to determine R and T. For this purpose (9.42) suggests that we define

$$Z = \rho c, \tag{9.43}$$

which is known as the *acoustic impedance*. This we observe to be the absolute value of the ratio of the pressure perturbation to the velocity perturbation. (It is analogous to the ratio of voltage to current in electrical problems—hence the use of the term impedance.)

If we solve (9.41) and (9.42), we obtain

$$T = \frac{2Z_1}{Z_1 + Z_2} U, \quad R = \frac{Z_1 - Z_2}{Z_1 + Z_2} U, \quad Z_1 = \rho_1 c_1, \quad Z_2 = \rho_2 c_2. \tag{9.44}$$

From these we see that if $Z_1 \ll Z_2$, the wave is almost completely reflected (the minus sign indicates a 180° phase change). We will have minimal reflection and almost complete transmission if $Z_1 \approx Z_2$. (This is referred to as *impedance matching*.) It is also of interest to observe that if $Z_1 \gg Z_2$, the wave is almost completely reflected and in addition there is a large transmitted wave (roughly double the amplitude).

As an example, consider the case of an air-water interface. The physical parameters, in cgs units, are

$$\text{Water}: \quad \rho_w = 1, \quad c_w = 1.5 \times 10^5, \quad Z_w = 1.5 \times 10^5;$$

$$\text{Air}: \quad \rho_a = 1.25 \times 10^{-3}, \quad c_a = 3.5 \times 10^4, \quad Z_a = 44.$$

It is evident that, in this case, the impedance mismatch is quite pronounced. This explains why we hear little outside noise when we are submerged in water.

Similar impedance considerations are important, for example, in assembling components of a stereo system. The tuner and speakers must have near identical impedances in order to communicate signals. In another context we mention that the cochlea of the inner ear contains a water-like fluid. As a result of this, impedance at the oval window of the middle ear is greatly different from that of the air in the outer ear, where soundwaves enter. It is the function of the middle ear, by means of a series of three bones (stapes, malleus, and incus) which act as levers, to remove some of this impedance mismatch.

To come back to the problem under consideration, we can now write the solution as follows:

In region 1

$$\phi_1 = \int_{-\infty}^{\infty} U(\omega)e^{2\pi i\omega(t-(x/c_1))}\,d\omega + \frac{Z_1 - Z_2}{Z_1 + Z_2}\int_{-\infty}^{\infty} U(\omega)e^{2\pi i\omega(t+(x/c_1))}$$

$$= u\left(t - \frac{x}{c_1}\right) + \frac{Z_1 - Z_2}{Z_1 + Z_2}u\left(t + \frac{x}{c_1}\right).$$

In region 2

$$\phi_2 = \frac{2Z_1}{Z_1 + Z_2}\int_{-\infty}^{\infty} e^{2\pi i\omega(t-(x/c_2))}U(\omega)\,d\omega$$

$$= \frac{2Z_1}{Z_1 + Z_2}u\left(t - \frac{x}{c_2}\right).$$

Thus the result is actually very simple.

Exercises

1. Solve $u_{tt} = u_{xx}$, $0 < t < \infty$ if

$$u(t = 0) = x, \quad u_t(t = 0) = x^2.$$

2. Find general solutions of

   (a)
   $$\frac{1}{c^2}\frac{\partial^2}{\partial t^2}\phi - \frac{\partial^2}{\partial x^2}\phi = x,$$

   (b)
   $$\frac{1}{c^2}\frac{\partial^2}{\partial t^2}\phi - \frac{\partial^2}{\partial x^2}\phi = xt + x$$

   for $\phi(t = 0) = f(x)$, $\phi_t(t = 0) = g(x)$.

3. Solve $\phi_{xx} = \phi_{tt}$, $-\infty < x < \infty$, $t > 0$; $\phi(x,0) = H$ if $|x| < L$, $\phi(x,0) = 0$ if $|x| > L$; $\phi_t(x,0) = 0$.

4. Solve
$$\phi_{tt} - c^2\phi_{xx} = x^2, \quad -a < x < a,$$
$$\phi(x,0) = x,$$
$$\phi_t(x,0) = 0$$

for (x,t) in the triangle $(-a,0)$, $(a,0)$, $(0,a/c)$.

5. Write
$$L = \frac{\partial^2}{\partial t^2} - c^2\frac{\partial^2}{\partial x^2}$$

and suppose
$$Lu = 0 = Lv.$$

Show
$$L(u_t v_t + c^2 u_x v_x) = 0.$$

6. The solution to the inhomogeneous wave equation (9.28) contains a line integral, over Γ^0, in which ϕ_t and ϕ_x appear—and not the given ϕ and ϕ_n. Suppose the curve Γ^0 is represented parametrically by
$$x = x(s), \quad t = t(s)$$

and on it
$$\phi = \alpha(s), \quad \frac{\partial\phi}{\partial n} = \beta(s).$$

Express
$$\int_{\Gamma^0}(\phi_x dt + \phi_t dx) = \int_{\Gamma^0}\left(\phi_x\frac{dt}{ds} + \phi_t\frac{dx}{ds}\right)ds$$

in terms of α and β (and their derivatives).

7. Find particular solutions of the following:

   (a) $\phi_{xx} - \phi_{yy} = x$,
   (b) $\phi_{xx} - 3\phi_{xy} + 2\phi_{yy} = \cos y$,
   (c) $\phi_{xx} + \phi_{yy} = x^2 + xy$,
   (d) $\phi_{xx} - \phi_{xy} + \phi_{yy} = x^2 + y$.

8. Solve
$$\phi_{xx} - \phi_{tt} = 16x^2, \quad 0 < x < 1, \quad t > 0,$$
$$\phi(x,0) = x(1-x),$$
$$\phi_t(x,0) = 0,$$
$$\phi(0,t) = \phi(1,t) = 0.$$

[Hint: Use the fact that $x^2 = [(x+t)+(x-t)]^2$ and use characteristic coordinates.]

9. Consider an infinite stretched string. The string to the left of the midpoint is of density per unit length ρ_1 and to the right, ρ_2. The tension can be assumed to be constant. What is the impedance at the midpoint? [Hint: Displacement and velocity must be continuous.]

10. Solve

$$\phi_{xx} - \frac{1}{c^2}\phi_{tt} = xt, \quad -\infty < x < \infty, \ t > 0$$

with

$$\phi_t(t = 0) = \phi(t = 0) = 0.$$

11. In two or three dimensions the analogue of (9.29) is

$$\frac{\partial^2}{\partial t^2}\phi = c^2 \nabla^2 \phi - k\phi.$$

Under what conditions does

$$\frac{dE}{dt} = 0$$

where

$$E = \frac{1}{2}\int_V (\phi_t^2 + c(\nabla\phi)^2 + k\phi^2)d\mathbf{x}?$$

12. Use the previous exercise to prove a uniqueness theorem for the wave equation in higher dimensions.

9.3 Parabolic Case—The Diffusion Equation

We recall from Section 9.1 that without loss of generality the parabolic case can be reduced to the diffusion equation.

Boundary and Initial Conditions

As we saw in our earlier discussion of the diffusion or heat equation, it seem appropriate to solve

$$\frac{\partial\theta}{\partial t} = \kappa\frac{\partial^2\theta}{\partial x^2} \tag{9.45}$$

by supplying one datum initially and, for a finite problem, one datum at each of the two endpoints. Thus in the sketch on the left hand of Figure 9.7, which corresponds to heat flow in a bar of length ℓ, we supply one datum on each of the heavily drawn lines. The solution is then determined in the shaded region. By analogy, we might suppose that the solution to (9.45) in the shaded region of the sketch on the right hand of Figure 9.7

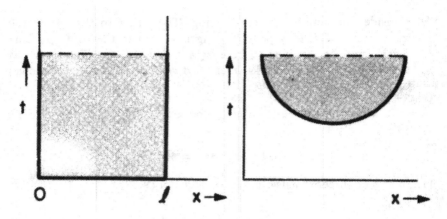

FIGURE 9.7.

is determined when one datum is given on the heavily drawn curve. This datum in the general case is of the form

$$a\theta + b\frac{\partial\theta}{\partial n} = f(s), \text{ on } x = x(s), \ t = t(s),$$

where $(x(s), t(s))$ is the curve.

If we imagine solving a heat flow problem, intuition tells us that the maximum or minimum temperature is achieved initially or at a boundary. This is the content of a theorem known as the *Maximum Principle.* Two other mathematical results which intuition suggests are uniqueness and energy conservation. We now prove each of these results.

Energy Conservation

Consider the three-dimensional form of the diffusion equation (the one- and two-dimensional cases are special cases)

$$\frac{\partial\theta}{\partial t} = \nabla \cdot (\kappa \nabla \theta), \tag{9.46}$$

where $\kappa > 0$ can be a function of position. If we integrate over a volume V in space and apply the Divergence Theorem, we obtain

$$\frac{\partial}{\partial t}\int_V \theta d\mathbf{x} = \int_{\partial V} \kappa(\mathbf{n} \cdot \nabla\theta)ds.$$

∂V refers to the surface of the volume V and $\mathbf{n}$ the outward normal to ∂V. The integral of the left-hand side gives the total physical energy in V (or the total amount of solute in V for a diffusion problem) and the right-hand side the outward flux from ∂V. Therefore, for a finite volume, if

$\mathbf{n} \cdot \nabla \theta$ vanishes on ∂V (i.e., no heat flow into the domain), or for an infinite volume, if $\mathbf{n} \cdot \nabla \theta \to 0$ sufficiently rapidly as infinity is approached we have

$$\frac{\partial}{\partial t} \int_V \theta \, d\mathbf{x} = 0$$

or

$$\int_V \theta \, d\mathbf{x} = \int_V \theta_0 d\mathbf{x}, \tag{9.47}$$

where θ_0 is the initial distribution and the integral of it is a conserved quantity.

Uniqueness

For mathematical reasons, another quantity is also referred to as energy, viz.,

$$\epsilon = \int_V \theta^2 / 2 \, d\mathbf{x}.$$

ϵ will be referred to as the *mathematical energy*.

In order to prove the uniqueness of solutions to (9.46), we begin with an identity. Multiply (9.46) by θ to obtain

$$\theta \frac{\partial \theta}{\partial t} = \frac{\partial}{\partial t} \frac{\theta^2}{2} = \theta \nabla \cdot \kappa \nabla \theta = \nabla \cdot (\theta \kappa \nabla \theta) - \kappa (\nabla \theta)^2.$$

If we integrate this over V and apply the Divergence Theorem to the first term on the right, we obtain

$$\frac{\partial}{\partial t} \int_V \frac{\theta^2}{2} d\mathbf{x} = \int_{\partial V} \theta \kappa \frac{\partial \theta}{\partial n} ds - \int_V \kappa (\nabla \theta)^2 d\mathbf{x}.$$

In the finite case, if $\partial \theta / \partial n = 0$ on ∂V, or in the infinite case, if $\theta(\partial \theta / \partial n) \to 0$ sufficiently rapidly at infinity, we obtain

$$\frac{\partial \epsilon}{\partial t} = \frac{\partial}{\partial t} \int_V \frac{\theta^2}{2} d\mathbf{x} = - \int_V \kappa (\nabla \theta)^2 d\mathbf{x}. \tag{9.48}$$

(*This last relation governing the energy ϵ is actually a form of the entropy principle of thermodynamics. It states that $\int_V (\theta^2/2) d\mathbf{x}$ continues to decrease if gradients are present.*)

Next we denote the initial distribution by θ_0 and integrate (9.48) over time to obtain

$$\int_V \frac{\theta^2}{2} d\mathbf{x} - \int_V \frac{\theta_0^2}{2} d\mathbf{x} = - \int_0^t \int_V \kappa (\nabla \theta)^2 d\mathbf{x} d\tau, \tag{9.49}$$

where it is understood that the first integral is evaluated at time t.

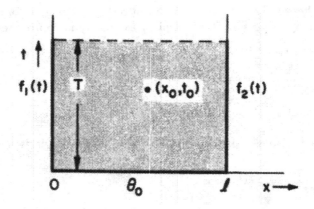

FIGURE 9.8.

To prove uniqueness, suppose θ_1 and θ_2 satisfy (9.46), have the same initial data, and are equal on ∂V (or have identical fluxes on ∂V). Then

$$\Theta = \theta_1 - \theta_2$$

satisfies (9.46), $\Theta(x,0) = 0$, and $\Theta = 0$ on ∂V (or $\Theta_n = 0$ on ∂V). It then follows from (9.49) that

$$\int_V \frac{\Theta^2}{2} dx = - \int_0^t \int_V \kappa(\nabla\Theta)^2 dx d\tau,$$

which, since the right-hand side is nonpositive and the left-hand side nonnegative, is only possible if $\Theta = 0$. Hence uniqueness has been proved.

Maximum Principle

It follows from the earlier intuitive derivation of the heat equation that heat is conducted away from regions of high temperature. From this it follows that a temperature maximum can only appear initially or on boundaries. This is the essence and the origin of the *Maximum Principle*, which states that the maximum of θ, (9.46), occurs either initially or on the boundaries.

We now prove this Maximum Principle in one space dimension. In particular, we consider (9.45), subject to the data

$$\theta(x,0) = \theta_0(x), \quad \theta(0,t) = f_1(t), \quad \theta(\ell,t) = f_2(t),$$

and show that a maximum *cannot* occur at an interior point (x_0, t_0) as depicted in Figure 9.8.

The proof is by contradiction and thus we start by assuming that M, the maximum of θ, occurs at (x_0, t_0), an interior point of the domain such

that
$$0 < x_0 < \ell, \quad 0 < t_0 \le T.$$
This is the shaded area of Figure 9.8. Consider
$$\Theta(x,t) = \theta(x,t) + \frac{M - m}{4\ell^2}(x - x_0)^2,$$
where m is the maximum of θ on the boundary, indicated by the heavily drawn lines in Figure 9.8. On this boundary we easily have that
$$\Theta < m + \frac{M - m}{4} = \epsilon M,$$
where $0 < \epsilon < 1$. This property, coupled with the fact that $\Theta(x_0, t_0) = M$, implies that Θ also has a maximum at an interior point, say, at the point (x_1, t_1). Therefore, at this point,
$$\Theta_{xx} \le 0, \quad \Theta_t \ge 0.$$
(If $t_0 < T$, $\Theta_t = 0$.) Hence
$$\Theta_t - \Theta_{xx} \ge 0.$$
But
$$\Theta_t - \Theta_{xx} = \theta_t - \theta_{xx} - \frac{M - m}{2\ell^2} < 0,$$
a contradiction, and thus the Maximum Principle is proven. By considering $-\theta$, we also prove that the minimum cannot occur at an interior point.

Transform Methods

We next consider a series of problems associated with the diffusion equation (9.45) that can be solved by transform methods. (Similar methods hold for the wave equation and this is left for the exercises.)

The following two-point boundary value problem
$$DT = \frac{\partial T}{\partial t} - \kappa \frac{\partial^2 T}{\partial x^2} = 0,$$
$$T(x,0) = h(x), \quad T(0,t) = f(t), \quad T(\ell,t) = g(t) \tag{9.50}$$
can be formatted as the sum
$$T = T_1 + T_2 + T_3$$
of the following three problems:

(I):
$$DT_1 = 0,$$
$$T_1(x,0) = h(x), \quad T_1(0,t) = T(\ell,t) = 0;$$

(II):
$$DT_2 = 0,$$
$$T_2(x,0) = 0, \quad T_2(0,t) = f(t), \quad T_2(\ell,t) = 0;$$

(III):
$$DT_3 = 0,$$
$$T_3(x,0) = 0, \quad T_3(0,t) = 0, \quad T_3(\ell,t) = g(t).$$

The first problem has already been solved by means of Fourier series in Section 7.2. Problems II and III are similar to one another and we focus on Problem II. To simplify the notation, we drop the subscript 2, and for later convenience we set $\kappa = 1/\beta^2$. We therefore restate problem II as

$$\frac{\partial}{\partial t}T = \frac{1}{\beta^2}\frac{\partial^2 T}{\partial x^2}, \quad T(x,0) = 0, \; T(0,t) = f(t), \; T(\ell,t) = 0. \qquad (9.51)$$

To begin, we introduce the Laplace transform in time:

$$\overline{T}(x,p) = \int_0^\infty e^{-pt} T(x,t) dt.$$

If this is applied to (9.51), we obtain

$$p\overline{T} = \frac{1}{\beta^2}\frac{\partial^2 \overline{T}}{\partial x^2}, \quad \overline{T}(0,p) = F(p), \; \overline{T}(\ell,p) = 0, \qquad (9.52)$$

where $F(p)$ is the transformed boundary condition

$$\overline{T}(x = 0) = F(p) = \int_0^\infty e^{-pt} f(t) dt.$$

Although the problem (9.52) only involves an ordinary differential equation which is easily integrated (see Exercise 7), to get some more practice with transforms we instead solve by introducing an additional Laplace transform, this time on the space variable. At the outset we mention two related difficulties to this approach. We have only one condition at $x = 0$, and $\overline{T}(x,p)$ is only defined for $0 < x < \ell$. We turn our back on these for the moment and set

$$\overline{\overline{T}}(s,p) = \int_0^\infty e^{-sx}\overline{T}(x,p) dx.$$

If this is introduced into (9.52), we obtain

$$p\overline{\overline{T}} = \frac{s^2\overline{\overline{T}}}{\beta^2} - \frac{\overline{T}'}{\beta^2} - \frac{s}{\beta^2}F(p), \qquad (9.53)$$

where the *constant* $\overline{T}'$ is unknown. From (9.53) we obtain

$$\overline{\overline{T}}(s^2 - \beta^2 p) = sF(p) + \overline{T}'(p),$$

$$\overline{T} = \frac{sF(p)}{s^2 - p\beta^2} + \frac{\overline{T}'(p)}{s^2 - p\beta^2}.$$

If we invert the s-transform first (since we still do not know $\overline{T}'(p)$), then

$$\overline{T}(x,p) = \frac{1}{2\pi i} \int_{\uparrow} e^{sx} \frac{sF(p)}{(s - \beta\sqrt{p})(s + \beta\sqrt{p})} ds,$$

$$+ \frac{1}{2\pi i} \int_{\uparrow} \frac{e^{sx}\overline{T}'(p)ds}{(s - \beta\sqrt{p})(s + \beta\sqrt{p})}.$$

This inversion is straightforward since we only pick up pole contributions as the path is moved to the left. We get

$$\overline{T}(x,p) = F(p)\cosh \beta x\sqrt{p} + \frac{\overline{T}'(p)}{\beta\sqrt{p}} \sinh \beta\sqrt{p}x. \qquad (9.54)$$

As a check we note that $\overline{T}(0,p) = F(p)$. (Equation (9.54) is what we would have obtained if we had directly solved the differential equation (9.52).)

At this point we are at liberty to impose the remaining condition, namely that

$$\overline{T}(\ell,p) = 0.$$

From this we obtain

$$\overline{T}' = -F(p)\beta\sqrt{p} \frac{\cosh \ell\beta\sqrt{p}}{\sinh \ell\beta\sqrt{p}},$$

and hence,

$$\overline{T}(x,p) = F(p)\left\{\cosh \beta x\sqrt{p} - \frac{\cosh \beta\ell\sqrt{p}}{\sinh \beta\ell\sqrt{p}} \sinh \beta\sqrt{p}x\right\}.$$

It is now necessary to invert this to obtain $T(x,t)$. However, individually each term in this expression *does not* possess an inverse since each term diverges along the Bromwich path. On the other hand, the terms added together do converge; in fact, we can easily combine these so that

$$\overline{T}(x,p) = F(p)\frac{\sinh \beta(\ell - x)\sqrt{p}}{\sinh \beta\ell\sqrt{p}}. \qquad (9.55)$$

Note that (9.55) could have been written down directly as the solution of (9.52).

Quarter-Space Problem

We first consider the special case when $\ell \uparrow \infty$, in which case (9.55) becomes

$$\overline{T}(x,p) = F(p)\lim_{\ell\uparrow\infty} \frac{\sinh \beta(\ell - x)\sqrt{p}}{\sinh \beta\sqrt{p}\ell} = F(p)e^{-x\beta\sqrt{p}}. \qquad (9.56)$$

The inversion of this transform is

$$T(x,t) = \frac{1}{2\pi i}\int_\uparrow e^{pt}F(p)e^{-x\beta\sqrt{p}}dp.$$

From the Convolution Theorem this can be written as

$$T(x,t) = G \star f(t) = \int_0^t f(t-\tau)G(x,\tau)d\tau,\tag{9.57}$$

where (see Exercise 1)

$$G(x,t) = \frac{1}{2\pi i}\int_\uparrow e^{pt}e^{-x\beta\sqrt{p}}dp$$

$$= \frac{\partial}{\partial x}\int_\uparrow e^{pt-x\beta\sqrt{p}}\frac{dp}{\sqrt{p}2\pi i(-\beta)} = \frac{x\beta}{2\sqrt{\pi}t^{3/2}}e^{-x^2\beta^2/4t}.\tag{9.58}$$

The quarter-space problem can therefore be considered a solved problem since (9.57) gives the solution everywhere in terms of the given data, $T(0,t) = f(t)$.

Finite Problem

We now return to the finite problem. From (9.55) the solution can still be expressed as a convolution product:

$$T = G \star f,\tag{9.59}$$

where

$$G = \frac{1}{2\pi i}\int_\uparrow e^{pt}\frac{\sinh(\ell-x)\beta\sqrt{p}}{\sinh \ell\beta\sqrt{p}}dp.\tag{9.60}$$

We evaluate G below by two different methods.

Method 1. We first note that $\sinh((\ell-x)\beta\sqrt{p})$ and $\sinh(\beta\ell\sqrt{p})$ are odd functions in the quantity $\sqrt{p}$. Therefore the ratio of the two is even in $\sqrt{p}$. From this it can be concluded that the integrand of G *does not* have a branch point at the origin of the p-plane since it is a function of the square of $\sqrt{p}$. In fact, we only have poles and their location is determined by

$$\sinh \ell\beta\sqrt{p} = 0$$

or

$$e^{\ell\beta\sqrt{p}} = e^{-\ell\beta\sqrt{p}}$$

or

$$2\ell\beta\sqrt{p} = 2n\pi i, \quad n = \pm 1, 2, \ldots.$$

The poles are therefore located at

$$p = -\frac{n^2\pi^2}{\ell^2\beta^2}, \quad n = 1, 2, \ldots.\tag{9.61}$$

$n = 0$ is deleted since the numerator also vanishes at this point. (Otherwise said, the origin is a removable singularity.)

To evaluate the integral in (9.60), we move the Bromwich path to the left indefinitely. We therefore pick up a pole contribution for each n in (9.61). The sum of the residues is

$$G(x,t) = \sum_{n=1}^{\infty} \frac{e^{-(n^2\pi^2/\ell^2\beta^2)t} \sinh(\ell - x)\beta(n\pi i/\ell\beta)}{((d/dp)(\sinh \ell\beta\sqrt{p})|_{\sqrt{p}=n\pi i/\ell\beta})}. \tag{9.62}$$

Since $\sinh iz = (e^{iz} - e^{-iz})/2 = i \sin z$, it follows that

$$\frac{d}{dp} \sinh \ell\beta\sqrt{p} = \frac{\ell\beta}{2\sqrt{p}} \cosh \ell\beta\sqrt{p},$$

which, evaluated at $p = -n^2\pi^2/(\ell^2\beta^2)$, gives

$$\frac{d}{dp} \sinh \ell\beta\sqrt{p} = \frac{\ell\beta}{2(n\pi i/\ell\beta)} \cos \ell\beta\frac{n\pi}{\ell\beta} = \frac{(\ell\beta)^2(-)^n}{2n\pi i}.$$

Putting all of this together, we end up with

$$G = \sum_{n=1}^{\infty} (-)^{n+1} \frac{2n\pi}{\ell^2\beta^2} \sin\left(\frac{\ell-x}{\ell}\right) n\pi \exp[-n^2\pi^2 t/(\ell^2\beta^2)]. \tag{9.63}$$

Method 2. The second method of evaluation comes from the observation that

$$\frac{\sinh(\ell - x)\beta\sqrt{p}}{\sinh \ell\beta\sqrt{p}} = \frac{e^{(\ell-x)\beta\sqrt{p}} - e^{-(\ell-x)\beta\sqrt{p}}}{e^{\ell\beta\sqrt{p}} - e^{-\ell\beta\sqrt{p}}}$$

$$= \frac{e^{-\beta x\sqrt{p}} - e^{-(2\ell-x)\beta\sqrt{p}}}{1 - e^{-2\ell\beta\sqrt{p}}}.$$

If we expand the denominator as a geometrical series, we get

$$\frac{\sinh(\ell - x)\beta\sqrt{p}}{\sinh \ell\beta\sqrt{p}} = \left\{e^{-\beta x\sqrt{p}} - e^{-(2\ell-x)\beta\sqrt{p}}\right\}\{1 + e^{-2\ell\beta\sqrt{p}} + \cdots\}$$

$$= \sum_{n=0}^{\infty} e^{-(2n\ell+x)\beta\sqrt{p}} - \sum_{n=0}^{\infty} e^{-\{2(n+1)\ell-x\}\beta\sqrt{p}}$$

$$= \sum_{n=0}^{\infty} e^{-(2n\ell+x)\beta\sqrt{p}} - \sum_{n=1}^{\infty} e^{-(2n\ell-x)\beta\sqrt{p}}.$$

The important feature to note in each of these series is that the real part in each exponent is negative for $0 < x < \ell$; hence, we have absolute and

uniform convergence of the series and these can be integrated term by term
to give

$$G(x,t) = \sum_{n=0}^{\infty} \frac{1}{2\pi i} \int_{\uparrow} e^{-(2n\ell+x)\beta\sqrt{p}} e^{pt} \, dp$$

$$- \sum_{n=1}^{\infty} \frac{1}{2\pi i} \int_{\uparrow} e^{-(2n\ell-x)\beta\sqrt{p}} e^{pt} \, dp. \tag{9.64}$$

Each term is of a type which has already been evaluated; in fact

$$G = \sum_{n=0}^{\infty} \frac{(2n\ell + x)\beta}{2\sqrt{\pi} t^{3/2}} e^{-(2n\ell+x)^2\beta^2/4t}$$

$$- \sum_{n=1}^{\infty} \frac{(2n\ell - x)\beta}{2\sqrt{\pi} t^{3/2}} e^{-(2n\ell-x)^2\beta^2/4t}. \tag{9.65}$$

From the way in which t appears, it is seen that Method 1 leads to a
representation which is useful for $t \uparrow \infty$, while Method 2 gives us a form
which is more valuable for short times.

Heat Flow on a Ring—Again

The appearance of the two different representations (9.63) and (9.65) can
be observed in a more elementary context. We have already considered the
problem of heat conduction on a ring (Section 7.2); namely,

$$\frac{\partial T}{\partial t} = \kappa \frac{\partial^2 T}{\partial x^2},$$

$$T(x,0) = T^0(x), \quad T(x+n,t) = T(x,t)$$

for all integers n. The solution in convolution form is given by

$$T = G \star T^0, \tag{9.66}$$

where

$$G = \sum_{n=-\infty}^{\infty} e^{-(2\pi n)^2 \kappa t + inx 2\pi} \tag{9.67}$$

(see (7.45)).

Another method of solution is afforded to us by considering $T^0(x)$, to
be defined for $-\infty < x < \infty$ as a one-periodic function. We can then use
(8.32) to write the solution as

$$T = U \star T^0 \tag{9.68}$$

with

$$U = \frac{e^{-x^2/4\kappa t}}{\sqrt{4\pi\kappa t}}.$$

Equations (9.66) and (9.68) cannot be directly compared with each other since (9.66) is a convolution over the unit interval while (9.68) is a convolution over the full real line. Let us consider the latter:

$$U \star T^0 = \int_{-\infty}^{\infty} \frac{e^{-(x-z)^2/4\kappa t}}{\sqrt{4\pi\kappa t}} T^0(z)dz$$

$$= \sum_{n=-\infty}^{\infty} \int_{n}^{n+1} \frac{e^{-(x-z)^2/4\kappa t}}{\sqrt{4\pi\kappa t}} T^0(z)dz$$

$$\xrightarrow{z-n=y} \sum_{n=-\infty}^{\infty} \int_{0}^{1} \frac{e^{-(x-y-n)^2/4\kappa t}}{\sqrt{4\pi\kappa t}} T^0(y)dy = G \star T^0, \qquad (9.69)$$

where

$$G = \sum_{n=-\infty}^{\infty} \frac{e^{-(x-n)^2/4\kappa t}}{\sqrt{4\pi\kappa t}}. \qquad (9.70)$$

From (9.66) it follows that $G = \mathcal{G}$ or

$$\sum_{n=-\infty}^{\infty} e^{-(2\pi n)^2 \kappa t + inx} = \sum_{n=-\infty}^{\infty} \frac{e^{-(x-n)^2/4\kappa t}}{\sqrt{4\pi\kappa t}}. \qquad (9.71)$$

The right-hand side of (9.71), $\mathcal{G}$, is useful for short times, since for $0 \le x \le 1$ the terms die out exponentially as $t \downarrow 0$. By contrast, the terms of the left-hand side die out exponentially (except for $n = 0$) as $t \uparrow \infty$.

An interesting relation results from (9.71) for $x = 0$, viz.,

$$\sum_{n=-\infty}^{\infty} e^{-(2\pi n)^2 \kappa t} = \frac{1}{\sqrt{4\pi\kappa t}} \cdot \sum_{n=-\infty}^{\infty} e^{-n^2/4\kappa t}, \qquad (9.72)$$

which is a form of the *Jacobi Relation* for the θ-function. In fact, the θ-function is defined to be

$$\theta(t) = \sum_{n=-\infty}^{\infty} e^{-\pi n^2 t}, \quad t > 0, \qquad (9.73)$$

and Jacobi's Relation, which we have just demonstrated, states

$$\theta(t) = \frac{1}{\sqrt{t}}\theta(1/t). \qquad (9.74)$$

Exercises

1. Show that

$$\frac{1}{2\pi i} \int_{\uparrow} e^{pt - x\beta\sqrt{p}} dp = \frac{x\beta}{2\sqrt{\pi}t^{3/2}} e^{-x^2\beta^2/4t}.$$

2. Solve the following problems using Laplace transforms:

   (a) $\partial_t T = \kappa \partial_x^2 T$, $T(x,0) = 0 = T(0,t)$, $T(1,t) = g(t)$.

   (b) $\partial_t T = \kappa \partial_x^2 T$, $T(0,t) = T(1,t) = 0$, $T(x,0) = 1$.

   (c) $\partial_t T = \kappa \partial_x^2 T$, $T(x,0) = 0 = \partial_x T(1,t) - T(1,t)$, $T(0,t) = f(t)$.

   [In the last problem just indicate the steps.]

3. Use the Laplace transform in time to solve the following:

   (a)
   $$\frac{\partial \theta}{\partial t} = \kappa \partial_x^2 \theta, \quad 0 < x < \infty,$$
   $$T(x,0) = 1, \quad T(0,t) = 0.$$

   (b)
   $$\frac{\partial T}{\partial t} = \kappa \partial_x^2 T,$$
   $$T(x,0) = 0, \quad T(0,t) = 0, \quad T(1,t) = \cos \omega t.$$

4. Solve the following by using the Laplace transform in time:

   (a)
   $$\frac{\partial^2 \phi}{\partial t^2} = c^2 \frac{\partial^2 \phi}{\partial x^2}, \quad 0 < x < \infty, \quad 0 < t < \infty,$$
   $$\phi(0,t) = 0, \quad \phi(x,0) = 0, \quad \frac{\partial \phi}{\partial t}(x,0) = 1.$$

   (b)
   $$\frac{\partial^2 \phi}{\partial x^2} = \frac{1}{c^2} \frac{\partial^2 \phi}{\partial t^2} - \cos \omega t, \quad -\infty < x < \infty, \quad 0 < t < \infty,$$
   $$\phi(0,t) = 0, \quad \phi(x,0) = 0, \quad \frac{\partial \phi}{\partial t}(x,0) = 0.$$

5. The *cable equations* are
   $$\frac{\partial e}{\partial x} + L \frac{\partial i}{\partial t} + Ri = 0,$$
   $$\frac{\partial i}{\partial x} + C \frac{\partial e}{\partial t} + Ge = 0.$$

   (a) Reduce these to a single second order equation in e.

   (b) Reduce this equation to canonical form.

6. Solve
   $$\frac{\partial \theta}{\partial t} = \frac{1}{2} \frac{\partial^2 \theta}{\partial x^2},$$
   $$\theta(0,t) = \theta(1,t) = 0, \quad \theta(x,0) = f(x)$$

(a) by eigenfunction expansion and

(b) by Method of Images and (8.32).

(c) Show that (a) and (b) yield

$$\frac{1}{\sqrt{2\pi t}} \sum_{n=-\infty}^{\infty} \{\exp[-(x-y-2n)^2/2t] - \exp[-(x+y-2n)^2/2t]\}$$

$$= \sum_{n=1}^{\infty} \exp(-n^2\pi^2 t) \sin n\pi x \sin n\pi y.$$

7. Obtain (9.55) by directly integrating the problem posed by (9.52).

8. Solve

$$\frac{\partial \theta}{\partial t} = \frac{\partial^2 \theta}{\partial x^2}, \quad 0 < x < 1, \quad t > 0,$$

with

$$\theta(t=0) = 0, \quad \theta(0,t) = 1 - e^{-t}, \quad \theta(1,t) = 0.$$

[Hint: As $t \uparrow \infty$, $\theta \to 1 - x$.]

9. Solve

$$\frac{\partial \theta}{\partial t} = \frac{\partial^2 \theta}{\partial x^2}, \quad -\infty < x < \infty, \quad t > 0,$$

with

$$\theta(x, t=0) = f(x).$$

[Hint: Fourier transform in x.]

10. Solve

$$\frac{\partial \theta}{\partial t} + \frac{\partial \theta}{\partial x} = \frac{\partial^2 \theta}{\partial x^2}, \quad -\infty < x < \infty, \quad t > 0,$$

with

$$\theta(x, t=0) = f(x).$$

9.4 The Potential Equation

The remaining canonical form for a second order partial differential equation is the elliptic case. We study in particular the Laplace equation, $\nabla^2 \phi = 0$, which in two and three dimensions is given by

$$\left(\frac{\partial^2}{\partial x^2} + \frac{\partial^2}{\partial y^2} \right) \phi = 0 \tag{9.75}$$

and

$$\left(\frac{\partial^2}{\partial x^2} + \frac{\partial^2}{\partial y^2} + \frac{\partial^2}{\partial z^2} \right) \phi = 0. \tag{9.76}$$

A function which satisfies the Laplace equation is said to be *harmonic*.

Physical Models

Laplace's equation is also referred to as the *potential equation*, a name which arises from electrostatic theory. In that case the potential (or voltage) V is defined, in terms of the electric field vector $\mathbf{E}$, by

$$\mathbf{E} = \nabla V. \tag{9.77}$$

(For electrostatics $\nabla \wedge \mathbf{E} = 0$, which implies (9.77).) $\mathbf{E}$, by definition, is the force on a unit charge in an electric field. It therefore follows from (9.77) that

$$\int_{-\infty}^{\mathbf{x}} \mathbf{E} \cdot d\boldsymbol{\ell} = V$$

is the work done in bringing a unit charge from infinity, along an arbitrary path, to the point $\mathbf{x}$. (It is assumed that $V \to 0$ as $|\mathbf{x}| \uparrow \infty$.)

In electrostatics, the electric field arises from a distribution of charge. If the charge density is denoted by $\sigma(\mathbf{x})$, then

$$\nabla \cdot \mathbf{E} = \sigma(\mathbf{x});$$

and if the potential V is substituted, then

$$\nabla^2 V = \sigma. \tag{9.78}$$

This is the inhomogeneous form of the Laplace equation and is known as the *Poisson equation*. In regions where $\sigma = 0$, we obtain the Laplace equation.

In another context, we observed that the Laplace equation can result as the limiting form of the diffusion equation. If the conditions of a diffusion problem are such that an *equilibrium* or *steady state* is obtained as $t \uparrow \infty$, then the equilibrium is governed by the Laplace equation. In the most general case, the boundary condition for (9.78) is of the form

$$a\phi + b\frac{\partial \phi}{\partial n} = f(\mathbf{x}), \quad \mathbf{x} \in \partial V,$$

where a and b can also be functions of position. ∂V denotes the set of bounding surfaces.

An interesting case for which the Laplace equation arises is that of a soap film or *bubble*. In this situation a membrane (the soap film) is stretched over a loop of wire. If the displacement of the membrane (measured from the (x, y)-plane) is denoted by $h = h(x, y)$, then it is found from a simple balance of forces that

$$\frac{\partial^2 h}{\partial x^2} + \frac{\partial^2 h}{\partial y^2} = 0.$$

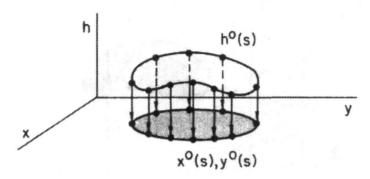

FIGURE 9.9.

This is to be solved subject to the conditions

$$h = h^0(s), \quad x = x^0(s), \quad y = y^0(s),$$

which specify the wire loop (see Figure 9.9). It is interesting to observe that the soap bubble problem is an *analogue* for solutions of Laplace's equation in two dimensions. For example, consider the *temperature* problem

$$\frac{\partial^2 T}{\partial x^2} + \frac{\partial^2 T}{\partial y^2} = 0,$$

$$T = T^0(x, y) \quad \text{on } \Gamma,$$

where Γ is a closed curve in the (x, y)-plane. To *solve* this problem in an analogue manner, we could construct a wire loop of elevation T^0 above each point of Γ in the (x, y)-plane. The height of the soap film stretched across the loop at each interior location then gives the solution to the corresponding temperature problem.

Another physical example which gives rise to the Laplace equation is *inviscid incompressible fluid flow*. In discussing fluid flow we focus on the fluid velocity

$$\mathbf{u} = \mathbf{u}(\mathbf{x}, t) = (u, v, w),$$

which gives the three components (in three space dimensions) of the fluid velocity at any point $\mathbf{x}$ at a time t. It follows from the *incompressibility* of the fluid that

$$\nabla \cdot \mathbf{u} = 0. \tag{9.79}$$

For common liquids such as air (air is incompressible, except at relatively high speed flow) and water, a useful approximation is that they flow without *friction*. Then under a wide and useful set of circumstances, we can prove

$$\nabla \wedge \mathbf{u} = \left(\frac{\partial w}{\partial y} - \frac{\partial v}{\partial z}, \frac{\partial u}{\partial z} - \frac{\partial w}{\partial x}, \frac{\partial v}{\partial x} - \frac{\partial u}{\partial y} \right) = 0. \tag{9.80}$$

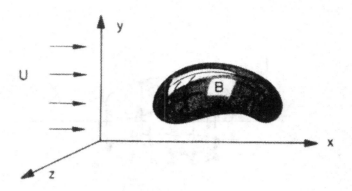

FIGURE 9.10.

From this last condition we have that there exists a scalar function ϕ, known as the *velocity potential*, such that

$$\mathbf{u} = \nabla\phi = \left(\frac{\partial\phi}{\partial x}, \frac{\partial\phi}{\partial y}, \frac{\partial\phi}{\partial z}\right),\tag{9.81}$$

and when this is substituted into (9.79), we obtain Laplace's equation

$$\nabla^2\phi = 0.$$

A fairly typical fluid problem is that of determining the fluid flow past a body B which is at rest relative to a fluid which is flowing with an upstream uniform velocity, say U in the x-direction (see Figure 9.10). The potential for the upstream *uniform* flow is

$$\phi^0 = Ux\tag{9.82}$$

since

$$\nabla\phi^0 = (U, 0, 0).$$

The boundary condition is that no fluid should penetrate the body. In symbols this states

$$\mathbf{n}\cdot\mathbf{u} = \mathbf{n}\cdot\nabla\phi = \frac{\partial\phi}{\partial n} = 0, \quad x \in \partial B,\tag{9.83}$$

where $\mathbf{n}$ is the outward normal at the body surface ∂B.

Boundary Value Problems

Two main boundary value problems are associated with Laplace's equation. These are

The Dirichlet Problem: $\nabla^2\phi = 0$, ϕ given on the boundary;

The Neumann Problem: $\nabla^2 \phi = 0$, $\partial\phi/\partial n$ given on the boundary.

A prototypical case for the Dirichlet Problem is that of the equilibrium temperature distribution in a volume V when the temperature is specified on ∂V. Both fluid flow and heat flow are typical of the Neumann Problem. In the case of heat flow, intuition tells us that we cannot arbitrarily prescribe the heat flowing at the boundaries (and hence $\partial\phi/\partial n$), but rather it must be specified in a manner such that no buildup or loss of heat occurs in the volume (since the time derivative is assumed to vanish). This can be seen analytically by integrating Laplace's equation over the volume V and by applying the Divergence Theorem:

$$0 = \int_V \nabla^2 \phi \, dx = \int_V \nabla \cdot (\nabla\phi) dx = \int_{\partial V} \frac{\partial\phi}{\partial n} ds.$$

Therefore a side condition on the data of a Neumann Problem is that

$$\int_{\partial V} \frac{\partial\phi}{\partial n} ds = 0, \tag{9.84}$$

where the integration is over all boundaries of V. In physical terms, (9.84) says that an equilibrium is impossible if there is a net heat flow into the volume.

Mean Value Property

In the case of two dimensions, we have seen that the real and imaginary parts of an analytic function are *harmonic*, and hence an analytic function is itself harmonic. We showed in Chapter 3 that an analytic function $f(z)$ at a point z_0 is the mean value of f over any circle (in which $f(z)$ is analytic) centered at z_0; viz.,

$$f(z_0) = \frac{1}{2\pi r} \int_0^{2\pi} f(z_0 + re^{i\theta}) r \, d\theta.$$

In general we have

Mean Value Theorem. *A harmonic function $\phi(\mathbf{x})$ at a point $\mathbf{x}_0$ is equal to the mean value of ϕ over any sphere in the domain of harmonicity, with center $\mathbf{x}_0$.*

This is now proven in three dimensions. Consider ϕ at a point on a sphere of radius r centered at $\mathbf{x}_0$:

$$\phi(\mathbf{x}_0 + \mathbf{r}) = \phi(x_0 + r\cos\theta, y_0 + r\sin\theta\sin\psi, z_0 + r\sin\theta\cos\psi)$$

$$= \phi(r, \theta, \psi; \mathbf{x}_0).$$

Here $\mathbf{r} = \mathbf{x} - \mathbf{x}_0$, $r = |\mathbf{x} - \mathbf{x}_0|$, and (θ, ψ) are the polar angles of $\mathbf{r}$ with $\mathbf{x}_0$ as the origin.

The average value of ϕ over a sphere of radius r is

$$\overline{\phi}(r; x_0) = \frac{1}{4\pi r^2} \int_0^\pi d\theta \int_0^{2\pi} d\psi \sin\theta \, r^2 \phi(r, \theta, \psi; x_0)$$

$$= \frac{1}{4\pi} \int_0^\pi d\theta \int_0^{2\pi} d\psi \sin\theta \, \phi(r, \theta, \psi; x_0).$$

If we take the derivative of $\overline{\phi}$ with respect to r, then

$$\frac{\partial \overline{\phi}}{\partial r} = \frac{1}{4\pi} \int_0^\pi d\theta \int_0^{2\pi} d\psi \sin\theta \frac{\partial \phi}{\partial r}$$

$$= \frac{1}{4\pi r^2} \int_{|x-x_0|=r} \mathbf{n} \cdot \nabla\phi \, ds = 0.$$

In this we have first transformed to a surface integral, with r a constant, and then made use of the condition (9.84). We therefore have that $\overline{\phi}$ is a constant and, in particular, that

$$\overline{\phi} = \overline{\phi}(0; x_0) = \phi(x_0).$$

A similar argument can be used to show that

$$\phi(x_0) = \frac{1}{(4/3)\pi R^3} \int_{|r| \le R} \phi(x_0 + r) dr, \qquad (9.85)$$

namely that $\phi(x_0)$ is equal to the average value over the volume of a sphere centered at x_0.

Maximum (Minimum) Principle

Both the temperature model and the soap film model intuitively indicate that the maximum and minimum of a harmonic function that is not constant are achieved at a boundary.

We prove this by contradiction. Suppose for example that ϕ, which is harmonic, has a local maximum at an interior point x_0. Then from continuity there exists an $\epsilon > 0$ such that

$$\phi(x_0) > \phi(x_0 + r), \quad |r| \le \epsilon.$$

If we integrate this inequality over the surface of the sphere $r = \epsilon$, we find

$$\int_{|x-x_0|=\epsilon} \phi(x_0) ds = 4\pi\epsilon^2 \phi_0(x_0) > \int_{|x-x_0|=\epsilon} \phi(x_0 + r) ds$$

$$= 4\pi\epsilon^2 \phi(x_0)$$

from the Mean Value Theorem; and hence there is a contradiction unless constant. If we consider $-\phi$, since it too is harmonic, it must achieve

a maximum on the boundary; therefore, ϕ takes on its minimum on the
boundary.

Uniqueness

It is now easy to prove uniqueness for the following somewhat more
general problem

$$\nabla^2 \phi = \sigma(x),$$

where ϕ is given on the boundaries and the source term $\sigma(x)$ is also a given
function. Suppose ϕ_1 and ϕ_2 are two solutions to this problem. Then $\phi_1 - \phi_2$
is harmonic and satisfies zero boundary conditions. But then $\phi_1 - \phi_2$, which
takes on its maximum and minimum at the boundary, must therefore be
identically zero everywhere, and uniqueness is proven.

Some Special Solutions

The Laplacian operator in three dimensions,

$$\nabla^2 = \frac{\partial^2}{\partial x^2} + \frac{\partial^2}{\partial y^2} + \frac{\partial^2}{\partial z^2},$$

expressed in spherical coordinates has the form

$$\nabla^2 = \frac{1}{r^2}\frac{\partial^2 r}{\partial r}\left(r^2\frac{\partial}{\partial r}\right) + \frac{1}{r^2\sin^2\theta}\frac{\partial^2}{\partial\psi^2} + \frac{1}{r^2\sin\theta}\frac{\partial}{\partial\theta}\left(\sin\theta\frac{\partial}{\partial\theta}\right). \quad (9.86)$$

Consider the *fundamental solution*

$$\hat{\phi} = -\frac{1}{4\pi r}, \quad (9.87)$$

which, when substituted into (9.86), is seen to be a solution provided that
$r \neq 0$. To consider the meaning of $\hat{\phi}$ in the neighborhood of the origin, we
consider the integral

$$\int_{|x|<\epsilon} f(x)\,\nabla^2\,\hat{\phi}dx,$$

where f is a smooth but otherwise arbitrary function and ϵ is small. (The
integral exists in a conditionally convergent sense, since $\nabla^2(1/r) = O(1/r^3)$
and $dx = O(r^3)$.)

We rewrite the above integral as follows:

$$\int_{|x|<\epsilon} f(x)\,\nabla^2\left(\frac{1}{-4\pi r}\right)dx = \int_{|x|<\epsilon}(f(x) - f(0))\,\nabla^2\left(-\frac{1}{4\pi r}\right)dx$$

$$+ f(0)\int_{|x|<\epsilon}\nabla\cdot\nabla\left(-\frac{1}{4\pi r}\right)dx.$$

Using the Divergence Theorem, the last integral can be written as

$$\int_{|\mathbf{x}|=\epsilon} \mathbf{n} \cdot \nabla \left(-\frac{1}{4\pi r} \right) ds;$$

but $\mathbf{n} \cdot \nabla$ on a sphere is simply $\partial/\partial r$, and therefore

$$\int_{|\mathbf{x}|=\epsilon} \frac{1}{4\pi r^2} ds = \int_{|\mathbf{x}|=\epsilon} \frac{1}{4\pi r^2} r^2 \sin\theta d\theta d\phi = 1.$$

From the smoothness of f and the above calculation, we can show that the first integral on the right-hand side is vanishingly small as $\epsilon \to 0$. We have thus demonstrated

$$\lim_{\epsilon \to 0} \int_{|\mathbf{x}|<\epsilon} f(\mathbf{x}) \nabla^2 \hat{\phi} \, d\mathbf{x} = f(0). \tag{9.88}$$

Since $\nabla^2 \hat{\phi} = 0$, $\mathbf{x} \neq 0$, (9.88) permits us to write

$$\nabla^2 \hat{\phi}(\mathbf{x}) = \delta(\mathbf{x}), \tag{9.89}$$

where $\delta(\mathbf{x})$ is the three-dimensional *Dirac delta* and has the property that

$$\int_V \delta(\mathbf{x}) f(\mathbf{x}) d\mathbf{x} = f(0),$$

if the origin is contained in V (and is zero otherwise). More generally we write

$$\int_V \delta(\mathbf{x} - \mathbf{y}) f(\mathbf{y}) d\mathbf{y} = f(\mathbf{x}), \quad \mathbf{x} \in V.$$

We also observe that since ∇^2 is *translationally invariant* in space,

$$\nabla_{\mathbf{x}}^2 \hat{\phi}(\mathbf{x} - \mathbf{y}) = \delta(\mathbf{x} - \mathbf{y}), \tag{9.90}$$

where

$$\hat{\phi} = -\frac{1}{4\pi|\mathbf{x} - \mathbf{y}|} = \frac{-1}{4\pi\sqrt{(x_1 - y_1)^2 + (x_2 - y_2)^2 + (x_3 - y_3)^2}}. \tag{9.91}$$

Next we multiply (9.90) by the *constant* $\rho(\mathbf{y})$. This gives us

$$\nabla_{\mathbf{x}}^2 \rho(\mathbf{y}) \hat{\phi}(\mathbf{x} - \mathbf{y}) = \rho(\mathbf{y}) \delta(\mathbf{x} - \mathbf{y}).$$

If we integrate over $\mathbf{y}$, then

$$\int_V \nabla_{\mathbf{x}}^2 \rho(\mathbf{y}) \hat{\phi}(\mathbf{x} - \mathbf{y}) d\mathbf{y} = \nabla_{\mathbf{x}}^2 \Phi = \rho(\mathbf{x}),$$

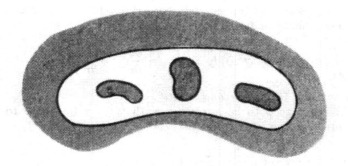

FIGURE 9.11.

where

$$\Phi(\mathbf{x}) = -\frac{1}{4\pi} \int_V \frac{\rho(\mathbf{y})}{|\mathbf{x} - \mathbf{y}|} d\mathbf{y}. \tag{9.92}$$

Equation (9.92) is known as *Poisson's Integral*.

We have therefore obtained a particular solution (9.92) to the Poisson equation

$$\nabla^2 \Phi = \rho(\mathbf{x}). \tag{9.93}$$

In general, a problem is fully posed by stating that Φ satisfies Poisson's equation, (9.93), with Φ given on the boundaries, say, $\Phi = f$ on ∂V (see Figure 9.11). We can now convert this to a problem involving just the Laplace equation. To see this write

$$\Phi = \phi + \int_V \rho(\mathbf{y})\hat{\phi}(\mathbf{x} - \mathbf{y})d\mathbf{y}.$$

Then

$$\nabla^2 \phi = 0$$

and

$$\phi = f - \int_V \rho(\mathbf{y})\hat{\phi}(\mathbf{x} - \mathbf{y})d\mathbf{y} \quad \text{on } \partial V.$$

Green's Identities

Consider a volume V with boundary ∂V; then for any two smooth functions ϕ and ψ, a simple application of the Divergence Theorem yields

$$\int_V \psi \nabla^2 \phi \, d\mathbf{y} = \int_V \nabla \cdot (\psi \nabla \phi) d\mathbf{y} - \int_V \nabla \psi \cdot (\nabla \phi) d\mathbf{y}$$

$$= \int_{\partial V} \psi \frac{\partial \phi}{\partial n} ds_{\mathbf{y}} - \int_V \nabla \psi \cdot (\nabla \phi) d\mathbf{y}. \tag{9.94}$$

This is known as *Green's First Identity*.

Next, from (9.94), we immediately have (interchange ϕ and ψ in (9.94) and subtract the result from (9.94))

$$\int_V (\psi \, \nabla^2 \, \phi - \phi \, \nabla^2 \, \psi) dy = \int_{\partial V} \left(\psi \frac{\partial \phi}{\partial n} - \phi \frac{\partial \psi}{\partial n} \right) ds_y. \qquad (9.95)$$

This is known as *Green's Second Identity*.

Then, if we take ϕ to be harmonic, with $\nabla^2 \phi = 0$, and ψ to be the fundamental solution, $\psi = -\hat{\phi}$,

$$\phi(\mathbf{x}) = \frac{1}{4\pi} \int_{\partial V} \left\{ \frac{1}{|\mathbf{x} - \mathbf{y}|} \frac{\partial}{\partial n} \phi(\mathbf{y}) - \phi(\mathbf{y}) \frac{\partial}{\partial n} \frac{1}{|\mathbf{x} - \mathbf{y}|} \right\} ds_y, \qquad (9.96)$$

which is known as *Green's Third Identity*.

This demonstrates that the potential ϕ is determined by ϕ and its normal derivative at the boundary. (Physical intuition tells us that only one piece of information is correct.) We mention in passing that the above representation is also said to replace boundaries by source and dipole (or doublet) distributions. To see the basis for this, recall that we saw above that $-1/4\pi r$ is a solution to the Laplace equation; and since it corresponds to a unit charge in the electrostatics model, it is referred to as a source solution. Further, since $\nabla^2 \phi = 0$ is linear with constant coefficients, any derivative of $\hat{\phi}$ is also a solution. In particular, for an arbitrary constant $\mathbf{a}$,

$$\mathbf{a} \cdot \nabla \left(-\frac{1}{4\pi r} \right)$$

is a solution and is referred to as a *dipole solution in the* $\mathbf{a}$ *direction*.

Green's Function

If in Green's Second Identity we write

$$\psi = G = \frac{1}{4\pi R} + U,$$

$$R = |\mathbf{x} - \mathbf{y}| = \left((x_1 - y_1)^2 + (x_2 - y_2)^2 + (x_3 - y_3)^2 \right)^{1/2}, \qquad (9.97)$$

where U is harmonic in $\mathbf{y}$ so that

$$\nabla_{\mathbf{y}}^2 U = 0 \qquad (9.98)$$

in the domain of interest, then

$$\phi = \int_{\partial V} \left(G \frac{\partial \phi}{\partial n} - \phi \frac{\partial G}{\partial n} \right) ds_y - \int_V G \, \nabla^2 \, \phi dy. \qquad (9.99)$$

Now if we can find a G such that $G = 0$ on ∂V, we have an explicit representation for the solution of the Dirichlet Problem. Alternatively, if we can find a G such that $\partial G/\partial n = 0$ on ∂V, then we have an explicit representation for the Neumann Problem. (Actually since $\nabla^2 \phi$ has not been specified, we could also be solving the corresponding Poisson problems.) Functions, G, having such properties are referred to as *Green's functions.*

Examples of Green's Functions

Consider the following half-space problem ($x_1 > 0$):

$$\left(\frac{\partial^2}{\partial x_1^2} + \frac{\partial^2}{\partial x_2^2} + \frac{\partial^2}{\partial x_3^2} \right) \phi = \rho(x_1, x_2, x_3), \quad -\infty < x_2, x_3 < \infty, \quad x_1 > 0,$$

$$\phi(x_1 = 0) = f(x_2, x_3).$$

It is also assumed that $\phi \to 0$ for $x_1 \uparrow \infty$. In order to make use of the above Green's representation, we must find a function U such that (9.98) is satisfied and such that

$$\left(\frac{1}{4\pi R} + U \right)_{y_1 = 0} = 0,$$

no matter what the value of $\mathbf{x}$. An immediate answer is obtained through the *Method of Images.* For $1/(4\pi R) = 1/(4\pi|\mathbf{x} - \mathbf{y}|)$ represents a unit negative charge at $\mathbf{x} = \mathbf{y}$. If we place an equal strength but opposite charge at $(-y_1, y_2, y_3)$, the potential should vanish at $x_1 = 0$. Specifically, consider

$$G = \frac{1}{4\pi\{(x_1 - y_1)^2 + (x_2 - y_2)^2 + (x_3 - y_3)^2\}^{1/2}}$$

$$- \frac{1}{4\pi\{(x_1 + y_1)^2 + (x_2 - y_2)^2 + (x_3 - y_3)^2\}^{1/2}},$$

which certainly vanishes at $y_1 = 0$ independently of $\mathbf{x}$. This represents a positive charge at $(-y_1, y_2, y_3)$ and a negative charge at the mirror image point (y_1, y_2, y_3). Clearly

$$U = \frac{-1}{4\pi\{(x_1 + y_1)^2 + (x_2 - y_2)^2 + (x_3 - y_3)^2\}^{1/2}}$$

is such that $\nabla_{\mathbf{y}}^2 U = 0$ for $y_1 > 0$. The solution to the above problem is therefore given by

$$\phi = \int_{-\infty}^{\infty} \int_{-\infty}^{\infty} dy_2 dy_3 f(y_2, y_3) \left(\frac{\partial}{\partial y_1} G \right)_{y_1 = 0}$$

$$- \int_0^{\infty} dy_1 \int \int_{-\infty}^{\infty} dy_2 dy_3 G\rho(\mathbf{y}).$$

The Green's function for the corresponding Neumann Problem is also easily obtained by means of images. In this case we want

$$\left.\frac{\partial G}{\partial y_1}\right|_{y_1=0} = 0.$$

This is obtained by taking a positive source at the image point; specifically,

$$G = +\frac{1}{4\pi}\left[\frac{1}{\{(x_1-y_1)^2+(x_2-y_2)^2+(x_3-y_3)^2\}^{1/2}}\right.$$
$$\left. + \frac{1}{\{(x_1+y_1)^2+(x_2-y_2)^2+(x_3-y_3)^2\}^{1/2}}\right],$$

which clearly satisfies the above condition at $y_1 = 0$. (Note that G is now an even function of y_1, and hence its derivative with respect to y_1 is odd.)

Exercises

1. Consider a membrane stretched across a closed wire loop: $x = x(\theta)$, $y = y(\theta)$, $0 \le \theta \le 2\pi$. Denote the pointwise elevation of the membrane by $h(x,y)$ and assume that the tension in the membrane is constant. Show that if $|\nabla h| \ll 1$, then the equation of the membrane is

$$\nabla^2 h = 0.$$

   (See Figure 9.9.)

2. Prove that if $\phi(\mathbf{x})$ is harmonic, it is equal to the mean value over the interior of any sphere centered at $\mathbf{x}$.

3. Set $\psi = \phi$ in Green's First Identity to show

$$\int_V \phi \nabla^2 \phi \, dy = \int_{\partial V} \phi \frac{\partial \phi}{\partial n} d s_y - \int_V (\nabla \phi)^2 dy.$$

   Use this to prove the Uniqueness Theorem.

4. Consider the Laplace equation in two-space,

$$\nabla^2 \phi = 0, \quad 0 < x < 1, \quad 0 < y < 2,$$

   with boundary conditions

$$\phi(x,2) = x(1-x) \quad \text{and} \quad \phi = 0 \text{ on the other three sides.}$$

   Solve. [Hint: Use a Fourier series in x.]

5. In three-space prove Liouville's Theorem: $\nabla^2 \phi = 0$ and $|\phi|$ bounded for all $\mathbf{x}$ imply that $\phi = $ constant. [Hint: Use the Mean Value Theorem.]

6. Suppose $\phi(r, \theta, \mu)$, expressed in spherical coordinates, satisfies Laplace's equation inside a sphere of radius R centered at the origin. For $r > R$ define

$$\Phi(r, \theta, \mu) = \frac{1}{r} \phi \left(\frac{R^2}{r}, \theta, \mu \right)$$

and prove $\nabla^2 \Phi = 0$ for $r > R$.

7. Is the product of two harmonic functions harmonic?

8. $\phi_1 = 1$ and $\phi_2 = (x^2 + y^2 + z^2)^{-1/2}$ both satisfy $\nabla^2 \phi = 0$ and both are unity on $r = 1$. Is this a contradiction of the Uniqueness Theorem?

9. In two dimensions show that

$$\nabla^2 \phi(r, \theta) = 0, \quad 0 \le r \le 1, \quad 0 \le \theta \le 2\pi,$$

with

$$\phi_r(r = 1, \theta) = 1,$$

has no solution.

Series Solutions

Next we comment on series solutions of Laplace's equation in three dimensions. For this purpose we consider Green's Third Identity (9.96) for the case of finite boundaries ∂V, in an otherwise infinite domain. Further, we place the origin in the interior of one of the bodies and consider the form of the potential $\phi(\mathbf{x})$ at distances $|\mathbf{x}| = r = (x_1^2 + x_2^2 + x_3^2)^{1/2}$ which are large compared with a typical distance $|\mathbf{y}| = (y_1^2 + y_2^2 + y_3^2)^{1/2}$ to the boundary points (Figure 9.12). Under this limit the Taylor series

$$\frac{1}{|\mathbf{x} - \mathbf{y}|} = \frac{1}{r} + (-\mathbf{y} \cdot \nabla \mathbf{x})\frac{1}{r} + \frac{(-\mathbf{y} \cdot \nabla)^2}{2!}\frac{1}{r} + \frac{(-\mathbf{y} \cdot \nabla)^3}{3!}\frac{1}{r} + \cdots \quad (9.100)$$

is appropriate since the terms are powers of $|\mathbf{y}|/|\mathbf{x}| < 1$.

It is clear that the series is uniformly convergent. If (9.100) is substituted into Green's Third Identity (9.96), then

$$\phi = \sum_{k=0}^{\infty} \frac{1}{4\pi} \left\{ \int_{\partial V} \left[\frac{(-\mathbf{y} \cdot \nabla)^k}{k!} \frac{1}{r} \right] \frac{\partial}{\partial n} \phi(\mathbf{y}) ds_\mathbf{y} \right.$$

$$\left. - \int_{\partial V} \phi \mathbf{n} \cdot \nabla \mathbf{x} \frac{(-\mathbf{y} \cdot \nabla)^k}{k!} \frac{1}{r} ds_\mathbf{y} \right\}.$$

If we use the summation convention which was introduced earlier (i.e., repeated subscripts are summed), then

$$\phi = \sum_{n=0}^{\infty} \left(A^{(n)}_{j_1 j_2 \cdots j_n} \frac{\partial}{\partial x_{j_1}} \frac{\partial}{\partial x_{j_2}} \cdots \frac{\partial}{\partial x_{j_n}} \right) \frac{1}{r}. \quad (9.101)$$

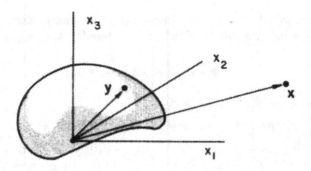

FIGURE 9.12.

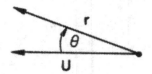

FIGURE 9.13.

Each of the *tensors* $A^{(n)}_{j_1 \ldots j_n}$ can be evaluated in terms of integrals over ∂V of ϕ and $\partial \phi / \partial n$; e.g.,

$$A^{(0)} = \frac{1}{4\pi} \int_{\partial V} \frac{\partial}{\partial n} \phi(\mathbf{y}) ds_{\mathbf{y}}. \tag{9.102}$$

These *tensors* can be given a physical interpretation. For example, in the case of fluid flow, $\partial \phi / \partial n$ is the normal velocity and (9.102) measures net outflow. Unless there are sources and sinks, this term is zero. (In spite of the side condition on the Neumann Problem (9.84), this integral is not necessarily zero—why?)

The above expression is referred to as a *multipole expansion*. Each term of the expansion is itself a solution of the Laplace equation. For any constant tensor $A^{(n)}_{j_1 \ldots j_n}$,

$$\mathbf{A}^{(n)} \cdot (\nabla)^n \frac{1}{r}$$

is called a *multipole solution of order n* and $\mathbf{A}^{(n)}$ is the nth *order multipole moment*. In particular,

$$\mathbf{A} \cdot \nabla \frac{1}{r}$$

is called a *dipole of dipole moment A*.

Consider as an illustration the dipole solution

$$\phi = \alpha \mathbf{U} \cdot \nabla \frac{1}{r} = -\alpha \frac{\mathbf{U} \cdot \mathbf{r}}{r^3}$$

for an arbitrary vector U and constant α. In the context of irrotational fluid flow this has an interesting interpretation. For this purpose a convenient coordinate system to choose is one in which U is the polar axis with θ the angle between r and U (see Figure 9.13). Then

$$\phi = -\alpha \frac{U \cos \theta}{r^2}.$$

The radial component of velocity is given by

$$\frac{\partial \phi}{\partial r} = 2\alpha \frac{U \cos \theta}{r^3}.$$

For the case of fluid flow we recall that the boundary condition states that a boundary is impermeable. This leads to the Neumann boundary condition

$$\mathbf{n} \cdot \nabla \phi|_{\partial V} = \mathbf{U}_B \cdot \mathbf{n},$$

where $\mathbf{U}_B$ is the velocity of the body itself and $\mathbf{n}$ is its normal at the point of interest. If the body under study is a sphere, the above discussion immediately solves the problem, for in this case $\mathbf{n} \cdot \nabla = \partial/\partial r$ and the forms agree. To fix up matters, we need only take

$$\alpha = a^3/2$$

and write U for $\mathbf{U}_B$. In summary, the claim is that

$$\phi = \frac{a^3}{2} \mathbf{U} \cdot \nabla \frac{1}{r}$$

solves the problem of the motion of a sphere of radius a moving with a velocity U into an otherwise motionless fluid. To verify these remarks, note first that

$$\nabla^2 \left(\frac{a^3}{2} \mathbf{U} \cdot \nabla \frac{1}{r} \right) = 0$$

and secondly that

$$(\mathbf{n} \cdot \nabla)\phi|_{r=a} = \left(\frac{\partial}{\partial r} \frac{a^3}{2} \mathbf{U} \cdot \nabla \frac{1}{r} \right)_{r=a}$$

$$= \left(\frac{a^3}{2} \mathbf{U} \cdot \mathbf{n} \frac{\partial^2}{\partial r^2} \frac{1}{r} \right)_{r=a} = \frac{a^3}{2} \mathbf{U} \cdot \mathbf{n} \frac{2}{r^3}|_{r=a} = \mathbf{U} \cdot \mathbf{n},$$

which is the appropriate boundary condition.

So far we have assumed that the origin is at the center of the sphere, and in order to allow the sphere to move with respect to a fixed frame we write instead

$$\phi = \frac{a^3}{2} \mathbf{U}_0(t) \cdot \nabla \frac{1}{|\mathbf{x} - \mathbf{x}_0(t)|},$$

where $x_0(t)$ is the trajectory of the sphere and $U_0(t) = (d/dt)x_0(t)$ is its velocity. Another representation follows if we choose a coordinate system situated in the body. In this instance we see a uniform flow coming toward the sphere from infinity (upstream). As is seen immediately, the velocity potential of the uniform flow is

$$\overline{\phi} = -U \cdot x,$$

and hence the potential to consider is

$$\Phi = \overline{\phi} + \phi = -U \cdot x + \frac{a^3}{2} U \cdot \nabla \frac{1}{|x|}.$$

Here the appropriate boundary condition is

$$n \cdot \nabla \Phi|_{r=a} = 0,$$

which is obviously satisfied by our construction.

Spherical Harmonics

The set of solutions to the Laplace equation generated by

$$(\nabla)^n \frac{1}{r}, \quad n = 0, 1, \ldots$$

are called *spherical harmonics*, a partial list of which includes

$$\frac{1}{r}, \quad \frac{\partial}{\partial x_j} \frac{1}{r}, \quad \frac{\partial^2}{\partial x_{j_1} \partial x_{j_2}} \frac{1}{r}.$$

These are referred to as spherical harmonics of degree $-1, -2, -3$, and so forth. It is clear that

$$S^{(k)} = \frac{\partial^k}{\partial x_{j_1} \cdots \partial x_{j_k}} \frac{1}{r}$$

is homogeneous of degree $-(k + 1)$; i.e.,

$$S^{(k)}(\alpha r) = S^{(k)}(r)/\alpha^{k+1}$$

for any α. Therefore

$$S^{(k)} = r^{k+1} S^{(k+1)} = r^{k+1} \frac{\partial^k}{\partial x_{j_1} \cdots \partial x_{j_k}} \frac{1}{r}, \tag{9.103}$$

known as a *spherical surface harmonic of order k*, is homogeneous of degree zero. As will be shown in the exercises,

$$s^{(k)} = r^k S^{(k)} \tag{9.104}$$

are in fact also solutions of the Laplace equation. In principle, we can solve the Laplace equation by an expansion in spherical harmonics of all degrees. It is clear that only spherical harmonics of negative degree, $S^{(k)}$, are required for an external problem and of positive degree, $s^{(k)}$, for an internal problem.

An entirely different approach to spherical harmonics is obtained by considering the Laplace equation in spherical coordinates,

$$0 = \frac{1}{r^2}\left\{ \frac{\partial}{\partial r} r^2 \frac{\partial \phi}{\partial r} + \frac{1}{\sin\theta}\frac{\partial}{\partial\theta}\sin\theta\frac{\partial\phi}{\partial\theta} + \frac{1}{\sin^2\theta}\frac{\partial^2\phi}{\partial\psi^2}\right\}, \qquad (9.105)$$

and seeking solutions in the *separated* form

$$\phi = R(r)\Theta(\theta)M(\psi). \qquad (9.106)$$

This is left for the exercises.

Exercises

1. Verify that $s^{(k)}$, (9.104), satisfies Laplace's equation.

2. If (9.105) is substituted into (9.106) show that

   (a) $M'' = -m^2 M$ where m is an integer;

   (b) $r^{-2}(r^2 R')' - cR/r^2 = 0$, c a constant.

   (c) Find the equation which Θ satisfies.

3. Obtain the form of multipole tensors $A_j^{(1)}$ and $A_{jk}^{(2)}$. (See (9.101).)

4. Express $S^{(k)}$ and $s^{(k)}$ for $k = 1$ and 2 in spherical coordinates.

9.5 Laplace's Equation—Two-Dimensional Problems

We return to the two-dimensional form of the Laplace equation and again consider the Dirichlet and Neumann problems (or more general boundary value problems). In general, two-dimensional problems have the physical interpretation of being cross sections of infinite cylinders in three dimensions.

We recall from our study of complex variable theory that if

$$f(z) = \phi + i\psi = \phi(x,y) + i\psi(x,y)$$

is an analytic function of z on some domain R, then ϕ and ψ are each harmonic functions of x and y in the same domain of (x,y)-space (and thus $f(z)$ is also harmonic). Alternately, if $\phi(x,y)$ is harmonic in R, then we can define ψ by

$$\phi_x = \psi_y, \quad \phi_y = -\psi_x$$

so that up to a constant

$$\psi = \int d\psi = \int (\psi_x dx + \psi_y dy) = -\int \phi_y dx + \int \phi_x dy, \qquad (9.107)$$

where any convenient path will do. Therefore the combination $\phi + i\psi$ is an...

In this way, the problem

$$\nabla^2 \phi = 0, \quad x, y \in R,$$

$$\phi \quad \text{given on } \partial R$$

can be embedded in the problem of finding a function, $f(z) = \phi + i\psi$, analytic on R and such that its real part, ϕ, is specified on the boundary ∂R of R. It is then natural to ask to what problem is ψ the solution. To answer this question we first observe that

$$\frac{\partial \phi}{\partial \ell} = \frac{\partial \phi}{\partial x}\frac{\partial x}{\partial \ell} + \frac{\partial \phi}{\partial y}\frac{\partial y}{\partial \ell}$$

is the derivative of ϕ in the direction $(\partial x/\partial \ell, \partial y/\partial \ell)$. ($\ell$ is the arc length along a curve $x = x(\ell)$, $y = y(\ell)$). If we substitute the Cauchy–Riemann equations into this expression, then

$$\frac{\partial \phi}{\partial \ell} = \frac{\partial \psi}{\partial y}\frac{\partial x}{\partial \ell} - \frac{\partial \psi}{\partial x}\frac{\partial y}{\partial \ell} = \frac{\partial \psi}{\partial n}, \qquad (9.108)$$

which, as indicated in Figure 9.14, is the derivative in the normal direction $(-\partial y/\partial \ell, \partial x/\partial \ell)$ of ψ. Next we observe that the specification of ϕ on the boundaries ∂R is equivalent to giving $\partial \phi/\partial \ell$, where ℓ is the arc length, on ∂R. Therefore, if ϕ solves a Dirichlet Problem, its harmonic conjugate, ψ, solves a Neumann Problem for which $\partial \psi/\partial n = \partial \phi/\partial \ell$.

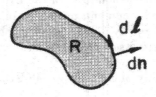

FIGURE 9.14.

Some Fluid Flows

We digress to reconsider fluid flows. In two dimensions, if the fluid velocity components are denoted by (u, v), then (9.79) and (9.80) become

$$\frac{\partial u}{\partial x} = -\frac{\partial v}{\partial y}, \quad \frac{\partial u}{\partial y} = \frac{\partial v}{\partial x}, \tag{9.109}$$

which are just the Cauchy–Riemann equations. This says that $iu(x, y) + v(x, y)$ is analytic (or if we multiply by $-i$, $u - iv$ is analytic). We have already introduced, via the second equation of (9.109), the velocity potential ϕ, with the property

$$u = \frac{\partial \phi}{\partial x}, \quad v = \frac{\partial \phi}{\partial y}. \tag{9.110}$$

When (9.110) is substituted in the second equation of (9.109), it is identically satisfied, and when substituted into the first it gives Laplace's equation. By the same token, the first equation of (9.109) implies the existence of a function ψ such that

$$u = \frac{\partial \psi}{\partial y}, \quad v = -\frac{\partial \psi}{\partial x}. \tag{9.111}$$

If (9.110) is substituted into this, then

$$\frac{\partial \phi}{\partial x} = \frac{\partial \psi}{\partial y}, \quad \frac{\partial \phi}{\partial y} = -\frac{\partial \psi}{\partial x}, \tag{9.112}$$

and again we obtain the Cauchy–Riemann equations. Therefore we have shown that the *complex potential*, $\phi + i\psi$, is an analytic function,

$$F(z) = \phi + i\psi. \tag{9.113}$$

Since the derivative of an analytic function can be computed by differentiating in any direction, we have, in particular, by differentiating in the x-direction,

$$\frac{dF}{dz} = \phi_x + i\psi_x = u - iv. \tag{9.114}$$

This is known as the *complex velocity*.

The function ψ is known as the *stream function* and the lines $\psi = $ constant are known as *streamlines*. To see the basis for this terminology, consider

$$\psi(x, y) = \text{constant}.$$

Then

$$d\psi = 0 = \frac{\partial \psi}{\partial x} dx + \frac{\partial \psi}{\partial y} dy = -v\, dx + u\, dy.$$

Thus on $\psi = $ constant

$$\frac{dy}{dx} = \frac{v}{u},$$

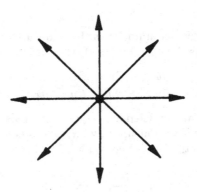

FIGURE 9.15.

or the flow is parallel to streamlines at each point. Since flow is parallel
to a material body, boundaries will appear as streamlines. The lines $\phi =$
constant are perpendicular to the streamlines and are known as *potential
lines*.

The above discussion implies that any analytic function can be regarded
as an incompressible irrotational flow field, and it is of interest to examine
some elementary cases.

Uniform Flow:
$$F = (U - iV)z,$$

$$\frac{dF}{dz} = u - iv = U - iV \quad (U \text{ and } V \text{ constant}).$$

Simple Source (strength m):

$$F = \frac{m}{2\pi} \ln z = \frac{m}{2\pi} \ln re^{i\theta} = \frac{m}{2\pi} \ln r + \frac{im}{2\pi}\theta,$$

$$\frac{dF}{dz} = u - iv = \frac{m}{2\pi z} = \frac{m}{2\pi}\left(\frac{x}{r^2} - \frac{iy}{r^2}\right).$$

In the last case we see that streamlines are $\theta =$ constant; hence, as
indicated in Figure 9.15, the interpretation is that of a source (or a sink).
The velocity in the radial direction is

$$\frac{\partial \phi}{\partial r} = \frac{1}{r}\frac{\partial \psi}{\partial \theta} = \frac{m}{2\pi r},$$

and as a result the outflow from the source is

$$\int_{r=\text{constant}} \mathbf{u} \cdot \mathbf{n}\,d\ell = \oint_{r=\text{constant}} u_r\, r\, d\theta = \int_0^{2\pi} \frac{m}{2\pi r} r\, d\theta = m$$

(since $d\ell = r\, d\theta$ on $r =$ constant). We can therefore understand why m is
designated as strength.

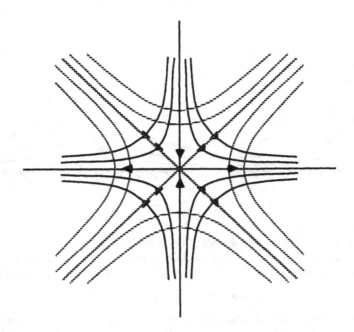

FIGURE 9.16.

Next consider

$$F = \alpha z^n$$

with, say, α real. If $n = 1$, we again get the case of uniform flow—in the x-direction. For $n = 2$,

$$F = \alpha(x + iy)^2 = \alpha(x^2 - y^2) + 2i\alpha xy,$$

$$\frac{dF}{dz} = u - iv = 2\alpha(x + iy).$$

Streamlines are given by the hyperbolas

$$2\alpha xy = \text{constant}$$

and are sketched as continuous lines in Figure 9.16. (The orthogonal equipotential lines $\phi = \text{constant}$ are sketched as dashed lines.) To obtain the flow directions, note, for example, that

$$u = \frac{\partial \psi}{\partial y} = 2\alpha x.$$

We observe that the positive and negative x- and y-axis can each be interpreted as material boundaries. This leads to two of the possible interpretations sketched in Figure 9.17. On the left we have flow in a corner, and on

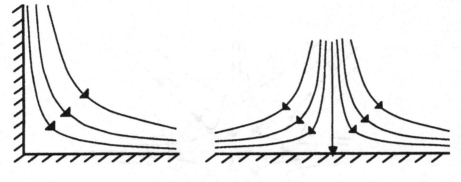

FIGURE 9.17.

the right flow impinging on a wall. The latter is known as *stagnation* point flow and the heavy dot as the *stagnation point*. It is a point at which a flow comes to rest. We can extend this discussion to other values of n; e.g.,

$$F = \alpha z^3 = \alpha r^3 e^{i3\theta} = \alpha r^3 \cos 3\theta + \alpha r^3 i \sin 3\theta,$$

so that the streamlines $\theta = 0,\ \pi/3$ can be regarded as material boundaries. This corresponds to flow in a $\pi/3 = 60°$ corner. (See Figure 9.18.) Next consider

$$F = \alpha z^{1/2} = \alpha r^{1/2}\left(\cos\frac{\Theta}{2} + i\sin\frac{\Theta}{2}\right),$$

for which $\theta = 0,\ 2\pi$ are zero streamlines and we have flow around an edge (see Figure 9.19). Note that $w = (1/2)(\alpha/\sqrt{z})$ so that infinite velocities are obtained at the origin and the flow is unrealistic in the neighborhood of the origin.

These simple flows can be added to construct more interesting cases since the potential equation is linear. For example, a source and a uniform flow is (taking the uniform flow in the x-direction)

$$F = Uz + \frac{m}{2\pi}\ln z \tag{9.115}$$

with the complex velocity

$$w = U + \frac{m}{2\pi z}. \tag{9.116}$$

Therefore, far from the origin the flow is uniform, and near the origin it resembles source flow. To get a better handle on the flow, we consider the stream function

$$\psi = Uy + \frac{m}{2\pi}\theta = Ur\sin\theta + \frac{m}{2\pi}\theta.$$

From the complex velocity (9.116) we see that

$$z_s = -\frac{m}{2\pi U}$$

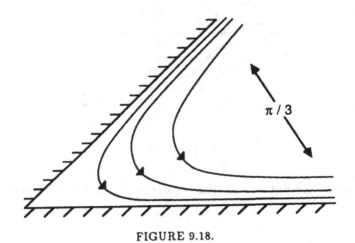

FIGURE 9.18.

FIGURE 9.19.

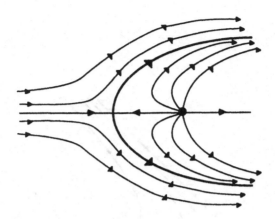

FIGURE 9.20.

is a stagnation point, $w = 0$. The negative real axis $\theta = \pi$ is the streamline

$$\psi = \frac{m}{2}.$$

From our discussion of the stagnation point in the case when $F = \alpha z^2$, we know there is a second orthogonal streamline along which $\psi = m/2$; and we therefore consider

$$U r \sin \theta + \frac{m}{2\pi} \theta = \frac{m}{2}$$

or

$$r = \frac{m}{2} \left(1 - \frac{\theta}{\pi} \right) / U \sin \theta,$$

which if $r(\theta)$ is a solution, then also $r(2\pi - \theta)$ is a solution. As $\theta \to 0$, $r \uparrow \infty$, and the sketch is as shown in Figure 9.20 so that this can be interpreted as flow past the infinite body that is indicated in the figure.

Another flow of interest is given by

$$F = Uz + \frac{m}{2\pi} \ln(z + a) - \frac{m}{2\pi} \ln(z - a), \qquad (9.117)$$

where $a > 0$ is real. This is the superposition of uniform flow, a source at $z = -a$, and an equal sink at $z = a$. A little thought shows that this can be considered as flow past the finite closed body sketched in Figure 9.21.

Some Heat Flow Problems

We will return to potential flow later and now consider some relatively simple heat flow problems in two dimensions. Suppose we have two boundaries fixed at two temperatures, T_1 and T_2, as indicated in Figure 9.22. What is the temperature in the intervening material (the region that is

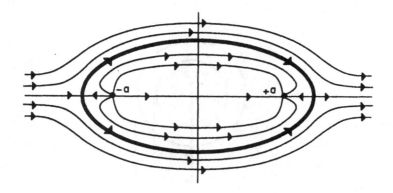

FIGURE 9.21.

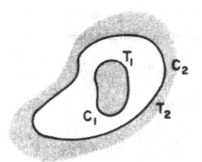

FIGURE 9.22.

not shaded)? The temperature distribution, we know, satisfies the Laplace equation

$$\frac{\partial^2 \phi}{\partial x^2} + \frac{\partial^2 \phi}{\partial y^2} = 0 \tag{9.118}$$

with $\phi = T_1$ on C_1 and $\phi = T_2$ on C_2.

We first consider the very simple case of two concentric circles, say, with radii R_1 and R_2 and, as above, at temperatures T_1 and T_2 (see Figure 9.23). The solution is immediate if we write the Laplace equation in cylindrical coordinates:

$$\nabla^2 \phi = \frac{1}{r^2} \frac{\partial^2 \phi}{\partial \theta^2} + \frac{1}{r} \frac{\partial}{\partial r} r \frac{\partial \phi}{\partial r}. \tag{9.119}$$

Then, from the angular symmetry of the problem, we can assume that $\partial/\partial \theta = 0$ so that (9.119) becomes

$$\frac{\partial}{\partial r} r \frac{\partial \phi}{\partial r} = 0.$$

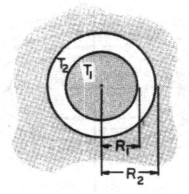

FIGURE 9.23.

This has the solution

$$A + B \ln r.$$

If we apply the boundary conditions, then

$$\phi = T_2 + \frac{T_1 - T_2}{\ln R_1/R_2} \ln(r/R_2), \tag{9.120}$$

which in complex form is

$$F(z) = T_2 + \frac{T_1 - T_2}{\ln R_1/R_2} \ln(z/R_2). \tag{9.121}$$

Next we consider the more complicated heat flow problem indicated in Figure 9.24. Without loss of generality we have set the outer circle temperature to zero, placed the origin at its center, and taken its radius to be unity. This problem is not as trivial as the previous one—especially since an angular dependence in the temperature is now to be expected. (The *queer* choices for a and b lead to simple forms later.)

The means by which we solve this problem come from a seemingly trivial observation. Suppose

$$w = w(z)$$

is an analytic function for $z \in D_1$. We will also write $w(D_1) = D_2$; i.e., the domain D_1 maps to a domain D_2. Therefore if F is analytic in D_2, then $F(w(z))$ is analytic for $z \in D_1$—which is a consequence of the chain rule of differentiation.

To be specific we write

$$w(z) = \alpha(x,y) + i\beta(x,y),$$

$$F = \phi(\alpha,\beta) + i\psi(\alpha,\beta) = \Phi(x,y) + i\Psi(x,y). \tag{9.122}$$

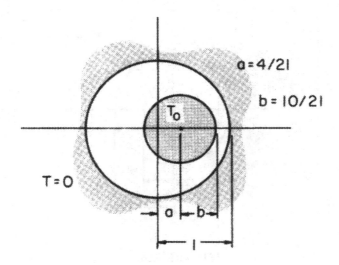

FIGURE 9.24.

Since w is analytic, α and β are harmonic with

$$\left(\frac{\partial^2}{\partial x^2} + \frac{\partial^2}{\partial y^2}\right)\alpha = \left(\frac{\partial^2}{\partial x^2} + \frac{\partial^2}{\partial y^2}\right)\beta = 0; \qquad (9.123)$$

and since F is analytic in $\alpha + i\beta$, then

$$\left(\frac{\partial^2}{\partial \alpha^2} + \frac{\partial^2}{\partial \beta^2}\right)\phi = \left(\frac{\partial^2}{\partial \alpha^2} + \frac{\partial^2}{\partial \beta^2}\right)\psi = 0. \qquad (9.124)$$

Finally, because F is also analytic in z, we have

$$\left(\frac{\partial^2}{\partial x^2} + \frac{\partial^2}{\partial y^2}\right)\Phi = \left(\frac{\partial^2}{\partial x^2} + \frac{\partial^2}{\partial y^2}\right)\Psi = 0. \qquad (9.125)$$

To solve the problem, consider the function

$$w = \frac{4z - 1}{4 - z}. \qquad (9.126)$$

This maps the unshaded region shown in Figure 9.24 into the annular region in Figure 9.25. We will show this below. For the time being, assume it to be true. The case of heat flow between concentric circles was solved by (9.120). For the case under study, this is

$$\phi = \frac{T_0}{\ln\frac{1}{2}}\ln|w| = \frac{T_0}{\ln\frac{1}{2}}\ln\left|\frac{4z - 1}{4 - z}\right|. \qquad (9.127)$$

This last substitution solves the original problem, as is clear from the above discussion.

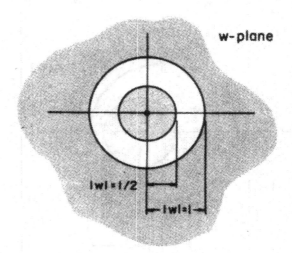

FIGURE 9.25.

We see that the original problem was in effect solved by the transformation (9.126). Some other examples in which problems are solved simply by
transformation will be given later. The question arises whether the above
transformation was fortuitous or could be derived. The latter is of course
true, and we consider this next.

Fractional Linear Transformations

Equation (9.126) is an example of a fractional linear transformation.
More generally this is defined by

$$w = \frac{az + b}{cz + d} = \frac{az + ad/c - ad/c + b}{cz + d}. \qquad (9.128)$$

After adding and subtracting as indicated in the second form this can be
written as

$$w = \frac{a}{c} + \frac{1}{c}\frac{bc - ad}{cz + d}. \qquad (9.129)$$

Thus for $c \neq 0$ the condition $bc - ad \neq 0$ leads to a nondegenerate transformation in the sense that it is not a constant. We characterize the above
mapping by the symbol L:

$$w = Lz, \quad L = \begin{pmatrix} a & b \\ c & d \end{pmatrix},$$

where the matrix is associated with L in the way indicated in (9.128). It
can be shown (see Exercise 6) that two successive fractional linear transformations L_1 and L_2 result in a fractional linear transformation given
by

$$L = L_2 L_1.$$

That is, it is given by ordinary matrix multiplication

$$L = \begin{pmatrix} a_2 & b_2 \\ c_2 & d_2 \end{pmatrix} \begin{pmatrix} a_1 & b_1 \\ c_1 & d_1 \end{pmatrix} = \begin{pmatrix} a_2 a_1 + b_2 c_1 & a_2 b_1 + b_2 d_1 \\ c_2 a_1 + d_2 c_1 & c_2 b_1 + d_2 d_1 \end{pmatrix}.$$

Therefore, since

$$bc - ad = -\det \begin{pmatrix} a & b \\ c & d \end{pmatrix},$$

the nondegenerate property is preserved if L_1 and L_2 have it.

From (9.129) the most general fractional linear transformation can be regarded as taking place in three steps:

$$z_1 = cz + d, \quad z_2 = \frac{1}{z_1}, \quad w = \frac{a}{c} + \frac{bc - ad}{c} z_2. \tag{9.130}$$

Alternatively, it is a product of the above three fractional linear transformations—the first and last of which are of the same type.

To understand this transformation, first consider

$$cz + d.$$

Set $d = 0$ and write $c = |c|e^{i\phi}$, $z = re^{i\theta}$. Then note that $cz = |c|e^{i\phi}z = |c|re^{i(\phi+\theta)}$ so that a point $z = re^{i\theta}$ is magnified by the factor $|c|$ and its angle is rotated by the angle ϕ. Next observe that a straight line of orientation θ_0 has the equation

$$z = x_0 + te^{i\theta_0}, \quad -\infty < t < \infty, \quad \theta_0 \text{ fixed}, \quad x_0 \text{ fixed}$$

and maps into

$$cz = cx_0 + tce^{i\theta_0} = cx_0 + t|c|re^{i(\theta+\phi)},$$

which is also a straight line, but rotated by ϕ.

Also observe that the circle $z = Re^{i\theta} + a$, $0 < \theta \le 2\pi$, of radius R and center a maps into

$$c(Re^{i\theta} + a) = |c|Re^{i(\theta+\phi)} + ca,$$

which is also a circle. Finally, the addition of the constant d to cz is simply a displacement by the *vector* d. In summary, $cz + d$ is a rotation, a stretching, and a translation, all of which preserve circles and straight lines.

Next consider $1/z$. This is called an *inversion in the unit circle* since all points such that $|z| > 1$ obviously map inside the unit circle, while $|z| < 1$ maps outside the unit circle. (Recall Figure 3.6.) In particular, $(z = 0) \rightarrow \infty$ and $(z = \infty) \rightarrow 0$. Under this transformation the circle $z = a + re^{i\phi}$, $0 \le \phi \le 2\pi$, becomes

$$\frac{1}{a + Re^{i\phi}}.$$

In this form this is somewhat opaque. To understand the significance of this form, observe that the equation of a circle is

$$x^2 + y^2 + Ax + By + C = 0$$

or, in polar coordinates $x = r \cos \theta$, $y = r \sin \theta$,

$$r^2 + (A \cos \theta + B \sin \theta)r + C = 0.$$

The transformation $1/z$ is characterized by $r \rightarrow 1/\rho$ and $\theta \rightarrow -\psi$, which on being substituted into the equation of a circle gives

$$\frac{1}{\rho^2} + (A \cos \psi - B \sin \psi)\frac{1}{\rho} + C = 0.$$

If $C \neq 0$, then it can also be written as

$$\rho^2 + \rho \left(\frac{A}{C} \cos \psi - \frac{B}{C} \sin \psi \right) + \frac{1}{C} = 0,$$

which is a circle. If $C = 0$, then

$$\rho(A \cos \psi - B \sin \psi) + 1 = 0,$$

and this is just a straight line. If $C = 0$, the original circle passes through the origin and therefore the transformed circle passes through infinity. Straight lines are regarded as special cases of circles, namely those which pass through infinity. Alternatively, if we consider a straight line

$$0 = Ax + By + C = Ar \cos \theta + Br \sin \theta + C,$$

then under inversion, $1/z$, this becomes

$$\frac{A}{\rho} \cos \psi - \frac{B}{\rho} \sin \psi + C = 0.$$

If $C \neq 0$, this is

$$\frac{A}{C} \cos \psi - \frac{B}{C} \sin \psi + \rho = 0,$$

which, on squaring, gives the equation of a circle passing through the origin. (If $C = 0$, multiply by ρ^2 to obtain the equation of a straight line passing through the origin.)

As mentioned before, the last of the three transformations (9.130) is the same type as the first. We have therefore demonstrated that (9.129) maps circles and straight lines into circles and straight lines.

Further properties may be noted:

$$z = -\frac{d}{c} \rightarrow w = \infty$$

so that circles (and straight lines) passing through $-d/c$ become straight lines

$$z = \infty \to w = \frac{a}{c}$$

so that straight lines (not passing through $-d/c$) become circles passing through $w = a/c$.

Although four constants appear—a, b, c, d —the multiplication of the numerator and denominator (9.128) by any constant leaves the transformation invariant. Therefore only three constants determine the fractional map. These can be determined by the solution of the three equations, giving the mapping

$$(z_1, z_2, z_3) \to (w_1, w_2, w_3).$$

Another way to effect this is through

$$\frac{(w_1 - w)(w_3 - w_2)}{(w_1 - w_2)(w_3 - w)} = \frac{(z_1 - z)(z_3 - z_2)}{(z_1 - z_2)(z_3 - z)} \tag{9.131}$$

which implies the following:

$$z \to z_1 \Rightarrow w \to w_1$$

$$z \to z_2 \Rightarrow w \to w_2$$

$$z \to z_3 \Rightarrow w \to w_3.$$

Incidentally, it should be remarked that

$$C = \frac{(z_1 - z_4)(z_3 - z_2)}{(z_1 - z_2)(z_3 - z_4)} \tag{9.132}$$

is called the *cross-ratio* and it can be seen to be an invariant under fractional linear transformations.

As is suggested by our remarks, the mapping is one-to-one. To see this, note that for any pair $(z_1, z_2) \to (w_1, w_2)$,

$$w_2 - w_1 = \frac{(bc - ad)}{c(cz_2 + d)} - \frac{(bc - ad)}{c(cz_1 + d)} = \frac{(bc - ad)(z_1 - z_2)}{(cz_2 + d)(cz_1 + d)},$$

which shows $z_1 \neq z_2$, $w_1 \neq w_2$ and vice versa.

As an application of these deliberations, we consider a transformation which maps the interior of the unit circle $|z| < 1$ into the interior of the unit circle $|w| < 1$. We take $(z_1, z_2, z_3) = (1, -1, \alpha)$ and $(w_1, w_2, w_3) = (1, -1, \beta)$ with $|\alpha| = |\beta| = 1$. We do this so that we are indeed mapping $|z| = 1$ into $|w| = 1$. Thus if we apply (9.131) we obtain

$$\frac{(1 - w)(\beta + 1)}{2(\beta - w)} = \frac{(1 - z)(\alpha + 1)}{2(\alpha - z)}. \tag{9.133}$$

This reduces to

$$w = \frac{z - \gamma}{1 - \gamma z} \tag{9.134}$$

where

$$\gamma = \frac{\alpha - \beta}{1 - \alpha\beta}. \tag{9.135}$$

If we divide the numerator and denominator by $\alpha\beta$, we obtain

$$\gamma = \frac{(1/\alpha) - (1/\beta)}{1 - (1/\alpha\beta)} = \frac{\bar{\alpha} - \bar{\beta}}{1 - \bar{\alpha}\bar{\beta}} = \bar{\gamma}$$

(since $|\alpha| = 1$, $\alpha^{-1} = \bar{\alpha}$). Thus γ is real.

Finally, since the interior of the disc $|z| < 1$ maps into the interior of the disc $|w| < 1$, the point $z = \gamma$ which maps into $w = 0$ must be such that $|\gamma| < 1$.

More generally,

$$w = C\frac{z - \mu}{1 - \bar{\mu}z}, \tag{9.136}$$

where $|C| = 1$ and $|\mu| < 1$ performs the required mapping. To see this, note that

$$|w| = \frac{|z - \mu|}{|1 - \bar{\mu}z|}.$$

On the unit circle $z = e^{i\theta}$, we have

$$|z - \mu| = |e^{i\theta} - \mu| = |e^{i\theta}||1 - \mu e^{-i\theta}|$$

$$= |1 - \mu e^{-i\theta}| = |1 - \bar{\mu}e^{i\theta}| = |1 - \bar{\mu}z|$$

and $|w| = 1$. The only way in which the condition $|\mu| < 1$ enters is by fixing that the interior of $|z| = 1$ goes into $|w| < 1$. Clearly if $|\mu| > 1$, then $|z| > 1$ maps into $|w| < 1$.

We consider the above special case (cf. (9.126))

$$w = \frac{z - \gamma}{1 - \gamma z}$$

with γ real and $|\gamma| < 1$. Then

$$\bar{w} = \frac{\bar{z} - \gamma}{1 - \gamma\bar{z}}$$

so that conjugate points map into conjugate points. Next we use the transformation to map the indicated pairs of circles in the z-plane to the indicated concentric circles in the w-plane (see Figure 9.26). From symmetry it will suffice if we map $(z_1, z_2) = (a - r, a + r)$ into $(w_1, w_2) = (-R, R)$, so that

$$R = \frac{a - r - \gamma}{1 - \gamma(a - r)}, \qquad -R = \frac{a + r - \gamma}{1 - \gamma(a + r)}.$$

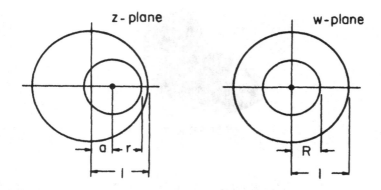

FIGURE 9.26.

Adding and cleaning up, the determination of γ comes from

$$\gamma^2 - \frac{\gamma}{a}(1 + a^2 - r^2) + 1 = 0,$$

and we get R from above. For example,

$$a = \frac{4}{21}, \quad r = \frac{10}{21}$$

give rise to

$$\gamma = \frac{1}{4}, \quad R = \frac{1}{2},$$

which furnish the values used earlier in the discussion of the heat flow problem, (9.126).

Flow Past a Body

We return to the problem of flow past a body. There is one particular case—and a very important one—in which we can compute flow fields with relative ease. This is the case of a circular cylinder. Consider a circular cylinder centered at the origin, $|z| = a$; then $F(z)$ can be considered a complex potential of flow past $|z| = a$ if $\mathrm{Im}(F(ae^{i\theta}))$ is constant, i.e., if the stream function is a constant on the circular cylinder. Suppose that $f(z)$ represents a *flow* and that it has no singularities in the region $|z| \leq a$. We then have that

$$F(z) = f(z) + \overline{f}\left(\frac{a^2}{z}\right) \tag{9.137}$$

represents that flow past the cylinder $|z| = a$. To see this, observe that on $z = ae^{i\theta}$, F is pure real, so that $\mathrm{Im}\, F = \psi = 0$ on $z = ae^{i\theta}$ and $F \to f(z)$ for $|z| \uparrow \infty$. As an example, consider uniform flow

$$f(z) = (U - iV)z. \tag{9.138}$$

FIGURE 9.27.

Then

$$F(z) = (U - iV)z + \frac{a^2}{z}(U + iV),\qquad(9.139)$$

which, ... is easily verified, has the circle $z = ae^{i\theta}$ as a streamline.

To deal with flow past an arbitrary body, let us denote the flow, i.e., the complex potential, by $F(z)$ and the body by Γ (see Figure 9.27). Denote by

$$w = g(z)$$

a mapping which takes $(z \in \Gamma)$ into the unit circle and the exterior of the body into the exterior of the unit circle in such a way so that

$$w \approx z \quad \text{as} \quad |z| \uparrow \infty.$$

Then, since

$$G(w) = F(w) + \overline{F}\left(\frac{1}{w}\right)\qquad(9.140)$$

gives flow past a circle in the w-plane,

$$\mathcal{G}(z) = F(g(z)) + \overline{F}\left(\frac{1}{g(z)}\right)\qquad(9.141)$$

is the sought-after complex potential.

As an illustration, consider uniform flow past a flat plate of breadth $2b$. This is sketched in Figure 9.28.

Uniform flow is expressed by

$$F = Uz.$$

Next consider the mapping

$$z = \frac{b}{2}\left(w - \frac{1}{w}\right).\qquad(9.142)$$

Then, if $w = e^{i\theta}$, i.e., the unit circle in the w-plane, it follows that

$$z = \frac{b}{2}2i\sin\theta,\qquad(9.143)$$

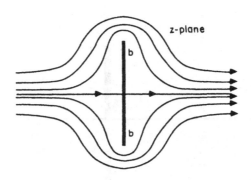

FIGURE 9.28.

whose locus for $0 < \theta < 2\pi$ is as shown in Figure 9.29 (bold line). Therefore $w(z)$ maps the flat plate onto a circle. Specifically,

$$w^2 - \frac{2z}{b}w - 1 = 0, \quad w = \frac{z}{b} + (1 + (z^2/b^2))^{1/2}.$$

Note that the positive branch was chosen so that as $|z| \uparrow \infty$, $w \to 2z/b$. The solution to the problem of flow past a flat plate is therefore given by

$$G(z) = \frac{b}{2}\left(wU + \frac{U}{w}\right) = \frac{b}{2}U\left(\frac{z}{b} + (1 + (z^2/b^2))^{1/2}\right.$$

$$\left. + \frac{1}{(z/b) + (1 + (z^2/b^2))^{1/2}}\right) = bU(1 + (z^2/b^2))^{1/2}. \tag{9.144}$$

The factor $b/2$ is inserted so that the complex potential approaches the correct limit of Uz as $|z| \uparrow \infty$.

A class of shapes of great interest arises from consideration of the transformation

$$z = \frac{1}{2}\left(w + \frac{1}{w}\right), \tag{9.145}$$

or

$$w = w(z). \tag{9.146}$$

This, as we have discussed, maps the unit circle of the w-plane into the real line segment $(-1, 1)$ and, with the proper choice of branches, the exterior of $(-1, 1)$ of the z-plane to the exterior of the unit circle in the w-plane. If, instead of considering $|w| = 1$ we consider circles passing through $w = 1$, we obtain the so-called *Joukowski airfoil* shapes indicated in Figure 9.30. In the first instance a symmetric airfoil is generated by choosing a center on the real axis. In particular, consider

$$|w + a| = 1 + a, \quad a > 0. \tag{9.147}$$

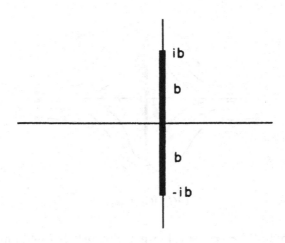

FIGURE 9.29.

Uniform flow past this circle has the complex potential

$$G(z) = U \left(w + a + \frac{(1+a)^2}{w+a} \right)$$

$$= U \left(1 + a + (z^2 - 1)^{1/2} + \frac{(1+a)^2}{1 + a + (z^2 - 1)^{1/2}} \right). \qquad (9.148)$$

A similar construction holds in the asymmetric case. Although we do not consider it here, these constructions can be used to determine lifting properties of airfoil shapes.

Exercises

1. Verify that ψ as defined by (9.107) and ϕ satisfy the Cauchy–Riemann equations.

2. Show that (9.121) is the complex form of (9.120). Find the imaginary part of (9.121) and verify that it satisfies (9.119). What problem does it solve?

3. Prove that potential and streamlines are orthogonal.

4. What *flow field* is described by

$$F(z) = \frac{i\kappa}{2\pi} \ln z$$

(compare with source flow)?

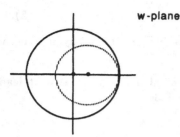

w-plane

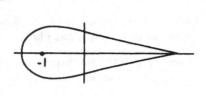

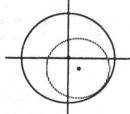

w-plane

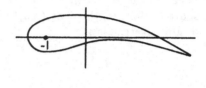

FIGURE 9.30.

5. What *flow field* is described by

$$f(z) = U\left(z + \frac{b^2}{z}\right)?$$

U and b are real.

6. Show that if

$$z_2(z_1) = \frac{a_1 z_1 + b_1}{c_1 z_1 + d_1}, \quad L_1 = \begin{pmatrix} a_1 & b_1 \\ c_1 & d_1 \end{pmatrix},$$

$$w(z_2) = \frac{a_2 z_2 + b_2}{c_2 z_2 + d_2}, \quad L_2 = \begin{pmatrix} a_2 & b_2 \\ c_2 & d_2 \end{pmatrix},$$

then

$$w(z_2(z_1)) = W(z_1) = \frac{\alpha z + \beta}{\gamma z + \delta}$$

with

$$\begin{pmatrix} \alpha & \beta \\ \gamma & \delta \end{pmatrix} = L_2 L_1.$$

7. Demonstrate that (9.132) is invariant under a general fractional linear transformation.

8. Verify that (9.133) reduces to (9.134) with γ given by (9.135).

9. Sketch the streamlines of the flow (9.144) and verify that it indeed describes flow past a flat plate.

10. Consider (9.148) for various values of a and show that it leads to flows past shapes shown in Figure 9.30.

11. Show that the transformation

$$w = z + b + \frac{(a-b)^2}{z+b}?$$

$0 < b < a$, leads to a symmetric airfoil.

12. What is the equation of the streamlines for the flow generated by sources of strength m at $(\pm a, 0)$ and a sink of strength $2m$ at the origin? Sketch these.

References

1. G. Arfken, *Mathematical Methods for Physicists* (1970) Academic, New York.

2. G. Birkhoff and G.-C. Rota, *Ordinary Differential Equations* (1969) Blaisdell, New York.

3. M. Braun, *Differential Equations and Their Applications* (1978), Springer-Verlag, New York.

4. W. Boyce and R.C. DiPrima, *Elementary Differential Equations and Boundary Value Problems* (1965) Wiley, New York.

5. R.V. Churchill, *Introduction to Complex Variables and Applications* (1948) McGraw-Hill, New York.

6. J.H. Curtiss, *Introduction to Functions of a Complex Variable* (1978) Dekker, New York.

7. R. Courant, *Differential and Integral Calculus* (1937) Nordemann, New York.

8. R. Courant and D. Hilbert, *Methods of Mathematical Physics.* Vol. 1 (1958) Interscience Publishers, New York.

9. A. Erdelyi, *Asymptotic Expansions* (1956) Dover, New York.

10. R. Haberman, *Mathematical Models* (1977) Prentice Hall, Englewood Cliffs, New Jersey.

11. P.R. Halmos, *Finite Dimensional Vector Spaces* (1958) van Nostrand, Princeton, New Jersey.

12. H. Jeffrys and B.S. Jeffrys, *Methods of Mathematical Physics* (1966) Cambridge University Press, Cambridge.

13. J. Mathews and R.L. Walker, *Mathematical Methods of Physics* (1970) Benjamin/Cummings, Menlo Park, California.

14. A. Naylor and G.R. Sell, *Linear Operator Theory in Engineering and Science* (1982) Springer-Verlag, New York.

15. Z. Nehari, *Introduction to Complex Analysis* (1964) Allyn and Bacon, Boston.

16. E.D. Nering, *Linear Algebra and Matrix Theory* (1970) Wiley, New York.

17. M.R. Spiegel, *Complex Variables (Schaum's Outline Series)* (1964) McGraw-Hill, New York.

Index